Lean Six Sigma
Using SigmaXL
and Minitab

D1472170

ABOUT THE AUTHORS

ISSA BASS is a Master Black Belt and senior consultant with Manor House and Associates. He is the founding editor of SixSigmaFirst.com. Bass has extensive experience in quality and operations management, and is the author of *Six Sigma Statistics with Minitab and Excel.*

BARBARA LAWTON, Ph.D., is a Six Sigma Black Belt and has been improving manufacturing processes using Lean techniques and Six Sigma for more than 15 years, in various industries on three continents. Dr. Lawton specializes in data analysis, and is a keen experimentalist and problem solver, using a wide range of statistical tools. She currently works for one of the world's leading aerospace organizations in the UK.

Lean Six Sigma
Using SigmaXL
and Minitab

Issa Bass

Barbara Lawton, Ph.D.

New York   Chicago San Francisco Lisbon   London Madrid
Mexico City   Milan   New Delhi San Juan Seoul
Singapore Sydney Toronto

The McGraw·Hill Companies

Cataloging-in-Publication Data is on file with the Library of Congress.

McGraw-Hill books are available at special quantity discounts to use as premiums and sales promotions, or for use in corporate training programs. To contact a special sales representative, please visit the Contact Us page at www.mhprofessional.com.

Lean Six Sigma Using SigmaXL and Minitab

1 2 3 4 5 6 7 8 9 0   DOC/DOC   0 1 4 3 2 1 0 9

ISBN 978-0-07-162130-4
MHID 0-07-162130-X

The pages within this book were printed on acid-free paper containing 15% post-consumer fiber.

Sponsoring Editor
Judy Bass

Editorial Supervisor
David E. Fogarty

Project Manager
Somya Rustagi, International
Typesetting and Composition

Copy Editor
Susan Fox-Greenberg

Proofreader
Surendra Nath Shivam,
International Typesetting and Composition

Indexer
Broccoli Information Management

Production Supervisor
Richard C. Ruzycka

Composition
International Typesetting and
Composition

Art Director, Cover
Jeff Weeks

To my mother Dema Diallo and my pretty niece Sira Basse

—Issa Bass

To my brother Peter and my late parents, Barbara and Derek.
To Professor Ron Pethig, Russel Johnson and Dr. Harry Watts for support and driving me toward a successful career.
To Issa Bass who has sent me on a new journey.
Finally, to Derek Richards, who taught me the mathematical foundation that has led to a successful engineering career. Thank you.

—Dr. Barbara Lawton

Contents

Preface

Business production methodologies have never stopped improving since Frederick W. Taylor, the inventor of scientific management, devised techniques for factory management and time and motion studies. Eli Whitney created the methods for interchangeable parts and Henry Ford developed the modern assembly lines used in mass production today. These inventions were the precursors of the modern day quality and productivity improvement methodologies. Over the past three decades, many managerial methodologies aimed at improving production processes have been introduced to businesses throughout the world. Some have resisted skepticism and have prevailed and are still being used, while others, such as the total quality management (TQM) or company wide quality control (CWQC), have been deemed to be nothing but fads and have disappeared almost immediately after they appeared. In fact, all the process improvement strategies (Six Sigma, TQM, CWQC, Lean, TOC, etc.) have the same underlying philosophy; they are all geared toward customer satisfaction and insist on the necessity for all sections of a company to cooperate in order to improve all aspects of its operations. They all insist on producing high-quality products at the lowest possible cost through a reduction of waste and continuous improvement.

Some companies have deployed TQM and failed because the deployment was conducted badly, their employees were poorly trained, or the areas they insisted on improving were areas that did not require improvement because their improvement would not have had a positive impact on the overall performance of the business. This costs money and does not generate any significant return on investment.

In most cases, TQM did not fail because it was in itself a bad methodology or that its application was conducive to poor performance and failure. In fact, the name of the methodology that a company uses to improve its processes should not be the most germane aspect of its management strategy. Currently, the most widespread methodologies used in management are Six Sigma and Lean, also known as the Toyota Production System (TPS).

Indeed, most of the tools that were used by TQM have been refined and are still being used in Six Sigma. Six Sigma and Lean have withstood skepticism largely because of the success some major corporations have seen as a direct result of their application. A careful observation of those corporations would reveal that Six Sigma and Lean are not partially used and, in most cases, they have become a culture, a way of managing for those companies instead of auxiliary instruments temporarily used to solve a circumstantial problem.

Six Sigma is a data-driven business strategy that seeks to streamline production processes to constantly generate quasi-perfect products and services in order to achieve breakthrough return on investment. One of the pillars of Six Sigma is the pursuit of the reduction of production process variation to an infinitesimal level.

Lean manufacturing or TPS is a management methodology originated in Japan and more often associated with Toyota Motor Company; it was introduced to the American public by James Womack and Daniel T. Jones in the 1990s. It is about doing things right the first time and every time at a steady pace. It is also about reducing cycle time and inventory by eliminating waste. The underlying foundation of Lean manufacturing is the organizational strategy that constantly seeks a continuous improvement through the identification of the non-value-added activities (*Muda*) and their elimination along with the reduction of the time it takes to perform the value-added tasks.

Most companies use these two methodologies simultaneously for process improvement because, taken in isolation, each of these methodologies can yield good results. However, when they are combined, the probability for success is even greater.

This book is written as a practical introduction to Lean Six Sigma project execution and follows the DMAIC (define, measure, analyze, improve, and control) roadmap. It is written in such a way that it can be used as a training text for beginners and a reference for seasoned practitioners. Six Sigma is, by definition, analytical and profoundly rooted in statistical analysis. Therefore, ample statistical theory and development are provided to support the analyses. Both a theoretical analysis and the two most widely used statistical software suites, SigmaXL and Minitab, are used throughout the examples to help the reader better understand how to execute a Lean Six Sigma project.

The book is based on years of teaching the Lean and Six Sigma methodologies to a wide variety of audiences from different industries. We hope that the content of the book will be helpful in furthering the understanding of Lean Six Sigma project execution.

Acknowledgments

We would like to thank all the people who have been so helpful in this endeavor. We would like to thank Randle Hooks, the PFS-Web, Memphis, TN Distribution Center's general manager; John Bradley from Bank of America; Bill Anderson from HP; Harun Sanjay from FedEx; and Tom McGregor from Wal-Mart for their support and advice.

To the Manor House and Associates team and all our trainees, thank you for making us better.

Special thanks are also addressed to Aissata Basse, Fatoumata Basse, and our good friend Vera Kea from the Kenco group.

Lean Six Sigma
Using SigmaXL
and Minitab

Introduction

The use of Lean Six Sigma has proved to be a powerful and effective way for providing sustained positive operational results in organizations worldwide. Lean Six Sigma is in fact a hybrid philosophy for continuously improving organizations. Lean aims at eliminating waste by creating a culture of improvement where people learn powerful tools for solving problems and continuous improvement based on visual management and standardization that sustains enhancement. On the other hand, Six Sigma is a methodology that aims at reducing variations in production processes in order to improve quality and meet customers' expectations.

Lean, also known as the Toyota Production System, has enabled many organizations to reach unparalleled levels of excellence. According to Fortune Magazine, Toyota made 50% more cars in 2005 than it did in 2001. It earned $11.4 billion more than all the major manufacturers combined and out of the 10 highest quality rated cars that run in America, 7 were made by Toyota at a time when GM was deemed close to declaring bankruptcy and Ford was mired in financial problems and had to lay off 30,000 employees. These kinds of successful results have made Toyota Motor Company the most benchmarked organization in the world.

The primary premise for Lean is the focus on the creation of value for the customers. The value creation that enhances the organization's overall productivity is done by eliminating non-value added activities using specific sets of tools that optimize the utilization of the people and the processes. According to Jeffrey Liker, the leading Toyota Production System's expert in America, "Toyota Production System is a total system of people, processes, and tools that evolve and grow stronger over decades. It is not a toolkit or program that you can 'implement' as you would a computer system. To really get anywhere close to the level of excellence of Toyota, senior leaders have to understand that Lean is a way of thinking."

The Lean thinking process focuses on satisfying customers while improving productivity, reducing lead time, reducing manufacturing and product cost, increasing inventory floor space, reducing new product time to market, and improving the cost of quality. This is done by implementing a strategy that constantly seeks a continuous improvement through the identification of the non-value-added activities (*Muda*) and their elimination along with the reduction of the time it takes to perform the value added tasks.

Defects within a production process are considered to be the results of deviations from the predefined targets. Six Sigma is a methodology that uses statistical and nonstatistical tools to define the optimal quality target and the tolerance around the target for a production process. It also seeks to identify and remove the causes of defects and errors in production processes by reducing variations around the target and containing them within the tolerance.

The Six Sigma approach to process improvements is project-driven; in other words, areas that show opportunities for improvements are identified and projects are selected to proceed with the necessary improvements.

The integration of Lean with Six Sigma came to be known as Lean Six Sigma. Lean is used to reduce waste but it does not monitor production processes to determine if they are in control. It does not use statistical tools to measure the processes' capabilities, i.e., their ability to generate reproducible products or services that meet or exceed customers' expectations. Because Six Sigma is project-driven, it is less flexible when it comes to addressing practical issues that occur daily and would not require even small projects to fix. Issues such as total preventive management (TPM), changeover time, labor and equipment efficiency, inventory reduction, and overproduction are better addressed using Lean techniques.

An integration of Lean and Six Sigma offers the possibility for reducing defects through a control of process variations and a reduction of waste using Lean techniques.

Lean Six Sigma process improvements are conducted through projects or Kaizen events. The project executions follow a rigorous pattern called the DMAIC (Define, Measure, Analyze, Improve, and Control). At every step in the DMAIC roadmap, specific tools are used to measure, analyze data, find root causes of problems, and determine the best options for their resolution.

Since Lean Six Sigma is data-driven, any project conducted using this methodology will require the use of some software. We elected to use SigmaXL and Minitab.

Most organizations use Microsoft Excel to organize and analyze their data. Excel is equipped with a substantial amount of tools for descriptive statistics and probability calculations but it still lacks capabilities for more complex data analyses. SigmaXL is a powerful statistics software suite that adds those capabilities to Microsoft Excel. It is very reliable and easy to use and because it is embedded in Excel spreadsheets, it makes data readily available for manipulation easy to access and analyze. Besides helping evaluate data through statistical analysis, SigmaXL also makes it easy to perform faster routine data organization and manipulation with Excel. Columns and rows are better manipulated and pivot tables are created much faster. Beyond statistical analysis, SigmaXL also contains a great deal of tools that help analyze quality, Lean, and Six Sigma related issues much faster through its many templates. SigmaXL is becoming a popular tool for Six Sigma Green Belts, Black Belts, quality and business professionals, engineers, and managers around the world.

Minitab software suit is widely used in many corporations and universities.

This book contains many examples and exercises to enable the reader to practice. The files required to complete the examples can be downloaded from www.mhprofressional.com/bass/. Minitab and SigmaXL will be needed in some cases to use those exercises. Trial versions of both SigmaXL and Minitab can be downloaded from their respective websites: www.sigmaXL.com and www.minitab.com.

An Overview of SigmaXL

To have SigmaXL always automatically appear on the menu bar whenever Microsoft Excel is open, from the **SigmaXL** menu select **Help**, then click on **Automatically load SigmaXL**.

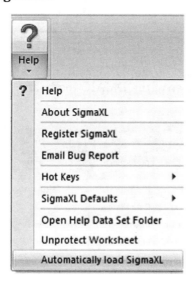

When the **Automatically load SigmaXL** box appears, press the **OK>>** button. Every time Microsoft Excel is open, the SigmaXL menu will appear on the menu bar.

SigmaXL menu bar

SigmaXL offers the possibility to organize the menu bar according to the preferences of the user. To change the menu from the default format to the DMAIC format, click on **Help,** then click on **SigmaXL defaults,** and then select **Menu Options–Set SigmaXL's Menu to Classical or DMAIC.**

| Data Selection Default |
| Clear Saved Defaults |
| Menu Options - Set SigmaXL's Menu to Classical or DMAIC |

When the Set **Menu** appears, make the selection before pressing the **OK>>** button.

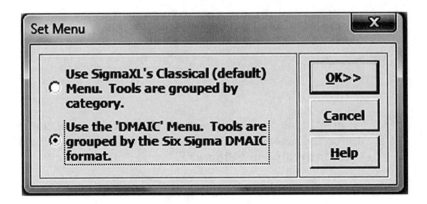

Selecting the DMAIC format makes it easier to see all the tools that are used at every step of the DMAIC roadmap.

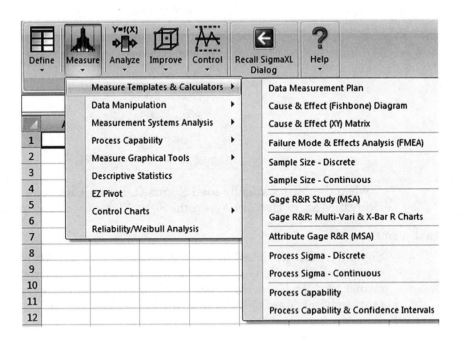

The data manipulation menu helps prepare and organize the data for further analysis.

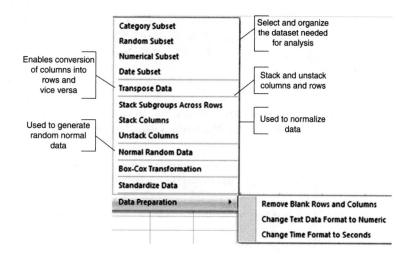

SigmaXL templates

The templates make basic computations such as the calculation of takt time and sample sizes very easy.

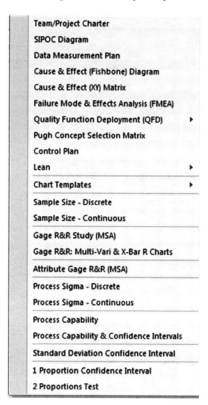

Generate two columns with 20 rows of normal random numbers with a mean equal to 10 and a standard deviation equal to 1.

Display the summary of the descriptive statistics.

Solution

From the **Data Manipulation** menu, click on **Random normal data.** When the **Normal Random Number Generator** box appears, fill it out as shown in the figure.

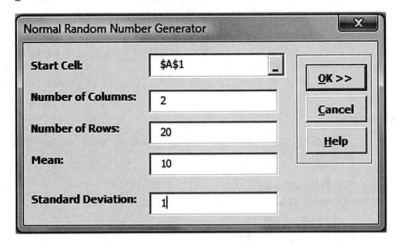

Press the **OK>>** button. Once the random numbers appear on a new sheet with the fields containing the data already selected, from the statistical tools menu click on **Descriptive Statistics**. The Descriptive Statistics box should appear with the field *Please select your data* populated.

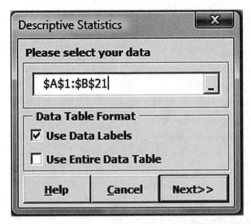

Press the **OK>>** button to show the Descriptive Statistics box. Using the **Numeric Data Variables (Y)>>** button, fill out the dialog box as shown in the figure.

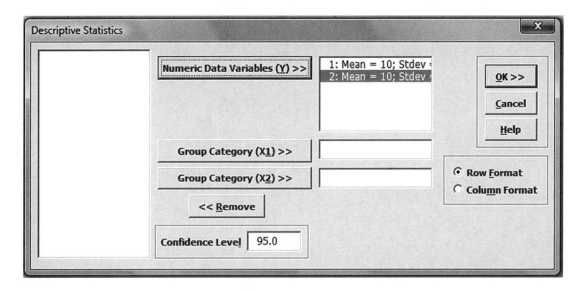

Press the **OK>>** button to obtain the results shown in the figure.

Descriptive Statistics	1: Mean = 10; Stdev = 1	2: Mean = 10; Stdev = 1
Count	20	20
Mean	9.526	10.401
Stdev	1.287	1.199
Range	4.819	5.193
Minimum	7.269	7.131
25th Percentile (Q1)	8.700	9.818
50th Percentile (Median)	9.428	10.566
75th Percentile (Q3)	10.508	11.218
Maximum	12.089	12.324
95.0% CI Mean	8.923 to 10.128	9.840 to 10.962
95.0% CI Sigma	0.978406 to 1.879	0.911737 to 1.751
Anderson-Darling Normality Test	A-Squared = 0.285917; P-value = 0.5876	A-Squared = 0.300528; P-value = 0.5472
Skewness	Skew = 0.249015; P-value = 0.6268	Skew = -0.899142; P-value = 0.0791
Kurtosis	Kurt = -0.422166; P-value = 0.6705	Kurt = 1.639; P-value = 0.0987

A 30-day trial version of SigmaXL can be downloaded from www. sigmaXL.com.

1

Define

Project Planning

A problem is defined as a contradiction, a gap between a concrete reality that is being faced and a desired situation. The first step in resolving a problem is to clearly understand and articulate both the current reality and the desired situation. A clear definition of a problem is the first step of a Lean Six Sigma problem solving roadmap. Lean Six Sigma problem resolutions are performed through projects.

The objective of a project is either to solve an existing problem or to start a new venture. In either case, a carefully planned and organized strategy is needed to accomplish the specified objectives. The strategy includes developing a plan that will define the goals, explicitly setting the tasks to be accomplished, determining how they will be accomplished, and estimating the time and the resources (both human and material) needed for their completion.

The way projects are planned and managed will seriously affect the profitability of the ventures for which they are intended and the quality of the products or services that they generate. The strategy used to plan the resources needed for the changes is called project management. It includes the specification of the tasks to be accomplished, how the objectives are to be achieved, the planning of the resources to be allocated, and the budgeting and timing as well as the implementation of the project and the controls involved.

Since not all the tasks can be executed at the same time because of their interdependence, a methodic scheduling is necessary for a timely and cost-effective outcome.

The human resources remain the most important aspect of any endeavor; therefore, a clear identification of the people who have a stake in the project as well as those who might resist changes is vital to its success.

However, before a project even starts, there are some prerequisites that must be fulfilled to ensure its success. These are listed as follows:

- The first step in a project management is its specification, the definition of its goals and objectives. A lucid definition of a project provides a solid basis for its prompt completion.
- The management must be committed to the changes envisaged.
- The overall strategy must be defined and be in agreement with the company's business vision.
- The project manager must be selected.
- The key participants to the projects must be identified, their skills and abilities assessed, and their roles defined and well understood.
- The risk involved in the changes to be implemented must be evaluated and managed.

Most project management plans are subdivided into four major phases: (1) the feasibility study, (2) the project planning, (3) the project implementation, and (4) the verification or evaluation. Each of these phases requires strategic planning.

The three major tools that are used for the purpose of planning and scheduling the different tasks in project management are the Gantt chart, the Critical Path Analysis (or Method), and the Program Evaluation and Review Technique.

However, before any scheduling starts, it is essential to estimate accurately the time that each task might require. A good scheduling must take into account the possible unexpected events and the complexity involved in the tasks themselves. This requires a thorough understanding of every aspect of the task before developing a list.

One way of creating a list of tasks is a process known as the Work Breakdown Structure (WBS). It consists of creating a tree of activities that take into account their lengths and contingence. The WBS starts with the project to be achieved and goes down to the different steps necessary for its completion. As the tree starts to grow, the list of the tasks grows. Once the list of all the tasks involved is known, based on experience or good wit, an estimation of the time required can be made and milestones determined. Milestones are the critical steps in a project that are used to help measure progress. Knowing the milestones of a project with certainty is extremely important because they can affect the timeliness of the project completion as a whole. Delays in project completions can have serious financial consequences and can cost companies market shares.

In global competitive markets, innovation is the driving force that keeps businesses alive, and this is more obvious in high-tech industries. Most companies have several lines of products and each one of them is required to put out a new product every year or every 6 months. If, for instance, Dell's Inspiron or Latitude fails to put out new products on time, it is likely to lose profit from the forgone sales (the loss is proportional to the products' time-to-live) and market shares to its competitors.

The Gantt chart

Gantt charts (named after its inventor, the American social scientist, Henry L. Gantt) are effective for scheduling complex tasks. They help arrange the different events in synchronism and associate each task with its owner and its estimated beginning and ending time. The charts also allow the project's team to visualize the resources needed to complete the project and the timing for each task. It therefore shows where the task owners must be at any given time in the execution of the projects. The team working on the project should know whether it is on schedule just by looking at the chart.

The chart itself is divided in two parts. The first part shows the different tasks, the tasks owners, the timing, and the resources needed for their completion; the second part graphically visualizes the sequence of the events.

The Gantt chart in Fig. 1.1 summarizes a project scheduling for a small IT project. The horizontal bars on the calendar side of the chart depict the beginnings and the ends of the scheduled tasks. Some tasks cannot start until the preceding ones are finished.

Critical path analysis.   The critical path analysis (CPA) is a tool used for complex projects. Not only does the CPA take into account the interdependence of the critical tasks, but it also considers the possibility of performing different tasks in parallel, at the same time, or at different times. It also helps monitor the execution of the tasks as they are being implemented. The CPA identifies the tasks that need to be completed on time and the ones that can be delayed without preventing the whole project from meeting its deadline. It helps estimate the critical path, the project duration, and the slack time for every activity. Since all the tasks included in a project cannot be executed at the same time because of their interdependence, a critical path needs to be determined when scheduling the activities. The critical path is the sequence of activities that cannot be delayed because any delay in any one of those activities will result in a delay on the whole project.

The CPA resembles the tree in WBS with the difference that it takes into account the timing of the tasks. It is a web of activities linked by arrows between every two nodes.

The first step in creating a CPA diagram is to list the tasks including their duration and the order in which they have to be completed. In some cases, the project team itself might need to complete the project before the time estimated by the CPA, which creates a need to reduce the time spent on some activities.

Example 1.1   Table 1.1 contains the information needed to create and display the critical path for a fictitious project.

TABLE 1.1

	Tasks summary	
Activity	Predecessor	Duration
A	NONE	3
B	A	5
C	A	3
D	B	5
E	C	6
F	D	7
G	E	4
H	G	2

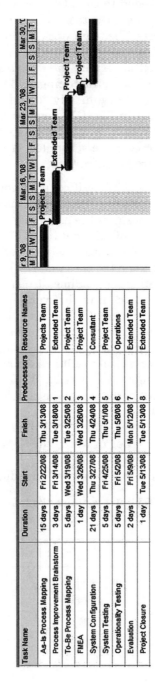

Task Name	Duration	Start	Finish	Predecessors	Resource Names
As-Is Process Mapping	15 days	Fri 2/22/08	Thu 3/13/08		Projects Team
Process Improvement Brainstorm	3 days	Fri 3/14/08	Tue 3/18/08	1	Extended Team
To-Be Process Mapping	5 days	Wed 3/19/08	Tue 3/25/08	2	Project Team
FMEA	1 day	Wed 3/26/08	Wed 3/26/08	3	Project Team
System Configuration	21 days	Thu 3/27/08	Thu 4/24/08	4	Consultant
System Testing	5 days	Fri 4/25/08	Thu 5/1/08	5	Project Team
Operationality Testing	5 days	Fri 5/2/08	Thu 5/8/08	6	Operations
Evaluation	2 days	Fri 5/9/08	Mon 5/12/08	7	Extended Team
Project Closure	1 day	Tue 5/13/08	Tue 5/13/08	8	Extended Team

Figure 1.1

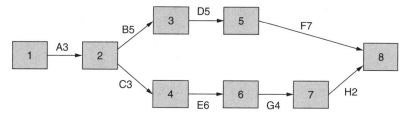

Figure 1.2

Based on this information, we can determine the critical path, the project duration, and the slack time for H.

Task A is the first on the list; no other task can start until it is completed. Tasks B and C come next; they are contingent on task A. Tasks E, G, and H are on the same path as C, while tasks D and F are on the same path and depend on B.

The letters on the diagram (Fig. 1.2) represent the different activities and the numbers beside them represent the time that it will take to accomplish those tasks. The diagram shows that there are two paths to the project: ABDF and ACEGH. The duration for ABDF is 20 days and the duration for ACEGH is 18 days. Since ABDF is the longest path, it is also the critical path; any delay on that path will result in a delay for the whole project. The earliest that task H can start is within 16 days.

The advantage of the Gantt chart over the CPA is the graphical visualization of the tasks along with their timing, the task owners, the start time, and the end time. The advantage of the CPA over the Gantt chart is the sequence of events that takes into account the interdependence of the tasks.

The CPA is a deterministic model because it does not take into account the probability for the tasks to be completed sooner or later than expected; the time variation is not considered.

Program evaluation and review technique (PERT)

The PERT is just a variation of the CPA with the difference that it follows a probabilistic approach while the CPA is a deterministic model. Once the critical tasks have been identified, their timing estimated, the sequence of events determined, and a list of activities established, we could evaluate the probability for the different tasks to be accomplished on time and the shortest possible time for each of them.

The completion of each task is said to follow a Beta distribution with the expected length of the project being

$$E(p) = \frac{Lt + St + 4lt}{6}$$

where Lt stands for longest expected time, St stands for shortest expected time, and lt stands for likely time.

The variance of the critical path will be

$$\sigma^2 = \frac{Lt - St}{6}$$

The estimated standard deviation is

$$\sigma = \sqrt{\sigma^2} = \sqrt{\frac{Lt - St}{6}}$$

The completion of the whole project follows a normal distribution.

Example 1.2 Based on the information in Table 1.2, find the critical path for the project, the project completion time, the probability of finishing it on time, and the probability of finishing it at least 1 day earlier.

TABLE 1.2

Activity	Predecessor	Most likely time	Shortest time	Longest time
A	NONE	3	2	4
B	A	5	4	6
C	A	3	2	4
D	B	5	3	6
E	C	6	5	6
F	D	7	5	8
G	E	4	3	4
H	G	2	1	3

Solution:

TABLE 1.3

Activity	Predecessor	Most likely time	Shortest time	Longest time	Estimated mean	Variance	Standard deviation
A	NONE	3	2	4	3	0.33	0.57
B	A	5	4	6	5	0.33	0.57
C	A	3	2	4	3	0.33	0.57
D	B	5	3	6	4.67	0.50	0.71
E	C	6	5	6	5.67	0.17	0.41
F	D	7	5	8	6.67	0.50	0.71
G	E	4	3	4	3.67	0.17	0.41
H	G	2	1	3	2.00	0.33	0.57

The critical path has the longest duration. It is critical because any delay in any of its tasks will cause a delay for the whole project. In this case, we have two paths—ABDF which will last 20 days and ACEGH which will last 18 days. Therefore, the critical path is ABDF (see Fig. 1.3).

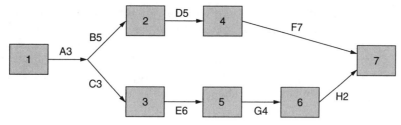

Figure 1.3

The estimated variance for the critical path is $0.029 + 0.11 + 0.25 + 0.11 = 0.499$ with a standard deviation of $\sqrt{0.499} = 0.706$

The probability for completing the project at least 1 day earlier means completing it in 19 days or less. That probability is found using a normal distribution:

$$Z = \frac{x - \mu}{\sigma} = \frac{19 - 20}{0.706} = -1.416$$

1.42 corresponds to 0.4222 on the normal table, therefore, $0.5 + 0.4222$ gives 0.9222 on the normal table; the area we are looking for will be on the right side of 0.9222 under the normal curve, which is $1 - 0.9222$ equal to 0.0778 (Refer to Z transformations in Chapter 2.).

Project Charter

Any initiative aimed at generating qualitative changes in a workplace requires a strategic planning of the resources needed for its realization. The resources involved are human, material, and financial, as well as the time needed for both the planning and the execution of the plans. However, before a project even starts, there are some prerequisites that must be fulfilled for its success.

- Management must be committed to the changes envisaged.
- The overall strategy must be defined and be in agreement with the company's business vision.
- The project champion should be nominated.
- The key participants to the projects must be identified, their skills and abilities assessed, and their roles defined and well understood.
- The risk involved in the changes to be implemented must be evaluated and managed.

Since most of the cost savings are contingent upon the planning and execution of projects, before any action is taken it is necessary to have a clear common understanding of all the aspects of the project, its extent, the key stakeholders, its goals, and its objectives. A good definition provides a clear appreciation of every stakeholder's role and what is expected of him or her. It also provides a tacit agreement between the parties. The definition of the project is displayed on a document called the project charter. The project charter is a written document that embodies the understanding between the different parties involved in the project; it specifies the overall mission, goals, and objectives of the project and the roles and responsibilities of each participant.

The project charter is written either by the project sponsor or by the project champion with the approval of the sponsor. Upper management issues the project charter to make the project official. The project charter gives the project champion and his or her team the power to use the organization's resources for its purpose (see Fig. 1.4). The charter is then made public and distributed to all the stakeholders. Among others, advantages for a charter are as follows:

- Provides a clear understanding among all the parties involved about the expectations.
- Defines the participant's role and responsibility.
- Clearly defines the scope of the project and the exclusions.
- Defines the deliverables and their timelines.
- Enables the team to have access to data that it may not have been able to access.
- Defines and provides the resources needed for a prompt competition.
- Releases the team members from their regular responsibilities.

Each organization has a standardized way to present its project charters but some elements are found on all project charters.

Project number

The project number depends on the organization's standards and structures. It is just a document reference.

Project champion

The project champion or sponsor ensures that the resources are available and that the project is executed in a timely and cost-effective manner.

Project definition

The first step in a project management is the specification of the mission of the team, the definition of its goals and objectives. A lucid definition of a project provides a solid basis for its prompt completion. The project definition generally starts with a background statement. The background statement explains the reason for the project and the context that lead to its need.

Project description

This defines the project, gives a clear reason for its purpose, and provides the expected measurable results. It sometimes includes the project background and relates it to our present conditions. It makes clear the expectations from every stakeholder. It also defines the success of the project and addresses the consequences of a failure. The project description defines the constraints and the factors that can affect the project.

Value statement

The value statement clearly sets out the positive impact of the project on the rest of the organization.

Stakeholders

The stakeholders start with the project sponsor, the team working on the project, the internal customers (who can be the project sponsor), and the external customers. This identifies the key people involved in the project and their roles.

Project monitoring

This entails how and when to conduct meetings, who participates, and for what purpose, as well as how the progress of the different aspects of the project are measured.

Scheduling

This includes important events and dates that affect the project, as well as different deadlines for subparts of the project.

Alternative plans

This includes what to do if things do not go according to plans.

Risk analysis

What risks are involved in the project for the rest of the organization?

Capturing the Voice of the Customer

Since all production processes aim at satisfying some customers (whether they are internal or external), their needs (explicit as well as implicit) need to be determined and integrated in the design of the production processes. Any new product development or product or process improvement should start with capturing the voice of the customer; in other words, what the customers expect to find in the product needs to be precisely assessed and integrated in the design of the products. Several techniques are used to capture the voice of the customer. Surveys,

TEAM/PROJECT CHARTER

Project Name:	
Date (Last Revision):	
Prepared By:	
Approved By:	

Business Case:	Opportunity Statement (High Level Problem Statement):
	Defect Definition:
Goal Statement:	**Project Scope:**
	Process Start Point:
	Process End Point:
Expected Savings/Benefits:	In Scope:
	Out of Scope:

Project Plan:				Team:		
Task/Phase	Start Date	End Date	Actual End	Name	Role	Commitment (%)

Figure 1.4 SigmaXL project charter template.

customer service data collection, interviews, and focus groups are among the tools used to identify the customer needs.

The first step in capturing the voice of the customer is the determination of who the customer is. Customers are generally divided in two groups; they are either internal or external. Internal customers are the next step in the process when one is dealing with production in progress. Every business is composed of several processes and a process is defined as a sequence of events, a chain of tasks. When the production process is started, the materials and information flow through the chain of tasks to generate the products or services needed and every task becomes a supplier to the tasks downstream and a customer to the tasks upstream.

Other types of internal customers are the process owners and the stakeholders.

External customers are the users of the final product or service.

The importance of the customers remains the same under all circumstances, whether the customers are internal or external, because every customer can make or break a company. Yet the methods used to capture their needs are different.

Capturing the voice of the external customer

Defining the external customer is not always easy. If a company sells a product to retailers, both the end users and the retailers are customers, but who should be considered the primary customer? Is it the end user or the retail store? The way these two groups of customers are treated is not the same because their needs are different. It is important to determine who should be considered the primary customer because their expectations should affect the production processes the most.

To better define the customers, it is good to divide them into actual and potential. Potential customers are those who are not currently buying the company's products but could possibly do so if changes are made in the company's operations. Most companies only sell a small proportion of products to the customers of the market that they serve. Potential customers are competitors' customers, lost customers, and prospective customers.

Since all external customers, be they primary or not, potential or actual, impact the process, their voices need to be captured and their expectations understood and integrated in the design of the products. The customers' expectations about a product are not necessarily homogeneous but it is always possible to segment the needs and find underlying common trends among the segments and manage the quintessential needs in ways that are conducive to the production of goods or services that meet their expectations at a reasonable cost for the producer. Customers' expectations can be collected in several ways, such as through surveys, market research, customer service records, focus groups, and interviews.

Survey.   A survey is a gathering of opinion about a product or service through a sample of randomly selected customers. It is generally based on a questionnaire with the idea of generating a well-constructed customer perception of the quality of products or services and identifying their weaknesses and their strengths. The pertinence of the survey is contingent upon how its statistical analysis was conducted, mainly on the sample size, the margin of error, and the confidence level. Surveys can be conducted in several ways. One way of conducting a survey is asking customers to rate some of the critical characteristics of a product in order to generate actionable data that can be geared toward improvement. These steps should be followed:

1. Clearly specify the goal of the survey, i.e., what is it that you want to learn about the product?
2. Determine the population to be addressed.
3. Determine your sample size. The determination of sample size follows some statistics rules.
4. Choose survey methodology. How will the customers be contacted?
5. Create the questionnaire. What questions will be asked?
6. Conduct interviews and gather data.
7. Analyze the data gathered and produce the report.

The Likert scale is a good example of how survey data can be organized and analyzed to generate objective results.

Likert scale.   A Likert scale is a metric used to measure customers' attitude or preferences about a product or service. A question related to an aspect of a product is asked but the responses to the question are not open and they are restricted. The responses are ordinal in the sense that they can be ranked from lowest to highest in value.

An example of a Likert scale ranking would be: "Not Relevant at all (0)," "Somewhat Relevant (1)," "Relevant (2)," "Very Relevant (3)," "Extremely Relevant (4)." The numbers in parenthesis are not additive; they are just codes that are not always necessary.

After the survey is completed, the responses to each question are summed up (how many "Very Relevant" did we have for question 1…?); and this is used to generate scores for every question.

Example 1.3   A charter company wanted to assess the relevance of television sets on its buses and ordered a survey. The results are summarized in Table 1.4.

TABLE 1.4

	Not relevant	Somewhat relevant	Relevant	Very relevant	Extremely relevant	Total
Group 1	15	22	9	33	16	95
Group 2	13	21	4	26	15	79
Group 3	17	21	5	23	17	83
Group 4	16	24	9	25	19	93
Group 5	15	25	9	24	17	90
Total	76	113	36	131	84	440

Once the questionnaire has been completed and the scores summarized in a table, the data gathered can be statistically tested. Because the data are generally ordinal, nonparametric statistics such as the Chi-square test, Mann-Whitney test, or Kruskal-Wallis test are used for the testing.

Customer service data collection.   Customer services (CSR) are, in general, the main point of contact between the customers and the business when poor quality products are sold. The CSR department becomes an invaluable resource for quality assessment because when customer complaints are recorded, the files obtained include the nature of the problems that the customers are encountering and in some cases what caused the problems. However, because customers very seldom call to complement businesses when they are satisfied with the products they buy, the data generated at CSR need to be viewed with caution.

Focus groups.   A focus group is a discussion group composed of a few chosen people (approximately five to nine) to talk about a selected subject. When used for marketing purposes or product development, the intention is generally to assess the customers' expectations about the product. It is particularly appropriate for obtaining numerous perspectives about the same matter.

Capturing the voice of the internal customer

Failure to clearly and precisely understand the needs of the internal customers and translate them into critical-to-quality characteristics of the products or services can seriously increase the cost of production through project overruns, product or service redesign or internal rework, product returns, and customer services expenses.

Capturing the voice of the customers of a project

A project is, in general, the work of a cross-functional team. Yet the members of the team that work on the design, development, or implementation of a project are seldom its customers. For instance, when a project for a new Warehouse Management System (WMS) is initiated, the people who will be managing the project implementation are hardly ever the ones who will actually be using the system in their daily activities. A group of project managers is gathered to set up the system and once it is ready, the operations department of the business is more likely to use it. Therefore, in this particular case, the primary customer for the project is operations.

One of the critical elements of a project is its clear definition. Since the main canon of communication between the steering committee (which oversees the project development and implementation) and the project team is the project charter, capturing the voice of the customer for a project becomes easier because it is clearly spelled out in a signed document, which should state the roles and responsibilities of each participant.

Capturing the voice of the next step in the process

As mentioned earlier, in the chain of events that constitutes a business process, the next step in the process is the customer of the current step when the production is in progress. To reduce the probability for internal rework and improve the quality level, efficiency, and productivity, the employee at the current step is expected to deliver a defect-free product to the next step. For that to happen, the employees will need to clearly understand what every step of the process expects from the previous ones. Several tools are used for that purpose.

Work instructions.   Work instructions or standard operating procedures (SOP) are official documents that contain instructions on how an employee should perform his or her tasks.

As is process mapping.   Process mapping is a graphical representation of a process flow. It is an effective tool to visualize a process in a very simple way. Flow charts are generally used to describe how material and information flow from step to step throughout the process. When unambiguous comments are added to every control of the flow chart, the chart becomes a good tool for understanding the requirements at every level of the process and for pinpointing potential sources of nonconformance.

Employee feedback.   Employees who are working "hands on" on the products or transferring information are an invaluable sources for understanding the materials on which they work and the conditions in which they move from task to task.

Quality assurance feedback.   In an enterprise that values organizational excellence, the role of the quality assurance (QA) department cannot be circumscribed to auditing products to prevent defective parts from reaching customers or monitoring process performance through control charts. The QA department should be associated with every step of the process, from concept design to development and implementation. Once production is in progress, QA is supposed to be the most knowledgeable entity about the quality of the products and the potential sources of their shortcomings. QA employees should be the resource that provides the feedback about where improvement efforts should concentrate within the chain of tasks.

Critical-to-Quality Tree

In order to be able to quickly solve customers' problems, it is necessary to not only understand what their requirements are but also to be able to translate the significant aspects of those requirements into measurable data that can be subjected to critical analysis. The purpose of the critical analysis is to determine what it takes to actually meet the customers' requirement. The critical-to-quality (CTQ) tree is one of the tools used in the Define phase of a Six Sigma project to capture the voice of the customers; it transforms requirements into quantifiable data.

Customers' demands are usually vague and they are a function of implicit factors that are not always clearly expressed when they place their orders. The critical part in satisfying the customers' requirements is the identification and

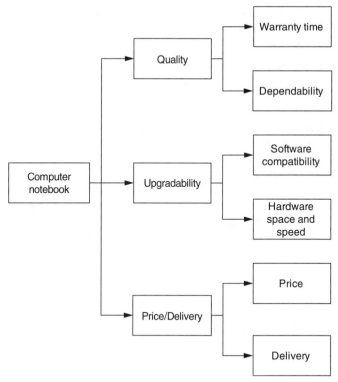

Figure 1.5

the analysis of those implicit factors; therefore, a strategic approach needs to be adopted to build the CTQ tree (see Fig. 1.5).

Building the CTQ tree can help to

- Identify implicit and vague customer needs and convert them into specific requirements.
- Make sure that all the critical requirements are inventoried.
- Speed up the understanding of those requirements by the project team while narrowing the scope to the critical few.

The following steps are used to create a CTQ tree:

1. Identify the product or service that is being analyzed.
2. Identify the key components of the product or service. These are the features that define the product. Without their presence, the product would not be what it is supposed to be.
3. Identify the critical customers' requirements for the products or services. The team identifies the basic requirements (quality, price, and delivery) that the customers expect from the products.

4. Identify the customers' first level of requirements. In this step, the team identifies critical requirements that satisfy the key customer need identified in the previous step.

5. Identify the customers' second level of requirements. In this step, the team identifies critical requirements that satisfy the key customer need identified in the previous step.

This step has to be repeated until quantifiable requirements are obtained.

The data gathering to determine the critical aspects of the products can be done through surveys, interviews, brainstorming session, or obtained from Customer Services data.

Once the data has been gathered and the tree built, the next step should consist of analyzing the tree to determine the aspects of the critical requirements that need improvements. The analysis is done in the Analyze phase of the project using the seven quality management tools or statistical analysis.

Kano analysis

Since the quality of a product or service is measured in terms of the satisfaction that the customer derives from using it, understanding the customers' needs and being able to quantify them becomes paramount. Customers' "needs" and "wants" are two distinct things, and the ways in which they react when their needs or wants are satisfied are not the same. The needs themselves are not appreciated at the same level; some needs are critical in a product, some are necessary but not as critical, and some are neither necessary nor critical but they do make customers feel delighted.

When a company chooses how to produce and sell its products, it needs to be able to measure what features of the product result in satisfying the expressed and implicit needs of the customers. Adding a multitude of features to a product or a service does not necessarily increase its appeal. Being able to know with accuracy what critical few features are demanded and how the customers react to their presence or nonpresence can help improve quality at a lower cost.

The Kano analysis (named after the inventor, Dr. Noriaki Kano) is a tool that helps determine what characteristics a producer might want to include in the product or service to increase customer satisfaction. The Kano model breaks down a product's features according to how they can meet customers' expectations, some of which are explicit and some latent.

Kano divides the products' features in three:

"The threshold (basic) features," which define the product, without them, the product is useless. These features are fundamental to the product. When you receive a call on your cellular phone, you expect to hear the person on the other end of the line and you expect that person to hear you. If you cannot find

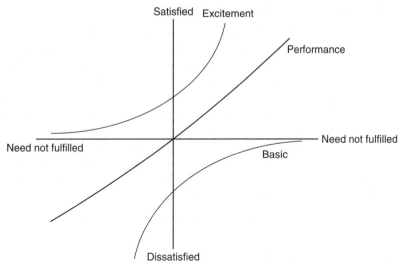

Figure 1.6

that feature on the phone, you would not buy it. However, the sheer presence of the basic features does lead to customer happiness.

"The performance features," which may not be as important as the threshold features but can increase the customers' satisfaction. The clarity of reception of a cellular phone can be put in that category.

"The delighter features," which the customers may not expect to find on the product but whose presence is exiting to them.

The ways in which the customers react to the presence or absence of these attributes on a product or service are summarized in Fig. 1.6.

The reasoning behind the Kano analysis thus far seems commonsensical, but since most competing products provide the same basic needs, the objective of the producer should be to identify and classify the essential attributes in order to satisfy the customers while not adding superfluous features that might end up being costly.

A producer can use surveys, brainstorming sessions, and focus groups to assess the customers' responses to the different features of a product or a service.

For a survey, the questionnaire must address all the possible features that can be found on a product because the objective is to eliminate the unnecessary attributes and retain the critical few.

The objective of the producer is not just to add the "delighter attributes" to a product but also to put them at an appropriate place as a distinguishable part of the product or service in order to rouse a "wow" effect.

Suppliers-Input-Process-Output-Customers (SIPOC)

In order to identify the voice of the customer in the Define phase of a Six Sigma improvement project, it is recommended to map the "as-is" processes through which information and products flow from suppliers to customers and identify the different components of the processes and how they contribute to bringing the products to the customers. Suppliers-Input-Process-Output-Customers (SIPOC) is a high-level diagram of the five key components of the process that contribute to creating and delivering the value demanded by the customers. It shows what those components are and how each one of them participates in the process. Having a visual map of the process handy makes it easier to link the different significant components of the process together, narrow the scope of the project with fewer resources.

The five key components of a SIPOC diagram are:

1. *Suppliers.* The providers of raw materials, services, and information used in the process to generate the value sold to customers.

2. *Inputs.* The actual services, raw materials, and information used to create the value sold to customers.

3. *Process.* The sequence of events used within the organization to transform the raw materials and serviced into value.

4. *Outputs.* The value created by the organization to satisfy customers' demands.

5. *Customers.* The users of the value created by the organization.

Since in the Define phase, what the project team is mapping is not the ideal state but the current state, the mapping process should start with the inventory of the customers, a high-level view of who the customers are. A nomenclature of the customers according to the kind of products or services that they expect from the organization and how those products are delivered to them should be created. The next step should consist of creating a classification of the products demanded by each group of customers under the Output grouping. Each product is created following a unique production process, so a high-level map of each product under the Output listing should be separately created under Process. The input (raw materials, services, and information) used in the production processes are unique to each process but some of the input will be used in several processes; therefore, the input can be listed in a way that shows which processes it is intended for without unnecessarily duplicating the items in the Input list. Finally, the list of the suppliers of the input should be created with each supplier tied to the product or service that it provides in the Input list. Both the suppliers and the customers can be either internal or external.

Example 1.4 A computer manufacturer uses a customer satisfaction index (CSI) to monitor how pleased its customers are with its products. A CSI of 99% or more is considered satisfactory. The company supplies computers to different customers classified as Homes, Governments, Schools, and Businesses. Lately, the CSI has dropped to 93.5% and a Six Sigma project has been initiated to investigate the problem and find a resolution. A distribution of the different customers with the models that they purchase and their level of satisfaction has been established and summarized in Table 1.5.

TABLE 1.5

Customers	Models	Satisfaction index (%)	Reason for dissatisfaction
Home	XYZ without Bluetooth WYZ with Linux	92	Product overheating
Government	XYZ with Bluetooth	99	Drops connections
Schools	AYZ without Bluetooth	90	Plastic breaks easily, overheats
Business	WYZ with Bluetooth	93	Fan not turning, locks up too often

Table 1.5 alone does not provide enough actionable data. It provides a glimpse of the extent of the problem but does not afford an understanding of the events that lead to the dissatisfaction. Therefore, in the Define phase of the project, the team decided to create a SIPOC diagram of the organization in order to isolate and organize the different models that might be causing the customers' dissatisfaction so that the scope of the project can be narrowed to only those models. Narrowing the project to only those models is expected to help reduce the resources needed for the project execution. Figure 1.7 shows the SIPOC created by the project team. Figure 1.7 is not a granular level map of the process; it is a high-level map that shows interactions between the different components at the highest level. Process mappings that are more detailed should be carried out in the Measure phase of the project.

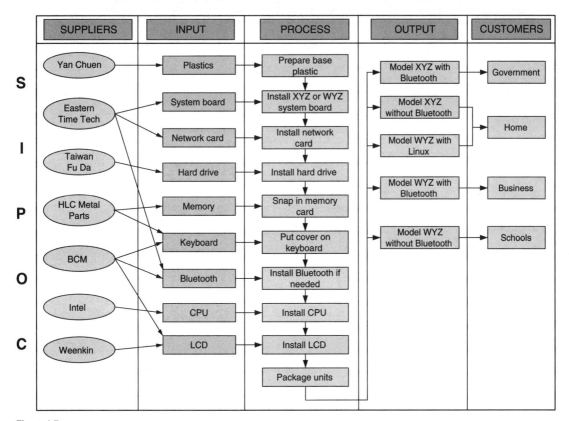

Figure 1.7

In addition to the list of the suppliers, the input that they provide, the output, and customers and the processes used to transform the input; it is also recommended to tabulate the metrics used in the organization's scorecards to track the performance of the different components of the process in terms of quality, delivery, cycle time, and cost. The matrix of the performance metrics helps to provide a quick view of where the problems might originate. Figure 1.8 is an example of the kinds of metrics that can be pulled from the scorecards to monitor performance at a SIPOC level. SigmaXL provides a good SIPOC template (Fig. 1.9).

Suppliers	Input	Process	Output	Customers	
• Expertise • Types or communication • Responsiveness	• Accuracy of order fill • Accuracy of database sync	• First time fix • Rework • Scrap • Return from customers	Order accuracy	• Satisfaction index • Rate of return	Quality
How fast do they process and fill orders ?	• Time to receive orders • Time to fill orders	• Order fill rate • Number of steps in a process • Time to complete a task	Order fill rate	• Retention rate • Acquisition rate • Attrition rate	Cycle time
How does the cost from suppliers compare to their competitors ?	• Order periodicity • Cost per order	• Labor cost • Cost per order • Machine cost	• Production cost per order • Shipping cost	• Marketing cost • Cost of acquiring new customers • Cost of losing customers	Cost

Figure 1.8

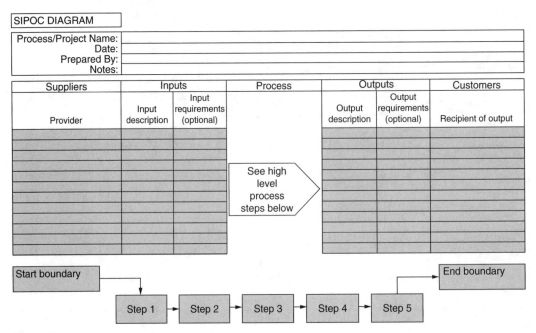

Figure 1.9

Cost of Quality

Assessing the cost of quality

The quality of a product is one of the most important factors that determine a company's sales and profit. In order to improve on sales, an organization needs to develop a strategic approach that relies on a standardized process to assess the impact of the cost of quality on profits and losses. Since in most organizations quality management is a silo separate from operations, the ability to quantify financially the impact of quality improvement can help motivate upper management to take action and help improve on production processes.

The definition of quality itself is not uniform. Dr. Joseph Juran defines quality in terms of its "fitness for use," while Philip Crosby defines it in terms of its conformance to requirements. Taguchi defines quality in terms of a degree of deviation from its predetermined target; the greater the deviation, the lower the quality, while Edward Deming approaches quality from a variability standpoint.

Nevertheless, one thing all quality practitioners agree on is that quality is measured in relation to the characteristics that customers expect to find in the product; therefore, the customers ultimately determine the quality level of the products. The customers' expectations about a product's performance, reliability, and attributes are translated into CTQ characteristics and integrated in the products' design by the design engineers. The CTQs that are expected by the customers become the standards, the reference by which the company goes to produce its products and services. A product is said to be of a poor quality every time it deviates from the predetermined CTQ target.

While designing the products, the engineers must also take into account the capabilities of the resources (machines, people, materials, etc.), that is, their ability to produce products that meet the customers' expectations. The production processes that use those resources to produce high-quality products and services come with a cost. When an enterprise measures the cost of quality, it considers the cost involved throughout the production process from when orders are received from customers to how orders are placed to suppliers and all the steps involved in meeting customers' demands. The cost of quality should be measured in terms of the overall productivity of the resources and the processes used to bring the products to the customers. If quality is assessed not only in terms of how many defects are produced or sold to customers but in terms of the capabilities of the production processes as well (in other words, in terms of the processes' abilities to meet or exceed customers' expectations), then poor quality would occur not only when poor quality goods are produced, but also whenever more resources than necessary are used to produce the (good) products. Situations that can cause a process to use more resources than necessary occur when a process does not perform at an optimal level and involves waste in the forms of rework, excessive testing, scrap, equipment rebuilding, etc.

Productivity measures the efficiency of the production processes. It determines how much input has been used for every unit of output. There is a positive correlation between quality and productivity. An improvement in the quality level of production processes will result in an improvement in the productivity of an enterprise. The relationship between productivity and quality can be expressed through the following equation:

$$P = \frac{R}{TC}$$

where P = productivity
R = revenue derived from the sales of the products
TC = total cost of production

A bad production process will increase waste in the form of rework and scrap, which will result in an increase in the costs of production. An increase in the costs of production will result in the decrease in the productivity of the resources used in the process.

The definition of the cost of quality is not unanimously accepted by all quality practitioners. Some practitioners consider the cost of quality as being the cost incurred for not doing things right, while others include the investment spent on getting the processes to produce good quality products and services.

The cost of quality can be assessed in different ways. It can be estimated by assessing the cost of meeting standards; in other words, the level of investment needed to improve the production processes to meet customers' expectations. That cost is called the cost of conformance. In the short run, a quality improvement might require an increase in the cost of conformance but in the end, the investments incurred to improve quality will be offset by the profits derived from an increase in sales.

The cost of quality can also be estimated in terms of the profit loss that occurs when nothing is done to improve on quality. Forrest Breyfogle calls that cost "the cost of doing nothing" (implementing Six Sigma). Some authors call that cost the cost of nonconformance. For instance, if the market standard for the longevity for car tires is 2 years and a tire manufacturer produces tires that last only 18 months, the cost of doing nothing would be measured in terms of the loss incurred when customers buy from the competitors because the company failed to meet market standards.

The cost of conformance includes the appraisal and preventive costs, while the cost of nonconformance includes the costs of internal and external defects.

Cost of conformance

The cost of conformance measures the investments incurred to produce goods and services that meet customers' expectations. It includes the preventive cost and the appraisal cost.

Preventive cost

The preventive cost is the cost incurred by the company to prevent nonconformance. That cost is incurred prior to the product being manufactured. It includes the costs of

- New process review
- Quality improvement meetings
- New quality improvement projects
- Process capability assessment and improvement
- The planning of new quality initiatives (process changes, quality improvement projects, etc.)
- Employee training

Appraisal cost

This is the cost incurred while assessing, auditing, and inspecting products and procedures to ensure that they conform to specifications. It is intended to detect quality related failures before the products are sent to customers. It includes

- Cost of process audits
- Inspection of products received from suppliers
- Process audit
- Testing on processing equipments
- Final inspection audit
- Design review
- Prerelease testing

Cost of nonconformance

The cost of nonconformance is the cost of having to rework products, process customers' complaints, and the loss of customers that results from selling poor quality products.

Internal failure

The cost of internal failure is incurred prior to the products being delivered to customers. It includes

- Cost of reworking products that failed audit
- Cost of bad marketing
- Scrap
- Non-value-adding activities

External failure

The cost of external failure is incurred after the products have been sent to the customers. It includes

- Cost of customer support
- Shipping cost of returned products
- Cost of reworking products returned from customers
- Cost of refunds
- Warranty claims
- Loss of customer goodwill
- Cost of discounts to recapture customers

Example 1.5 Haere-Lao Technologies manufactures computers sold directly to end-user customers. The cost of quality incurred during its first quarter is summarized in Table 1.6. The total revenue from sales for that quarter was $5,345,987.00 with a total operational expense of $4,232,897.00.

TABLE 1.6

	Absolute value	Relative value (%)
Cost of conformance		
Preventive cost		
Quality control labor cost	$95,678.00	12.07
Quality improvement special projects	$17,543.00	2.21
Training	$32,099.00	4.05
Total	**$145,320.00**	**18.34**
Appraisal cost		
Incoming product audit	$12,909.00	1.63
Process audit labor cost	$73,987.00	9.34
Down time due to equipment testing	$9,823.00	1.24
Product prerelease testing	$5,692.00	0.72
Total	**$102,411.00**	**12.92**
Total cost of conformance	**$247,731.00**	**31.26**
Cost of nonconformance		
Internal failure		
Scrap	$42,561.00	5.37
Rework	$135,671.00	17.12
Scrap collection	$12,908.00	1.63
Non-value-adding activities	$42,352.00	5.34
Total	**$233,492.00**	**29.47**
External failure		
Shipping cost from customers	$31,891.00	4.02
Scrap	$132,431.00	16.71
Customer services	$91,324.00	11.52
Rework	$23,132.00	2.92
Warranty claims	$11,092.00	1.40
Refunds	$21,345.00	2.69
Total	**$311,215.00**	**39.27**
Total cost of nonconformance	**$544,707.00**	**68.74**
Grand total	**$792,438.00**	**100.00**

1. What is the return on investment (ROI)?

2. What would the ROI have been if the total cost of quality were equal to $100,000.00?

3. If the tax rate were 33% of the gross profit, how much would have been spent on quality for each dollar made in profit after tax?

4. How much profit would have been made after tax if the cost of nonconformance were equal to $50,000.00?

Solution:

1. **The Return on Investment (ROI)**

   Total revenue from sales for that quarter = $5,345,987.00

   Total operational expense = $4,232,897.00

   $$\text{ROI} = \frac{\text{total revenue from sales} - \text{total operational expenses}}{\text{total operational expense}}$$

   $$= \frac{\$5,345,987.00 - \$4,232,897.00}{\$4,232,897.00} = \$0.263$$

   In other words, for every dollar invested, the company has made $0.263.

2. **What would the ROI have been if the total cost of quality were equal to $100,000.00?**

   If the total cost of quality were $100,000.00, then the total expenses would have been:

   $4,232,897.00 − ($792,438.00 − $100,000.00) = $3,540,459.00

   Therefore, the ROI would have been:

   $$\text{ROI} = \frac{\$5,345,987.00 - \$3,540,459.00}{\$3,540,459.00} = \$0.510$$

   For every dollar invested, the company would have made $0.510.

3. If the tax rate were 33% of the gross profit, how much would have been spent on quality for each dollar made in profit after tax?

   Profit before tax = $5,345,987.00 − $4,232,897.00 = $1,113,090.00

   The profit after tax = $1,113,090.00 × (1 − 0.33) = $745,770.30

   $$\frac{\$745,770.30}{\$792,438.00} = \$0.941$$

   Therefore, for each dollar invested in quality, only $0.941 was made in after tax profit.

4. How much profit after tax would have been made if the cost of nonconformance were equal to $50,000.00?

If the cost of nonconformance were equal to $50,000.00, the total operational expenses would have been:

$4,232,897.00 – ($544,707.00 – $50,000.00) = $3,738,190.00

The profit before tax would have been

$5,345,987.00 – $3,738,190.00 = $1,607,797.00

The profit after tax would have been

$1,607,797.00 \times (1 - 0.33) = $1,077,223.99

Optimal Cost of Quality

The ideal situation would be to have a production process that would enable every single item produced to meet the predefined CTQ target, but that would be impossible. Conversely, producing poor quality goods would result in customers' attrition. The question that management should ask itself is "What is the optimal cost of quality?"

In the short term, there is a positive correlation between quality improvement and the cost of conformance and a negative correlation between quality improvement and the cost of nonconformance. In other words, an improvement in the quality of the products will lead to an increase in the cost of conformance that generated it. This is because an improvement in the quality level of a product might require extra investment in R&D, more spending in appraisal cost, more investment in failure prevention, and so on.

However, a quality improvement will lead to a decrease in the cost of nonconformance because fewer products will be returned from the customers and, therefore, less operating cost will go to customer support and there will be less internal rework.

For instance, one of the CTQs for a liquid crystal display (LCD) is the number of pixels it contains. The brightness of each pixel is controlled by individual transistors that switch the backlights on and off. The manufacturing of LCDs is very complex and expensive and it is hard to determine the number of dead pixels on an LCD before the end of the manufacturing process. In order to reduce the number of scrapped units, if the number of dead pixels is infinitesimal or the dead pixels are almost invisible, the manufacturer would consider the LCDs as "good enough" to be sold. Otherwise, the cost of scrap or internal rework would be so prohibitive that it would jeopardize the cost of production. Improving the quality level of the LCDs to zero dead pixels would therefore increase the cost of conformance.

On the other hand, not improving the quality level of the LCDs will lead to an increase in the probability of having returned products from customers and internal rework, therefore increasing the cost of nonconformance.

The graph in Fig. 1.10 plots the relationship between quality improvement and the cost of conformance on one hand and the cost of nonconformance on the other hand.

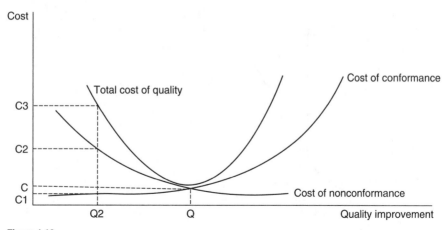

Figure 1.10

If the manufacturer determines the quality level at Q2, the cost of conformance would be low (C1), but the cost of nonconformance would be high (C2) because the probability for customer dissatisfaction will be high and more products will be returned for rework, therefore increasing the cost of rework, customer services, and shipping and handling. The total cost of quality would be the sum the cost of conformance and the cost of nonconformance, which would be C3 for a quality level of Q2.

$$C3 = C1 + C2$$

Should the manufacturer decide that the quality level would be at Q1, the cost of conformance (C2) would be higher than the cost of nonconformance (C1), and the total cost of quality would be at C3 (see Fig. 1.11).

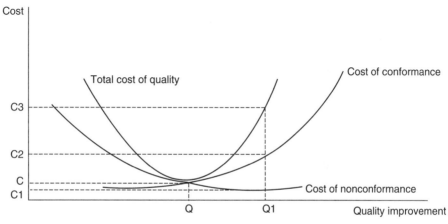

Figure 1.11

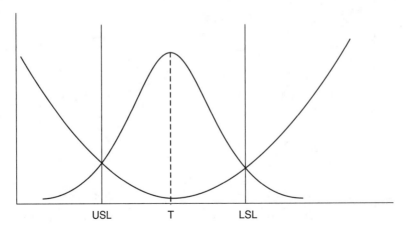

USL   T   LSL

Figure 1.12

The total cost of quality is minimized only when the cost of conformance and the cost of nonconformance are equal.

It is worth noting that, currently, the frequently used graph to represent the throughput yield in manufacturing is the normal curve. For a given target and specified limits, the normal curve helps estimate the volume of defects that should be expected. Therefore, while the normal curve estimates the volume of defects, the U curve estimates the cost incurred because of producing parts that do not match the target.

The graph in Fig. 1.12 represents both the volume of expected conforming and nonconforming parts and the costs associated with them at every level.

Cost of Quality According to Taguchi

In the now traditional quality management acceptance, engineers integrate all the CTQs in the design of their new products and clearly specify the target for their production processes as they define the characteristics of the products to be sent to the customers. However, because of unavoidable common causes of variation (variations that are inherent to the production process and that cannot be eliminated), they allow some variation or tolerance around the target. Any product that falls within the specified tolerance is considered as meeting the customers' expectations, and any product outside the specified limits would be considered as nonconforming.

However, according to Taguchi, the products that do not match the target do not operate as intended even if they are within the specified limits and any deviation from the target, be it within the specified limits or not, will generate financial loss to the customers, the company, and society. The loss incurred is proportional to the deviation from the target.

Suppose that a design engineer specifies the length and diameter of a certain bolt that needs to fit a given part of a machine. Even if the customers do not notice it, any deviation from the specified target will cause the machine to wear out faster, causing the company financial loss under the form of repair of the products under warranty or a loss of customers if the warranty has expired.

Taguchi constructed a loss function equation to determine how much society loses every time the parts produced do not match the specified target. The loss function determines the financial loss that occurs every time a CTQ of a product deviates from its target. The loss function is the square of the deviation multiplied by a constant k, with k being the ratio of the cost of defective product and the square of the tolerance.

The loss function quantifies the deviation from the target and assigns a financial value to the deviation.

$$l(y) = k(y - T)^2$$

where $k = \dfrac{\Delta}{m^2}$

Δ = cost of a defective product

$m = |LSL - T|$

According to Taguchi, the cost of quality in relation to the deviation from the target is not linear because the customers' frustrations increase (at a faster rate) as more defects are found in a product. That is why the loss function is quadratic.

The graph that depicts the financial loss to society that results from a deviation from the target resembles the total cost of quality U graph that we built earlier, but the premises that helped build them are not the same. While the total cost curve was built based on the costs of conformance and nonconformance, Taguchi's loss function is primarily based on the deviation from the target and measures the loss from the customers' expectation perspective (see Fig. 1.13).

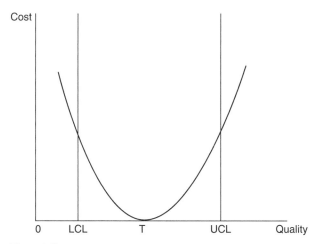

Figure 1.13

Example 1.6   Suppose a machine manufacturer specifies the target for the diameter of a given rivet to be 6 in. and the lower and upper limits to be 5.98 and 6.02 in., respectively. A bolt measuring 5.99 in. is inserted in its intended hole of a machine. Five months after the machine was sold, it breaks down as a result of loose parts. The cost of repair is estimated at $95. Find the loss to society incurred because of the part not matching its target.

Solution: We must first determine the value of the constant k

$$l(y) = k(y - T)^2$$

where $k = \dfrac{\Delta}{m^2}$

$T = 6$

$y = 5.99$

$USL = 6.02$

$m = |USL - T| = 6.02 - 6 = 0.02$

$\Delta = \$95$

$k = (\$95/0.0004) = \$237{,}500$

$(y - T)^2 = (5.99 - 6)^2 = 0.01^2 = 0.0001$

Therefore,

$$l(y) = k(y - T)^2 = \$237{,}500 \times 0.0001 = \$23.75$$

Not producing a bolt that matches the target would have resulted in a financial loss to society that would amount to $23.75.

Stakeholder Analysis

The stakeholders are the people who can affect or can be affected by a project. They can be department managers, customers, suppliers, or anyone among the employees who will contribute to its concretization. But the importance of all the stakeholders in a project is not the same. Therefore, a stratification of the stakeholders according to how they affect or are affected by the project is very important.

The technique used to identify the stakeholders, determine their relative importance for a project, and stratify them is called stakeholder analysis. The benefit of using this technique is that it helps anticipate how the different groups will influence the project and therefore develop the appropriate response strategies to remove obstacles and reduce negative impacts. Stakeholder analysis can be used to identify the most powerful stakeholders to the project and make it easier to anticipate their reactions. Better communication with those stakeholders can facilitate access to resources.

The first step is to determine who the stakeholders are and how they affect the project. This step is better achieved during a brainstorming session. After the list of the stakeholders is agreed upon, they are subdivided into groups according to their domain of interest and how they can benefit or hinder the project (a special tool called force field analysis [FFA] deals with how to approach the negative forces that might be opposed to changes).

The next step will consist of classifying them according to their importance to the project and then assessing the responsibilities and expectations for all (see Table 1.7).

- Assess every stakeholder's positive or negative impact on the project and grade that impact (from 1 to 5 or from A to E, for example).

- Determine what can be done to lessen the negative impacts and improve the positive contribution to the project.

TABLE 1.7

Stakeholders	How do they relate to the project?	How does the project impact them?	How do they impact the project?	How can we maximize the positive impacts on stakeholders?	How can we minimize negative impacts on stakeholders?	Total
Customers	5	5	2	2	5	**19**
Program managers	4	5	5	5	5	**24**
Operations	4	5	5	5	5	**24**
Inventory	3	2	3	4	4	**16**
Suppliers	1	1	2	2	2	**8**
Employees	3	2	2	3	3	**13**
Institutions outside the company	1	1	1	0	2	**5**

The following matrix is an example of a synthesis of a stakeholder analysis.

Once the importance of each stakeholder has been assessed, the next step should be determining the power that each stakeholder has on the project. The powers can range from resource allocations to the power to delay or block the execution of the project. Some stakeholders have high interest and power on the project and some have neither.

A power/interest matrix can be used to map out the interest and power of the stakeholders (see Fig. 1.14).

Force field analysis (FFA)

Qualitative change will always be opposed by restraining forces that are either too comfortable with the status quo or are afraid of the unknown. In an improvement project, identifying those forces in order to assess the risks involved and to better weigh the effectiveness of potential changes becomes an imperative.

The FFA is a managerial tool used for that purpose. FFA is a technique developed by Kurt Lewin (a 20th-century social scientist) as a tool for analyzing

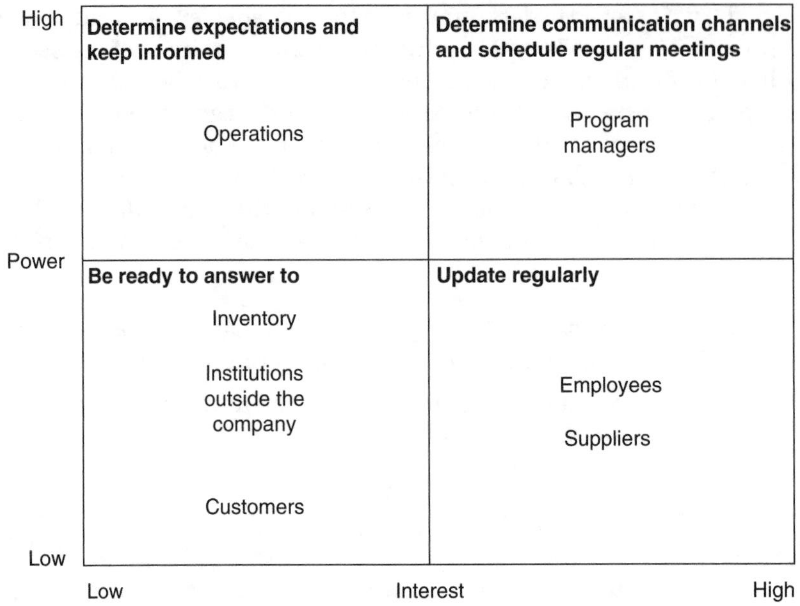

High	**Determine expectations and keep informed** Operations	**Determine communication channels and schedule regular meetings** Program managers
Power	**Be ready to answer to** Inventory Institutions outside the company Customers	**Update regularly** Employees Suppliers
Low		
	Low	Interest High

Figure 1.14

forces opposed to change. It rests on the premise that change is the result of a conflict between opposing forces; in order for it to take place, the driving forces must overcome the restraining forces.

Whenever changes are necessary, FFA can be used to determine the forces that oppose or stimulate the proposed changes. The opposing forces that are closely affected by the changes must be associated with the risk assessment and the decision making. The two groups are charted according to how important they can affect the changes, with the objective of abating the repulsive forces and invigorating the proponents of changes.

To conduct an FFA, a certain number of steps should be taken:

- First, describe the current and the ideal states to analyze how they compare and what will happen if changes are not made.

- Describe the problem to be solved and how to go about it. Brainstorming sessions can be an effective tool for that purpose.

- Identify and divide the stakeholders who are directly implicated in the decision making into two groups, the proponents for the changes and the restraining forces, and then select a facilitator to mend the fences.

- Each group should list the reasons why it is for or against the changes. The listing can be based on questionnaires for or against changes.

- The listing should classify the reasons according to their level of importance; a scale value can be used as a weight for each reason. Some of the issues to be considered are:

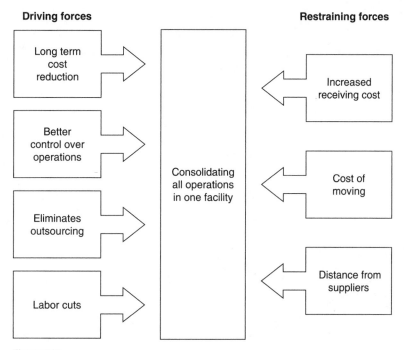

Figure 1.15

   Company's needs

   Cost of the changes

   Company's values

   Social environment (institutions, policies, etc.)

   Company's resources

   How the company usually operates

   Stakeholders' interests

   Stakeholders' attitudes

   The two lists are merged in the same chart to visualize the conflicting forces.

- Question every item on the lists to test their validity and determine how critical they are for the proposed changes.
- Add the scores to determine the feasibility of the changes. If the reasons for a change are overwhelming, take the appropriate course of action by strengthening the forces for change.

Example 1.7 An operation manager has suggested that all the operations of a fictitious company should be consolidated in one facility. The diagram in Fig. 1.15 depicts an example of an FFA.

2

Measure

Data Gathering

One of the first steps in assessing current process performance is data gathering. The process used to collect the data is very important because it ultimately determines their relevance. If wrong data are collected or if the correct data are collected in a wrong way, the results of their analysis would be misleading and would distort the decision-making process. Therefore, a consistent approach must be followed to ensure the reliability of the data. Several important steps need to be followed to guarantee that the collected data accurately reflect the process performance:

1. Clearly define the purpose for the data collection.
2. Decide on the critical to qualities (CTQs) to be considered.
3. Make sure that the measurement process for data collection is accurate and precise.
4. The appraisers must have a common understanding of the operational definitions for the collection process.
5. Decide on the appropriate sampling technique.

Once the data collection has been completed, a data analysis based on the data type is conducted.

Data Types

The more generally used data types are the attribute and the variable data, even though locational data are also used when analyzing data that describe particular positions.

Attribute data

Attribute data are discrete, countable, and can only take finite values. They are categorical data that cannot be broken into fractions. An example of attribute data is the number of defects at the end of a processing line—we cannot have 10.5 defective parts; we either have 10 or 11 defective parts.

Variable data

Variable data are measurable data on a continuum. They can assume any negative or positive value. Variable data measure length, time, distance, etc.

Locational data

Locational data indicate the position of a CTQ of interest. Locational data would, for instance, indicate what area in a manufacturing plant is the source of 95% of the defects.

Basic Probability

Uncertainty is an inherent part of business operations. No matter how well structured an organization is, no matter how good its processes are, it is impossible to predict with absolute certitude the outcome of every decision that is made within it. When an entrepreneur decides to invest in the acquisition of a new business, he does so based on the assumption of the likelihood that the venture would be profitable. Business decision makings are based on the probability of positive outcomes.

So what is probability?

Probability is the chance or the likelihood that something will happen. In statistics, the words chance and likelihood are seldom used to describe the possibilities for an event to take place; instead, the word probability is used along with some other basic concepts the meaning of which is inferred from our everyday use. Probability is the measure of the possibility for an event to take place. It is a number between 0 and 1. If there is a 100% chance that the event will take place, the probability will be 1 and if it is impossible for it to happen, the probability will be 0.

An event is the outcome of an experiment. Determining the number of defects out of a sample of 100 is an experiment and there are many possible events, the possible outcomes can be anywhere between 0 and 100.

An experiment is the study of the process by which an event occurs. An example of an experiment would be the sorting of defective parts from a production line.

Discrete versus continuous distributions

Suppose that an experiment is being conducted to determine the number of items that fail an audit during one shift. The number of outcome can be 0, 1, 2... n items; it ranges from 0 to the number of items produced during that shift.

The numbers 0, 1, 2 . . . n are values of the random variable. The outcome of an experiment is random. A random variable is a variable whose value can change in an experiment. Random variables are divided into two types: discrete random variables and continuous random variables.

Discrete random variables are variables whose values are either countable or finite as in the case of the number of items that fail the audit. Continuous variables can take any value within a continuum. For instance, a bottle of Coke is expected to contain 1 L, but the machine overfilled it and it ended up containing 1.09 L. In the case of the items that failed audit, we could not have 1.05 items that failed audit because the counts were based on increments of 1.

Briefly, discrete variables apply to counts while continuous variables apply to measurements.

A probability distribution shows the possible events and the associated probability for each of these events to occur.

Example 2.1   Table 2.1 depicts the distribution for the probability associated with numbers of defective items taken from a production line.

TABLE 2.1

Number of defective items	Probability
0	0.05
1	0.15
2	0.17
3	0.23
4	0.30
5	0.10

Expected value, variance, and standard deviation of discrete distribution

Measures of location and measures of variability can be derived from the discrete distribution.

Expected value or mean.   The values of the probability that are in Table 2.1 pertain to one trial for each event. If the trials are repeated long enough, chances are that the average of the outcomes will approach the expected value or long-term average.

The mean or expected value for discrete probability is

$$E(X) = \mu = \sum [X.P(X)]$$

where $E(X)$ = long-term average
X = an outcome
$P(X)$ = probability for that outcome

Example 2.2 What is the mean (expected value) for the distribution in Table 2.2?

Solution:

TABLE 2.2

Number of defects X	Probability $P(X)$	$X.P(X)$
0	0.05	0
1	0.15	0.15
2	0.17	0.34
3	0.23	0.69
4	0.30	1.2
5	0.10	0.5
Total	$\sum[X.P(X)]$	**2.88**

Since we are dealing with defective items, we round it up to three items.

Variance and standard deviation. The variance of a discrete distribution is the sum of the product of the squared deviations of each outcome from the mean outcome and the probability for its occurrence.

$$\sigma^2 = \sum[(X-\mu)^2.P(X)]$$

The standard deviation is the square root of the variance

$$\sigma = \sqrt{\sigma^2} = \sqrt{\sum[(X-\mu)^2.P(X)]}$$

Example 2.3 Based on the information in Table 2.3, find the standard deviation for the distribution.

Solution:

TABLE 2.3

Number of defects	Probability	$X.P(X)$	$(X-\mu)^2$	$(X-\mu)^2 P(X)$
0	0.05	0	$(0-2.88)^2 = 8.2944$	0.41472
1	0.15	0.15	$(1-2.88)^2 = 3.5344$	0.53016
2	0.17	0.34	$(2-2.88)^2 = 0.7744$	0.131648
3	0.23	0.69	$(3-2.88)^2 = 0.0144$	0.003312
4	0.3	1.2	$(4-2.88)^2 = 1.2544$	0.37632
5	0.1	0.5	$(5-2.88)^2 = 4.4944$	0.44944
		2.88	**18.366**	**1.906**

The variance is 1.906, and the standard deviation is $\sigma = \sqrt{1.906} = 1.380$.

Discrete probability distributions

A distribution is said to be discrete if it is built on discrete random variables. The four most used discrete probability distributions in business operations are the binomial, the Poisson, the geometric, and the hyper-geometric distributions.

Binomial distribution. The binomial distribution is one of the simplest probability distributions. It assumes an experiment with n identical trials, with each trial having only two possible outcomes considered as success or failure. Each trial is independent of the previous ones. The independence of the trials means that the outcome for each trial is the same as the rest of the trials.

For the remainder of this section, p will be considered as the probability for a success and q as the probability for a failure.

$$q = (1 - p)$$

The formula for a binomial distribution is as follows:

$$P(x) = C_x^n p^x q^{n-x}$$

where $P(x)$ is the probability for the event x to happen. x may take any value from 0 to n and

$$C_x^n = \frac{n!}{x!(n-x)!}$$

The mean, variance, and standard deviation for a binomial distribution are

$$\mu = E(x) = np$$

$$\sigma^2 = npq$$

$$\sigma = \sqrt{\sigma^2} = \sqrt{npq}$$

Example 2.4 The probability that a container is shipped to customers half-empty is 0.057. Twenty containers have been shipped this morning. What is the probability that five of them are half-empty?

Solution:

$$n = 20$$

$$x = 5$$

$$p = 0.057$$

$$q = 1 - 0.057 = 0.943$$

$$P(x = 5) = C_x^n p^x q^{n-x} = C_5^{20}(0.057)^5(0.943)^{15}$$

where

$$C_x^n = \frac{n!}{x!(n-x)!} = \frac{20!}{5!(15!)} = 15504$$

and

$$p^x = 0.057^5 = 6.017\text{E-}07$$

$$q^{n-x} = 0.943^{15} = 0.415$$

Therefore,

$$P(x=5) = C_x^n p^x q^{n-x} = 0.004$$

The probability that five containers are half-empty is 0.004.

Using Excel   From the Excel tool bar, click on the *Insert Function* button. When the **Insert Function** box appears, select Statistical for the Or select a category: option and select BINOMDIST before clicking on the **OK** button. Fill out the **Function Arguments** box as indicated in Fig. 2.1.

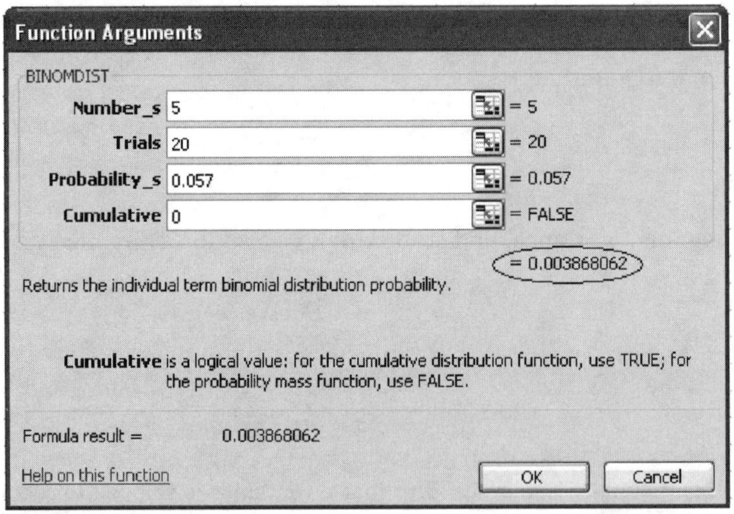

Figure 2.1

Example 2.5   An operations manager has 16 production lines that produce the exact same type of product. The production lines are supposed to be operational at 6:00 AM every day. The probability for a machine on a line to go down is 0.15. The operations manager wants to know how many lines will be operational at the start of the shift. He also wants to have the distribution for the probability of the lines not operating.

Solution: The probability for a machine to *not* go down is $1 - 0.15 = 0.85$.

$$\mu = E(X) = np = 16 \times 0.85 = 13.6$$

Since we cannot have 13.6 lines, we round that number up to 14. He should expect to have 14 lines operational.

$$\text{Variance} = npq = 16(0.85)(0.15) = 2.04$$

$$\text{Standard deviation} = \sqrt{npq} = \sqrt{2.04} = 1.428$$

To obtain the distribution, we can use Minitab or Excel.

Using Minitab   Open the file *ops line.MTW* and from the menu bar, click on **Calc,** then click on **Probability Distributions** and then on **Binomial**. When the **Binomial Distribution** box appears, fill it out as shown in Table 2.4, and then press on the **OK** button. Table 2.4 shows the probability distribution for the lines to go down.

TABLE 2.4   Probability Distribution for Lines to Go Down

Lines down	Probability
1	0.210
2	0.277
3	0.229
4	0.131
5	0.056
6	0.018
7	0.005
8	0.001
9	0.000
10	0.000
11	0.000
12	0.000
13	0.000
14	0.000
15	0.000
16	0.000

Example 2.6   The probability for a shipment to arrive late from a warehouse is 0.25 and 25 shipments have been sent from the warehouse. What is the probability that only five will arrive on time?

Solution: The probability that only five will arrive on time is the same as the probability that 20 arrived late.

$$n = 25$$

$$x = 20$$

$$p = 0.25$$

$$q = 1 - 0.25 = 0.75$$

$$P(x = 20) = C_x^n p^x q^{n-x} = C_{20}^{25} (0.25)^{20} (0.75)^5$$

$$C_x^n = \frac{n!}{x!(n-x)!} = \frac{25!}{20!(5!)} = 53130$$

$$C_x^n p^x q^{n-x} = C_{20}^{25} (0.25)^{20} (0.75)^5 = 53130(9.095E - 13)(0.237) \approx 0$$

The probability for having only five shipments arriving on time is equal to 0. Excel gives Fig. 2.2.

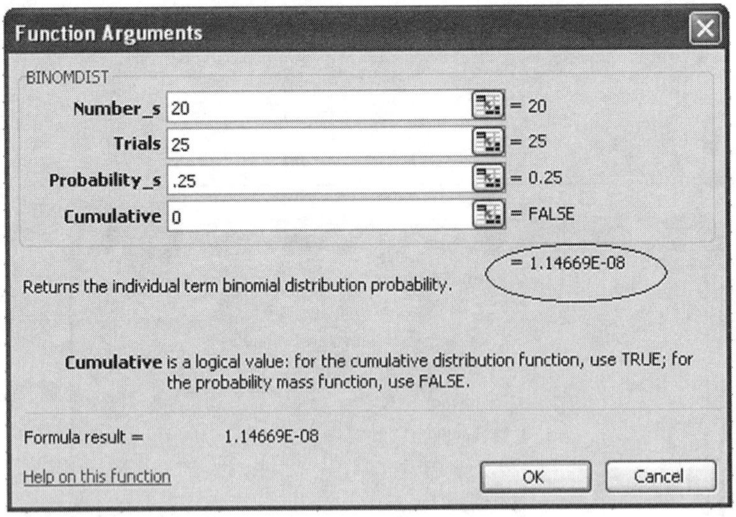

Figure 2.2

Example 2.7   Of the products sold, customers have returned 5.5%. A survey is being conducted over the phone to determine customer satisfaction. What is the probability of coming across four dissatisfied customers out of the first 15 customers called?

Solution:

$$n = 15$$

$$x = 4$$

$$p = 0.055$$

$$q = 1 - 0.055 = 0.945$$

$$P(x = 4) = C_4^{15}(0.055)^4(0.945)^{11}$$

$$C_x^n = \frac{n!}{x!(n-x)!} = \frac{15!}{4!(11!)} = 1365$$

$$P(x = 4) = C_x^n p^x q^{n-x} = C_4^{15}(0.055)^4(0.945)^{11} = 1365(0.000009151)(0.537) = 0.007$$

Minitab output

Binomial with n = 15 and p = 0.055

x P(X = x)
4 0.0067040

Hyper-geometric distribution.   The hyper-geometric distribution is very close to the binomial distribution. One of the conditions of a binomial distribution is the independence of the trials, which made the probability of a success to be the same for every trial. If successive trials are done without replacement and the sample size or population is small, the probability for each observation will vary. If a sample has 10 stones, the probability for taking a particular stone out of the 10 will be 1/10. If that stone is not replaced into the sample, the probability of taking another one will be 1/9. However, if the stones are replaced each time, the probability of taking a particular stone will remain the same, 1/10.

When the sampling is finite (relatively small and known) and the outcome changes from trial to trial, the hyper-geometric distribution is used instead of the binomial distribution.

The hyper-geometric distribution assumes

- A finite population of N

- Two outcomes described as success and failure with k successes in the population

- Samples of n items are taken without replacement

The probability distribution depends therefore on k, n, x, and N. The formula for the hyper-geometric distribution is as follows:

$$P(x) = \frac{C_x^k C_{n-x}^{N-k}}{C_n^N}$$

where x is an integer whose value is between 0 and n.

$$x \leq k$$

$$\mu = n\left(\frac{k}{N}\right)$$

$$\sigma^2 = n\left(\frac{k}{N}\right)\left(1 - \frac{k}{N}\right)\left(\frac{N-n}{N-1}\right)$$

Example 2.8   Of the 20 employees at HR, five are men. If three HR employees are randomly selected for a survey, what is the probability that two or more will be men?

Solution:

$$N = 20$$

$$n = 3$$

$$k = 5$$

$$x \geq 2$$

We have to find the probability for two selected employees to be men and the probability for three to be men.

Probability for selecting two men

$$P(x = 2) = \frac{C_2^5 C_{3-2}^{20-5}}{C_3^{20}} = \frac{C_2^5 C_1^{15}}{C_3^{20}} = \frac{10(15)}{1140} = 0.132$$

Minitab output

Probability Density Function
Hypergeometric with N = 20, M = 3, and n = 5
x P(X = x)
2 0.131579

The probability for selecting two men is 0.132.

Probability for selecting three men

$$P(x = 3) = \frac{C_3^5 C_{3-3}^{20-5}}{C_3^{20}} = \frac{C_3^5 C_0^{15}}{C_3^{20}} = \frac{10(1)}{1140} = 0.009$$

Minitab output

Probability Density Function
Hypergeometric with N = 20, M = 3, and n = 5
x P(X = x)
3 0.0087719

The probability for selecting three men is 0.009.

The probability for selecting two men or more out of the three selected is

$$0.132 + 0.009 = 0.141$$

Example 2.9 A veterinarian artificially inseminates 7 cows out of a lot of 21. He comes the next day to inseminate five more and realizes that his assistant has inadvertently released the seven inseminated cows back with the rest of the cows. What is the probability that out of the five that he has selected today, three have already been inseminated?

Solution:

$$N = 21$$

$$n = 7$$

$$k = 5$$

$$x = 3$$

$$P(x) = \frac{C_x^k C_{n-x}^{N-k}}{C_n^N}$$

$$P(x = 3) = \frac{C_3^5 C_{7-3}^{21-5}}{C_7^{21}} = \frac{C_3^5 C_4^{16}}{C_7^{21}} = \frac{10(1820)}{116280} = 0.157$$

See Excel output for Fig. 2.3.

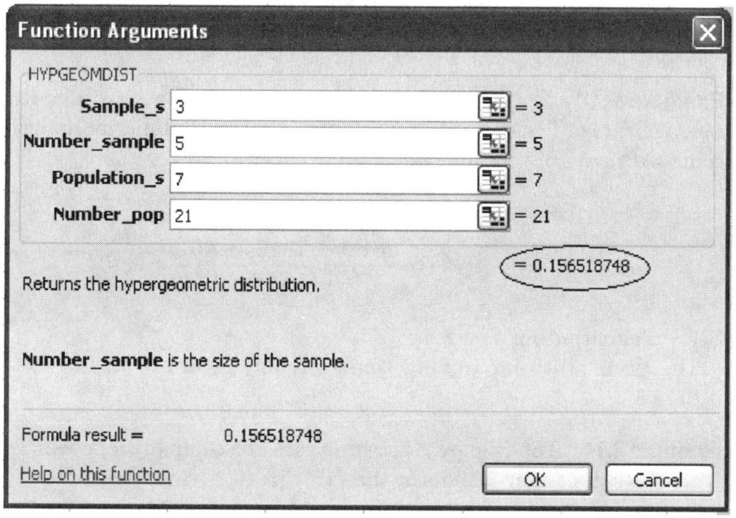

Function Arguments		
HYPGEOMDIST		
Sample_s	3	= 3
Number_sample	5	= 5
Population_s	7	= 7
Number_pop	21	= 21

= 0.156518748

Returns the hypergeometric distribution.

Number_sample is the size of the sample.

Formula result =   0.156518748

Help on this function

OK   Cancel

Figure 2.3

The probability that out of the five that he has inseminated today, three have already been inseminated is 0.157.

Poisson distribution.   The binomial and the hyper-geometric distributions are used to calculate the probability for one outcome out of two to occur. Not all situations exhibit only two alternatives. For instance, an engineer might want to know the number of defects on a machine; a doctor might be interested in the amount of sugar in the blood of a patient with diabetes, while a quality engineer may be interested in the number of calls received at a call center within a certain time frame. These examples do not state a number of trials nor do they state the number of alternative outcomes. They only describe the occurrence of an event within a time frame, a distance, an area, or a volume. These types of cases call for a Poisson distribution. The Poisson distribution applies to a situation that can be described by a discrete random variable that takes on integers (whole numbers like 1, 2, 3, etc.) with the events occurring at a known average rate.

The density function of the Poisson distribution is

$$P(x) = \frac{\mu^x e^{-\mu}}{x!}$$

where $P(x)$ is the probability for the event x to happen

μ is the arithmetic mean for the number of occurrences in a particular interval

e is the constant 2.718282

The mean and the variance of the Poisson distribution are the same and the standard deviation is the square root of the variance.

$$\mu = \sigma^2$$

$$\sigma = \sqrt{\mu} = \sqrt{\sigma^2}$$

Example 2.10   Employees come to pick up orders at an inventory location at an average rate of five every 15 min. What is the probability of having seven employees come within a 15-min interval?

Solution:

$$P(7) = \frac{5^7 e^{-5}}{7!} = \frac{78125(0.007)}{5040} = 0.104$$

See Excel output for Fig. 2.4.

The probability for having seven employees come within a 15-min interval is 0.104.

Example 2.11   The employee attrition rate at a company follows a Poisson distribution with a mean of four a month; the HR director wants to know the probability that between five and seven employees would leave the company next month.

Solution: The probability that between five and seven employees would leave the company next month will be the sum of the probabilities for having five, six, and seven employees leave.

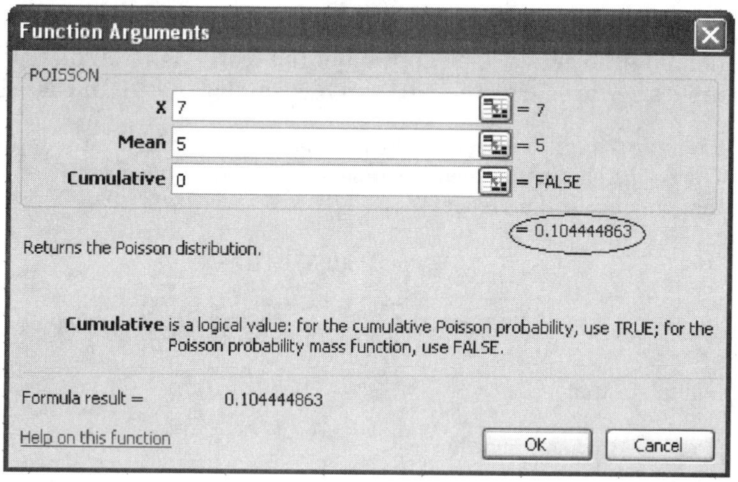

Figure 2.4

The probability of five employees leaving is

$$P(5) = \frac{4^5 e^{-4}}{5!} = \frac{1024(0.018)}{120} = 0.156$$

The probability of six employees leaving is

$$P(6) = \frac{4^6 e^{-4}}{6!} = \frac{4096(0.018)}{720} = 0.104$$

The probability of seven employees leaving is

$$P(7) = \frac{4^7 e^{-4}}{7!} = \frac{16384(0.018)}{5040} = 0.060$$

Minitab output

Probability Density Function
Poisson with mean = 4
x P(X = x)
5   0.156293
6   0.104196
7   0.059540

The probability of between five and seven employees leaving the company next month is equal to 0.156 + 0.104 + 0.06 = 0.32.

Approximating binomial problems by Poisson distribution

Binomial problems can be approximated by the Poisson distribution when the sample sizes are large $(n > 20)$ and p is small $(np \leq 7)$. In that case, $\mu = np$.

Example 2.12 The probability of receiving material from suppliers late is 0.05. Thirty-five shipments have been sent from the warehouse. What is the probability that only five will arrive late?

$$\mu = np = 35(0.05) = 1.75$$

$$P(5) = \frac{1.75^5 e^{-1.75}}{5!} = \frac{16.413(0.174)}{120} = 0.024$$

Minitab output

Probability Density Function
Poisson with mean = 1.75
x P(X = x)
5   0.0237681

The probability that only five will arrive late is 0.024.

Continuous distribution

A continuous random variable is a variable whose set of possible values is a whole interval of numbers and those values are generated from measurements as opposed to counts. The most widely used continuous distribution is the normal distribution.

Normal distribution.   The normal distribution is certainly the most important probability distribution because it is the most extensively used as the basis for inferential statistics.

Most of nature and human characteristics are normally distributed, and so are most production outputs. For instance, the height of adult male human beings is said to be normally distributed because the average adult male is about 5 ft and 9 in. tall and a significant percentage of adult males are close to that height. The characteristics of a production output of a manufacturing process are also generally normally distributed. Not all the items that come off a production process may be identical, but a high percentage of them are expected to be very close to the mean in order to have a stable and predictable process. When a population is normally distributed, most of the components of that population are very closely clustered around the mean. A characteristic of a population is said to be normally distributed if 68.27% of the population are within one standard deviation away from the mean, 95.47% are within two standard deviations away from the mean, and 99.73% are within three standard deviations away from the population mean.

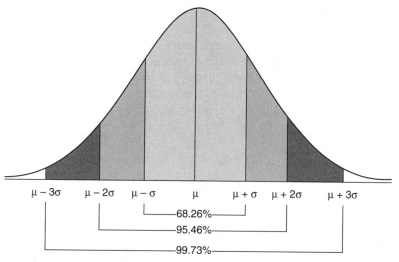

Figure 2.5

Normality in a production process is very important because it enables the producer to make predictions on future performances (Fig. 2.5).

The normal probability density function is expressed as

$$f(x) = \frac{e^{-\frac{(x-\mu)^2}{2\sigma^2}}}{\sigma\sqrt{2\pi}}$$

where $e \approx 2.7182828$ and $\pi \approx 3.1416$.

The curve associated with that function is bell shaped, has an apex at the center, and is symmetrical about the mean. The two tails of the curve extend indefinitely without ever touching the horizontal line (Fig. 2.6).

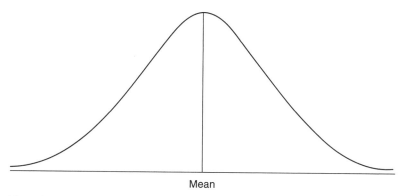

Figure 2.6

The shape of the curve depends only on two parameters, μ and σ. Every unique combination of μ and σ, $N(\mu, \sigma)$, is associated with a unique curve. If the standard deviation changes, the shape of the curve changes; if the mean changes, the whole curve shifts to the right or to the left while its shape remains the same.

Figure 2.7 shows that the mean has shifted from 4 to 8, and the curve has drifted while its shape remains the same.

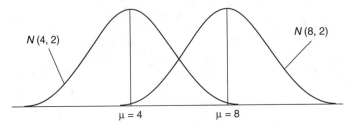

Figure 2.7

Figure 2.8 shows that the standard deviation has changed from 2 to 1, while the mean remained the same. The shape of the curve has changed as a result.

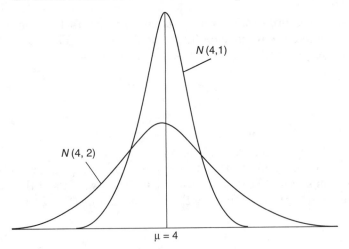

Figure 2.8

The area under the curve between a and b on Fig. 2.9 represents the probability that a random variable assumes a value within that interval.

Z transformation

The complexity associated with the computations using the normal probability density function makes its use very tedious, but fortunately, density functions can be converted using the Z distribution. The conversion of the normal probability generates a standardized Z distribution.

$$Z = \frac{X - \mu}{\sigma}$$

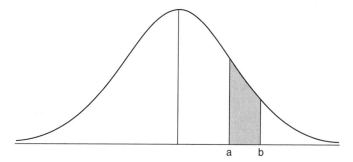

Figure 2.9

The standardized Z distribution also depends on the same two parameters as the normal density function, $N(\mu, \sigma)$. The Z distribution generates a Z score, which measures the number of standard deviations that X is away from the mean since the numerator is nothing but the deviation of X from the mean μ. In other words, the Z score indicates by how much X is to the left (when Z is negative) or to the right (when Z is positive) of the mean in terms of standard deviations.

The value of the Z score is used to locate the probability associated with it on the Z distribution table.

Example 2.13   The weekly number of turbo chargers ordered from a manufacturer is normally distributed with a mean of 170 and a standard deviation of 10. What is the probability that less than 175 turbo chargers will be ordered next week? What is the probability that more than 175 will be ordered?

Solution:

$$X = 175, \quad \mu = 170, \quad \sigma = 10$$

$$Z = \frac{X - \mu}{\sigma} = \frac{175 - 170}{10} = 0.5$$

From the Z score table in Table 2.5, 0.5 corresponds to 0.1915.

The probability that less than 175 turbo chargers will be ordered is equal to $0.5 + 0.1915 = 0.6915$ (see Fig. 2.10).

The probability that more than 175 will be ordered is $1 - 0.6915 = 0.3085$.

See the Excel output in Fig. 2.11.

Minitab output

```
Cumulative Distribution Function

Normal with mean = 170 and standard deviation = 10

  x  P( X <= x )
175    0.691462
```

TABLE 2.5

Z	.00	.01	.02	.03	.04	.05	.06	.07	.08	.09
0.0	.0000	.0040	.0080	.0120	.0160	.0199	.0239	.0279	.0319	.0359
0.1	.0398	.0438	.0478	.0517	.0557	.0596	.0636	.0675	.0714	.0753
0.2	.0793	.0832	.0871	.0910	.0948	.0987	.1026	.1064	.1103	.1141
0.3	.1179	.1217	.1255	.1293	.1331	.1368	.1406	.1443	.1480	.1517
0.4	.1554	.1591	.1628	.1664	.1700	.1736	.1772	.1808	.1844	.1879
0.5	.1915	.1950	.1985	.2019	.2054	.2088	.2123	.2157	.2190	.2224
0.6	.2257	.2291	.2324	.2357	.2389	.2422	.2454	.2486	.2517	.2549
0.7	.2580	.2611	.2642	.2673	.2704	.2734	.2764	.2794	.2823	.2852
0.8	.2881	.2910	.2939	.2967	.2995	.3023	.3051	.3078	.3106	.3133
0.9	.3159	.3186	.3212	.3238	.3264	.3289	.3315	.3340	.3365	.3389
1.0	.3413	.3438	.3461	.3485	.3508	.3531	.3554	.3577	.3599	.3621
1.1	.3643	.3665	.3686	.3708	.3729	.3749	.3770	.3790	.3810	.3830
1.2	.3849	.3869	.3888	.3907	.3925	.3944	.3962	.3980	.3997	.4015
1.3	.4032	.4049	.4066	.4082	.4099	.4115	.4131	.4147	.4162	.4177
1.4	.4192	.4207	.4222	.4236	.4251	.4265	.4279	.4292	.4306	.4319
1.5	.4332	.4345	.4357	.4370	.4382	.4394	.4406	.4418	.4429	.4441
1.6	.4452	.4463	.4474	.4484	.4495	.4505	.4515	.4525	.4535	.4545
1.7	.4554	.4564	.4573	.4582	.4591	.4599	.4608	.4616	.4625	.4633

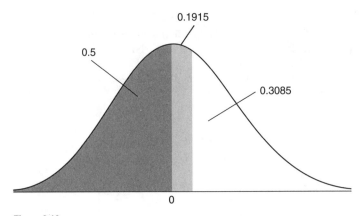

Figure 2.10

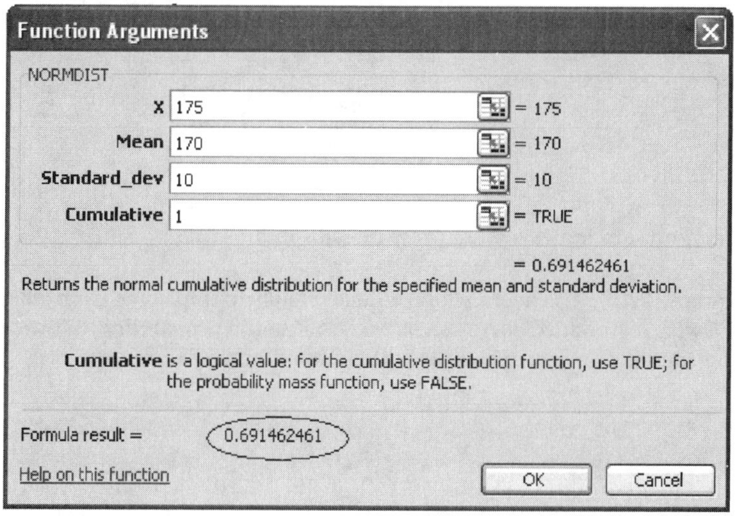

Figure 2.11

The probability that less than 175 turbo chargers will be ordered next week is 0.5 + 0.1915 = 0.6915.

Example 2.14   A manufacturer wants to set a minimum life expectancy on a newly manufactured light bulb. A test has revealed a mean $\mu = 250$ h and a standard deviation $\sigma = 15$. The production of light bulbs is normally distributed. The manufacturer wants to set the minimum life expectancy of the light bulbs so that less than 5% of the bulbs will have to be replaced. What minimum life expectancy should be put on the light bulb labels?

Solution: In Fig. 2.12, the area shaded under the curve between X and the end of the tail represents the 5% (or 0.05) of the light bulbs that might need to be replaced. The area between X and the mean μ (250) represents the 95% of good bulbs.

To find Z, we need to deduct 0.05 from 0.5 (0.5 represents half of the area under the curve)

$$0.5 - 0.05 = 0.45$$

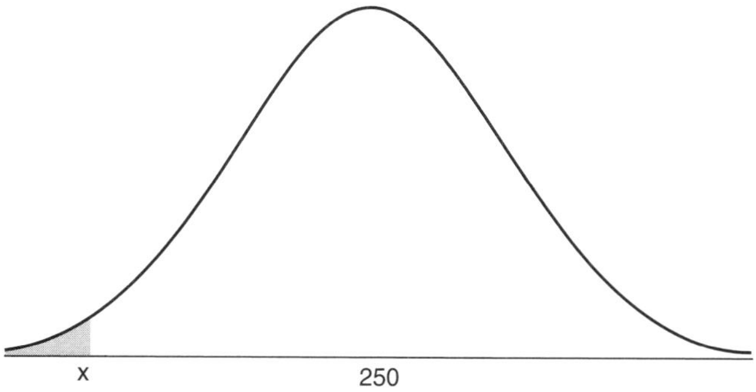

Figure 2.12

0.45 corresponds to 1.645 on the Z table. Because the value is to the left of μ,

$$Z = -1.645$$

$$Z = \frac{X - 250}{15} = -1.645$$

$$X = 225.325$$

The minimum life expectancy for the light bulb is 225 h.

Example 2.15   The average of the defective parts that come from a production line is 10.5 with a standard deviation of 2.5. What is the probability that the defective parts for a randomly selected sample will be less than 15?

Solution:

$$Z = \frac{15 - 10.5}{2.5} = 1.8$$

1.8 corresponds to 0.4641 on the Z table.

So the probability that the defective parts will be less than 15 is 0.9641 (0.5 + 0.4641) (see Fig. 2.13).

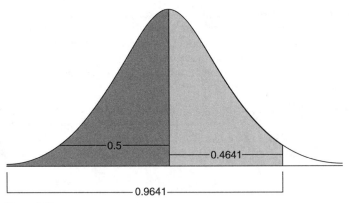

Figure 2.13

Exponential distribution.   The exponential distribution is close to the Poisson distribution. The Poisson distribution is built on discrete random variables and describes random occurrences over some intervals, while the exponential distribution is continuous and is used to determine the probability for events that occur at a constant average rate to happen between points of time.

Examples of an exponential distribution are the time between machine breakdowns, the time between inventory replenishment, and the waiting time on a line at a supermarket.

The exponential distribution is determined by the following formula:

$$P(x) = \lambda e^{-\lambda x}$$

The mean and the standard deviation are

$$\mu = \frac{1}{\lambda} \quad \text{and} \quad \sigma = \frac{1}{\lambda}$$

The shape of the exponential distribution is determined by only one parameter λ. Each value of λ determines a different shape of the curve. Figure 2.14 shows the graph of exponential distributions.

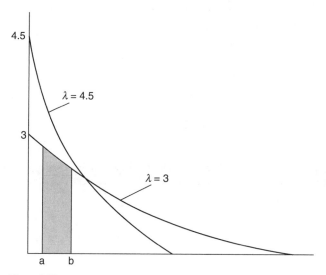

Figure 2.14

The area under the curve between two points determines the probabilities for an event to occur within that interval. The following formula can be used to calculate that probability.

$$P(x \geq a) = e^{-\lambda a} \quad \text{with} \quad a \geq 0$$

If the number of events taking place in an interval of time follows a Poisson distribution with a mean λ, then the interval between these events are exponentially distributed with the mean interval time equal to $\mu = 1/\lambda$.

Example 2.16   If the number of items arriving at inspection at the end of a production line follows a Poisson distribution with a mean of 10 per hour, then the arrivals follow an exponential distribution with a mean between arrival times of 6.

$$\mu = \frac{1}{\lambda} = \frac{1}{10} = 0.10$$

$$1 \text{ hour} = 60 \text{ min}$$

$$\mu = 0.10(60) = 6 \text{ min}$$

Example 2.17   Suppose that the time in months between line stoppages on a production line follows an exponential distribution with $\lambda = 0.5$.

1. What is the probability that the time until the line stops again will be more than 15 months?
2. What is the probability that the time until the line stops again will be less than 20 months?
3. What is the probability that the time until the line stops again will be between 10 and 15 months?
4. Find μ and σ. Find the probability that the time until the line stops will be between $(\mu - 3\sigma)$ and $(\mu + 3\sigma)$.

Solution:

1. $P(x > 15) = e^{-15\lambda} = e^{-15(0.5)} = 0.001$

   The probability that the time until the line stops again will be more than 15 months is 0.001.

   See the Excel outputs in Fig. 2.15.

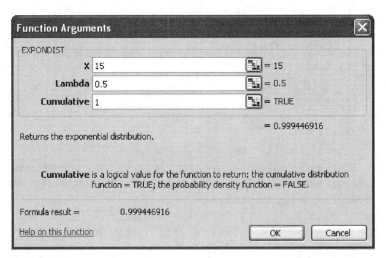

Figure 2.15

Excel gives the probability for "less than," what we were looking for is the probability for "more than"; therefore, we have to subtract the Excel results from 1. Therefore, $1 - 0.999 = 0.001$.

2. $P(x < 20) = 1 - P(x > 20) = 1 - e^{-20(0.5)} = 1 - e^{-10} \approx 1 - 0.000 \approx 1$

   The probability that the time until the line stops again will be less than 20 months is 1.

Using Minitab   From the menu bar, click on **Calc**, then select **Probability Distributions** and the from the menu list, select **Exponential**. Fill out the **Exponential**

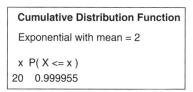

Figure 2.16

box as indicated in Fig. 2.16. Notice that we have "2" for **Scale**; this is because we choose to use the mean and set the threshold to 0. The mean is

$$\mu = \frac{1}{\lambda} = \frac{1}{0.5} = 2$$

Minitab output

Cumulative Distribution Function
Exponential with mean = 2
x P(X <= x)
20   0.999955

3. We have already found that $P(x > 15) = e^{-15\lambda} = e^{-15(0.5)} = 0.001$. We need to find the probability that the time until the line stops again will be more than 10 months.

$$P(x > 10) = e^{-10(0.5)} = e^{-5} = 0.007$$

The probability that the time until the line stops again will be between 10 and 15 months is the difference between 0.007 and 0.001.

$$P(10 < x < 15) = 0.007 - 0.001 = 0.006$$

4. The mean and the standard deviation are given by

$$\mu = \sigma = \frac{1}{\lambda} = \frac{1}{0.5} = 2$$

Therefore,

$$(\mu - 3\sigma) = 2 - 6 = -4$$

$$(\mu + 3\sigma) = 2 + 6 = 8$$

Therefore, we need to find $P(-4 \leq x \leq 8)$ which is equal to $P(0 \leq x \leq 8)$

$$P(0 \leq x \leq 8) = 1 - P(x \geq 8)$$

Therefore,

$$P(-4 \leq x \leq 8) = 1 - e^{-8(0.5)} = 1 - 0.018 = 0.982$$

The probability that the time until the line stops again will be between $(\mu - 3\sigma)$ and $(\mu + 3\sigma)$ is 0.982.

Minitab output

Cumulative Distribution Function
Exponential with mean = 2
x P(X <= x)
8   0.981684

Planning for Sampling

When studying data that pertain to mass production, it is often impossible to test every single produced item because of the cost, time, and space required for the testing; consequently, the appraisers usually take subsets of production yield over time and proceed with the analysis. Based on the results that they obtain and some statistics rules, they make an inference about the whole production process. For instance, a manufacturer of soft drinks produces millions of bottles every day; if every bottle had to be tested for quality, the cost of quality would be too exorbitant; therefore, samples are taken at preset time intervals for testing and the results of the tests are used for decision making for the whole production process. If the processes' yield is normally distributed and the processes are stable and under control, sample statistics can be used to interpret the production parameters.

The same idea is used when conducting a survey about a product or a situation or for polling a population to predict the results of an election. It is often impossible to consult every person of interest about the study and when it is possible to consult every person, the study is no longer a survey but a referendum.

The process used for sampling needs to be carefully decided based on the objective of the sampling, the cost associated with it, and the nature of the products being sampled.

Random Sampling versus Nonrandom Sampling

Random sampling and nonrandom sampling are the two main types of sampling. A random sampling is done in such a way that every single item in the population from which the sample is taken has the same chance of being considered, while in a nonrandom sampling the appraiser is deliberately selective about which item to favor for the study. Not all the items have the same probability for being selected. For instance, when auditing products in a given inventory location, if only products whose part numbers end with a given suffix among all homogeneous products are considered, the auditor would be performing a nonrandom sampling because only some parts have a chance of being selected.

Random sampling

Stratified sampling. Stratified sampling is a sampling process where the population from which the subsets are taken is divided into homogeneous strata and the experimenter takes random samples from each stratum. Stratified sampling is done to reduce the probability of making mistakes because each sample is taken from a subset of the population.

Example 2.18 Suppose that an audit of an inventory of products is being conducted and the inventory is spread over five different location aisles. If each aisle is considered as a stratum, then the samples of five items taken from each aisle are analyzed separately and the inference made from the results would only apply to the aisle from which they came (see Fig. 2.17).

Stratified sampling can be proportionate or disproportionate. It is proportionate when the samples taken from each stratum are equal as in the example above and it is disproportionate when the samples are unequal.

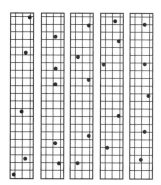

Figure 2.17

Systematic sampling.   In systematic sampling, the sample size n is predefined and every kth item is selected for sampling with k being predefined as well. The objective of systematic sampling is not to reduce the opportunities for error but to simplify the sampling process. For a population N with sample size n,

$$k = \frac{N}{n}$$

Example 2.19   A quality assessor wants to build control charts to monitor the quality of circuit board. He decides to select two boards every 20 min for inspection.

Cluster sampling.   In stratified sampling, the strata are homogeneous, i.e., the items within each block are identical. In cluster sampling, the appraiser divides the population being studied into strata as in the case of the stratified sampling, but in this case, the blocks called "clusters" are not homogeneous. The items within the clusters are not all identical. If the population being studied is made up of a wide variety of items, each cluster should be built in such a way that it mirrors the diversity of the population. After the clusters have been built, the appraiser randomly makes selections from each cluster to build his samples.

Nonrandom sampling

Nonrandom sampling is a technique of sampling that does not ensure that all the items in the population being sampled have the same probability of being selected. Convenient sampling and judgment sampling are examples of nonrandom sampling

Convenience sampling.   In convenience sampling, the appraiser selects the items based on their handiness or their willingness to participate in the study. If a researcher wants to test some items and only a few are available, he can choose to do so with those that are accessible.

Judgment sampling.   Based on experience, a researcher can know exactly what items in a population can make his experiment easy and select only those items.

Nonsampling errors

Nonsampling errors are errors that are not because the samples are not representative of the population. Nonsampling errors can be calculation errors or data collection errors.

Sampling error

The purpose of sampling is to study a subset of a population and use the results of the study to make an inference about the population. A sampling error happens when the subsets taken for the study are not representative of the entire population. For instance, suppose a population of 15 gaskets has thickness measures of 5.00, 5.11, 5.12, 5.12, 5.14, 5.10, 5.09, 5.08, 5.07, 5.90, 5.00, 5.10, 5.00, 5.12, and 6.00. The mean μ for that population would be 5.197. If a sample of the following three measures, 5.11, 5.12, and 5.10, is taken from the population, the mean of the sample would be

$$\overline{X} = \frac{5.11 + 5.12 + 5.10}{3} = 5.11$$

and the sampling error would be

$$E = \overline{X} - \mu = 5.11 - 5.197 = -0.087$$

Let us take another sample of three measures: 6.00, 5.10, and 5.11. This time the sample mean will be 5.403 and the sampling error will be $E = \overline{X} - \mu = 5.403 - 5.197 = 0.206$.

If another sample is taken and estimated, its sampling error might be different. These differences are said to be due to chance.

Therefore, if it is possible to make mistakes while estimating the population's parameters from a sample statistic, how can we be sure that sampling can help us obtain a good estimate? Why use sampling as a mean of estimating the population parameters?

The *central limit theorem* can help us answer these questions.

Central Limit Theorem

The central limit theorem states that for sufficiently large sample sizes ($n \geq 30$), regardless of the type of the distribution of the population, if samples of size $\boldsymbol{n}$ are randomly drawn from a population that has a mean μ and a standard deviation σ, the samples' means $\overline{X}$ are approximately normally distributed. If the population is normally distributed, the samples' means are normally distributed regardless of the sample sizes.

$$\mu_{\overline{x}} = \mu$$

$$\sigma_{\overline{x}} = \frac{\sigma}{\sqrt{n}}$$

where $\mu_{\overline{x}}$ is the mean of the samples' mean and $\sigma_{\overline{x}}$ is the standard deviation of the samples' means.

The implication of this theorem is that for sufficiently large populations, the normal distribution can be used to analyze samples drawn from populations that are not normally distributed or whose shapes are unknown.

When means are used as estimators to make inferences about a population's parameters, and $(n \geq 30)$, then the estimator will be approximately normally distributed in repeated sampling.

Sampling distribution of the mean $\overline{X}$

We have seen in the example of the gasket thickness that the mean of the first sample was 5.11; the mean of the second was 5.403. If the means of all possible samples are obtained and organized, we could derive the *sampling distribution of the means*.

In that example we had 15 gaskets. If all possible samples of three were to be computed, there would be 455 samples and means.

$$C_n^N = \frac{N!}{n!(N-n)!} = \frac{15!}{3!(15-3)!} = 455$$

The mean and standard deviation of that sampling distribution are given as

$$\mu_{\overline{x}} = \mu$$

$$\sigma_{\overline{x}} = \frac{\sigma}{\sqrt{n}}$$

Example 2.20   Gajaga Electronics is a company that manufactures circuit boards. The average imperfection on a board is $\mu = 5$ with a standard deviation of $\sigma = 2.34$ when the production process is under control.

A random sample of $n = 36$ circuit boards has been taken for inspection and a mean of $\overline{x} = 6$ defects per board was found. What is the probability of getting a value of $\overline{x} \leq 6$ if the process is under control?

Solution: Since the sample size is greater than 30, the central limit theorem can be used in this case even though the number of defects per board follows a Poisson distribution. Therefore, the distribution of the sample mean $\overline{x}$ is approximately normal with the standard deviation

$$\sigma_{\overline{x}} = \frac{\sigma}{\sqrt{n}} = \frac{2.34}{\sqrt{36}} 0.39$$

Therefore,

$$z = \frac{\overline{x} - \mu}{\sigma / \sqrt{n}} = \frac{6-5}{0.39} = \frac{1}{0.39} = 2.56$$

$z = 2.56$ corresponds to 0.4948 on the Z score table. The probability of getting a value of $\overline{x} \leq 6$ is $0.5 + 0.4948 = 0.9948$ (see Fig. 2.18).

$$P(\overline{x} \leq 6) = 0.9948$$

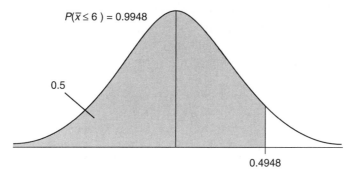

$P(\bar{x} \le 6) = 0.9948$

0.5

0.4948

Figure 2.18

The previous example is valid for an extremely large population. Sampling from a finite population will require some adjustment called the finite correction factor:

$$\sqrt{\dfrac{N-n}{N-1}}$$

Z will therefore become equal to

$$\dfrac{\bar{x} - \mu}{\dfrac{\sigma}{\sqrt{n}} \sqrt{\dfrac{N-n}{N-1}}}$$

Example 2.21   A city's 450 restaurant employees average \$35 tips a day with a standard deviation of 9. A sample of 50 employees is taken. What is the probability that the sample will have an average of less than \$37 tips a day?

Solution:

$$N = 450, \quad n = 50, \quad \sigma = 9, \quad \bar{x} = 37, \quad m\mu = 35$$

$$Z = \dfrac{37 - 35}{\dfrac{9}{\sqrt{50}} \sqrt{\dfrac{400}{449}}} = \dfrac{2}{1.27(0.944)} = \dfrac{2}{1.201} = 1.66$$

On the Z score table, 1.66 corresponds to 0.4515; therefore, the probability of getting an average daily tip of less than \$37 will be 0.4515 + 0.5 = 0.9515.

If the finite correction factor had not been taken into account, Z would have been 1.57, which corresponds to 0.4418 on the Z score table and therefore the probability of having a daily tip of less than \$37 would have been 0.9418.

Estimating the population mean with large sample sizes

Suppose a company has just developed a new process for prolonging the life of a light bulb. The engineers want to be able to date each bulb to determine its longevity, yet it is not possible to test each bulb in a production process that generates hundreds of thousands of bulbs a day. However, they can take a random sample and determine its average longevity and, from there, they can estimate the longevity of the whole population.

Using the central limit theorem, we have determined that the Z value for sample means can be used for large samples.

$$Z = \frac{\overline{X} - \mu}{\sigma / \sqrt{n}}$$

By rearranging this formula, we can derive the value of μ.

$$\mu = \overline{X} - Z \frac{\sigma}{\sqrt{n}}$$

Since Z can be positive or negative, the next formula would be more accurate

$$\mu = \overline{X} \pm Z \frac{\sigma}{\sqrt{n}}$$

In other words, μ will be within the following confidence interval:

$$\overline{X} - Z \frac{\sigma}{\sqrt{n}} \leq \mu \leq \overline{X} + Z \frac{\sigma}{\sqrt{n}}$$

However, a confidence interval presented as such does not take into account alpha (α), the area under the normal curve that is outside the confidence interval. We estimate with some confidence that the mean μ is within the interval:

$$\overline{X} - Z \frac{\sigma}{\sqrt{n}} \leq \mu \leq \overline{X} + Z \frac{\sigma}{\sqrt{n}}$$

Nevertheless, we cannot be certain that it is really within that interval unless the confidence level is 100%.

For a two-tailed normal curve, if we want to be 95% sure that μ is within that interval, then α would be equal to $100 - 95 = 0.05$, and then since the normal curve is two-tailed, the area under the tails will be $\alpha/2 = 0.05/2 = 0.025$; then $Z_{\alpha/2} = Z_{0.025}$, which corresponds to 1.96 on the Z table (see Fig. 2.19 and Table 2.6).

The confidence interval should be rewritten as

$$\overline{X} - Z_{\alpha/2} \frac{\sigma}{\sqrt{n}} \leq \mu \leq \overline{X} + Z_{\alpha/2} \frac{\sigma}{\sqrt{n}}$$

or

$$\overline{X} - Z_{0.025} \frac{\sigma}{\sqrt{n}} \leq \mu \leq \overline{X} + Z_{0.025} \frac{\sigma}{\sqrt{n}}$$

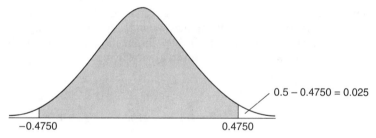

Figure 2.19

TABLE 2.6

z	.00	.01	.02	.03	.04	.05	.06	.07	.08	.09
0.0	.0000	.0040	.0080	.0120	.0160	.0199	.0239	.0279	.0319	.0359
0.1	.0398	.0438	.0478	.0517	.0557	.0596	.0636	.0675	.0714	.0753
0.2	.0793	.0832	.0871	.0910	.0948	.0987	.1026	.1064	.1103	.1141
0.3	.1179	.1217	.1255	.1293	.1331	.1368	.1406	.1443	.1480	.1517
0.4	.1554	.1591	.1628	.1664	.1700	.1736	.1772	.1808	.1844	.1879
0.5	.1915	.1950	.1985	.2019	.2054	.2088	.2123	.2157	.2190	.2224
0.6	.2257	.2291	.2324	.2357	.2389	.2422	.2454	.2486	.2517	.2549
0.7	.2580	.2611	.2642	.2673	.2704	.2734	.2764	.2794	.2823	.2852
0.8	.2881	.2910	.2939	.2967	.2995	.3023	.3051	.3078	.3106	.3133
0.9	.3159	.3186	.3212	.3238	.3264	.3289	.3315	.3340	.3365	.3389
1.0	.3413	.3438	.3461	.3485	.3508	.3531	.3554	.3577	.3599	.3621
1.1	.3643	.3665	.3686	.3708	.3729	.3749	.3770	.3790	.3810	.3830
1.2	.3849	.3869	.3888	.3907	.3925	.3944	.3962	.3980	.3997	.4015
1.3	.4032	.4049	.4066	.4082	.4099	.4115	.4131	.4147	.4162	.4177
1.4	.4192	.4207	.4222	.4236	.4251	.4265	.4279	.4292	.4306	.4319
1.5	.4332	.4345	.4357	.4370	.4382	.4394	.4406	.4418	.4429	.4441
1.6	.4452	.4463	.4474	.4484	.4495	.4505	.4515	.4525	.4535	.4545
1.7	.4554	.4564	.4573	.4582	.4591	.4599	.4608	.4616	.4625	.4633
1.8	.4641	.4649	.4656	.4664	.4671	.4678	.4686	.4693	.4699	.4706
1.9	.4713	.4719	.4726	.4732	.4738	.4744	.4750	.4756	.4761	.4767
2.0	.4772	.4778	.4783	.4788	.4793	.4798	.4803	.4808	.4812	.4817

or

$$\overline{X} - 1.96\frac{\sigma}{\sqrt{n}} \leq \mu \leq \overline{X} + 1.96\frac{\sigma}{\sqrt{n}}$$

Table 2.7 shows the most commonly used confidence coefficients and their Z score values.

TABLE 2.7

Confidence interval $(1 - \alpha)$	α	$Z_{\alpha/2}$
0.90	0.10	1.645
0.95	0.05	1.96
0.99	0.01	2.58

Example 2.22   A survey of companies was conducted to determine how many hours of overtime the employees perform every month. A random sample of 55 responses produced a mean of 45 h. Suppose the population standard deviation for this question is 15.5 h. Find the 95% confidence interval for the mean.

Solution:

$$n = 55, \quad \overline{X} = 45, \quad \sigma = 15.5, \quad Z_{\alpha/2} = 1.96$$

$$45 - 1.96\frac{15.5}{\sqrt{55}} \leq \mu \leq 45 + 1.96\frac{15.55}{\sqrt{55}}$$

$$40.90 \leq \mu \leq 49.1$$

Minitab output

One-Sample Z

The assumed standard deviation = 15.5

N	Mean	SE Mean	95% CI
55	45.0000	2.0900	(40.9036, 49.0964)

We can be 95% sure that the mean will be between 40.90 and 49.10 h; in other words, the probability for the mean to be between 40.9 and 49.1 will be 0.95.

$$prob[40.9 \leq \mu \leq 49.1] = 0.95$$

When the sample size is large $(n \geq 30)$, the sample's standard deviation can be used as an estimate of the population standard deviation.

Estimating the population mean with small sample sizes and σ unknown *t*-distribution

We have seen that when the population is normally distributed and the standard deviation is known, μ can be estimated to be within the interval

$$\mu = \overline{X} \pm Z_{\alpha/2}\frac{\sigma}{\sqrt{n}}$$

In fact, the Z formula has been determined not to always generate normal distributions for small sizes even if the population is normally distributed.

So in the case of small samples and when σ is not known, the *t*-distribution is used instead.

The formula for that distribution is given as:

$$t = \frac{\overline{X} - \mu}{s/\sqrt{n}}$$

The right side of this equation is identical to the one of the Z formula with the difference that σ is replaced with s but the tables used to determine the values are different from the ones used for the Z values.

Just as in the case of the Z formula, t can also be manipulated to estimate μ, but since the sample sizes are small, in order not to produce a biased result, the degrees of freedom (df) will be taken into account, $df = n - 1$.

So, the mean μ will be found within the interval

$$\overline{X} \pm t_{\alpha/2,n-1} \frac{s}{\sqrt{n}}$$

or

$$\overline{X} - t_{\alpha/2,n-1} \frac{s}{\sqrt{n}} \leq \mu \leq \overline{X} + t_{\alpha/2,n-1} \frac{s}{\sqrt{n}}$$

Example 2.23 The manager of a car rental company wants to know the number of times luxury cars are rented a month. She takes a random sample of 19 cars, which produces the following result:

$$3\ 7\ 12\ 5\ 9\ 13\ 2\ 8\ 6\ 14\ 6\ 1\ 2\ 3\ 2\ 5\ 11\ 13\ 5$$

She wants to use these data to construct a 95% confidence interval to estimate the average.

Solution:

$$3 + 7 + 12 + 5 + 9 + 13 + 2 + 8 + 6 + 14 + 6 + 1 + 2 + 3 + 2 + 5 + 11 + 13 + 5 = 127$$

$$\overline{X} = \frac{127}{19} = 6.68$$

$$s = 4.23, \quad n = 19, \quad df = n - 1 = 19 - 1 = 18$$

$$t_{0.025,18} = 2.101 \text{ (see Table 2.8)}$$

TABLE 2.8

df	0.10	0.05	0.025	0.01	0.005	0.001	0.0005	
13	1.350	1.771	2.160	2.650	3.012	3.852	4.221	13
14	1.345	1.761	2.145	2.624	2.977	3.787	4.140	14
15	1.341	1.753	2.131	2.602	2.947	3.733	4.073	15
16	1.337	1.746	2.120	2.583	2.921	3.686	4.015	16
17	1.333	1.740	2.110	2.567	2.898	3.646	3.965	17
18	1.330	1.734	2.101	2.552	2.878	3.610	3.922	18
19	1.328	1.729	2.093	2.539	2.861	3.579	3.883	19

$$6.68 - 2.101\frac{4.23}{\sqrt{19}} \le \mu \le 6.68 + 2.101\frac{4.23}{\sqrt{19}}$$

$$4.641 \le \mu \le 8.72$$

$$Prob[4.641 \le \mu \le 8.72] = 0.95$$

Using SigmaXL   Open the file CarRental.xl and select the area containing the data including the title. From the menu bar, click on **SigmaXL**, then on **Statistical Tools**, and then on **1 Sample t-Test & Confidence Intervals** as shown in Fig. 2.20

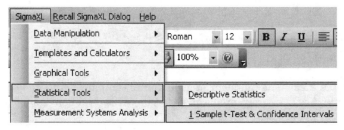

Figure 2.20

The 1 Sample t-Test should appear with the data already selected in the field "**Please select your data.**" Press on the **Next >>** button (Fig. 2.21).

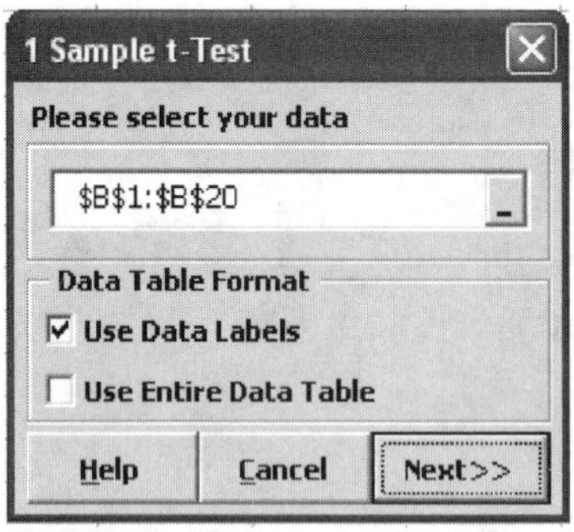

Figure 2.21

Select the Option "**Stacked Column Format,**" then select **Car Rental** for the field "**Numeric Data Variable (Y).**" Press the **OK >>** button (Fig. 2.22).

The results appear as shown in Table 2.9.

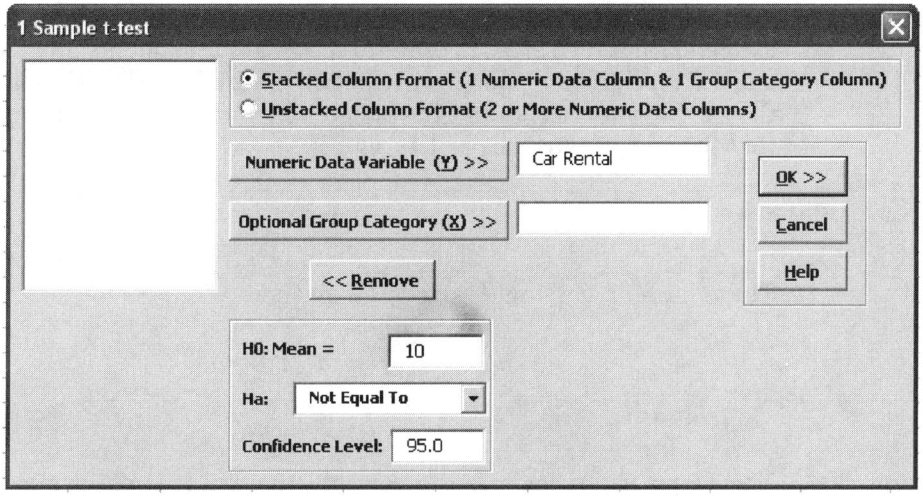

Figure 2.22

TABLE 2.9

1 Sample t-test	
H0: Mean (Mu) = 10	
Ha: Mean (Mu) Not Equal To 10	
	Car Rental
Count	19
Mean	6.684
StDev	4.230
SE Mean	0.970478
t	-3.417
P-value (2-sided)	0.0031
UC (2-sided, 95%)	8.723
LC (2-sided, 95%)	4.645

Minitab output

One-Sample T: Car Rental					
Variable	N	Mean	StDev	SE Mean	95% CI
Car Rental	19	6.68421	4.23022	0.97048	(4.64531, 8.72311)

The probability for μ, to be between 4.64 and 8.72, is 0.95.

χ^2 Distribution

In most cases, in quality control, the objective of the auditor is not to find the mean of a population but rather to determine the level of process variations. For instance, he would want to know how much variation the production process exhibits about the target in order to see what adjustments are needed to reach a defect-free process. The χ^2 distribution determines the relationship between the sample variance s^2 and the population variance σ^2. We have already seen that the sample variance is determined as

$$s^2 = \frac{\sum (x - \bar{x})^2}{n-1}$$

The χ^2 formula for single variance is given as

$$\chi^2 = (n-1)\frac{s^2}{\sigma^2}$$

$$(n-1) = df$$

The shape of χ^2 resembles the normal curve but it is not symmetrical and its shape depends on the degree of freedom (see Fig. 2.23).

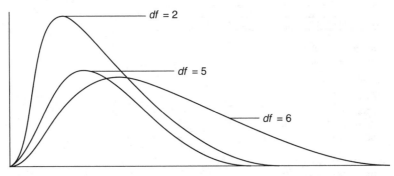

Figure 2.23

The χ^2 formula can be rearranged to find σ^2.

$$\chi^2 = (n-1)\frac{s^2}{\sigma^2}$$

σ^2 will be within the interval

$$\frac{(n-1)s^2}{\chi^2_{\alpha/2}} \leq \sigma^2 \leq \frac{(n-1)s^2}{\chi^2_{1-\alpha/2}}$$

Example 2.24   A sample of nine screws was taken out of a production line and their lengths are as follows:

13.00 mm

13.00 mm

12.00 mm

12.55 mm

12.99 mm

12.89 mm

12.88 mm

12.97 mm

12.99 mm

We are trying to estimate the population variance σ^2 with 95% confidence.

Solution: We need to determine the *point of estimate,* which is the sample's variance.

$$s^2 = 0.112$$

with a degree of freedom *df* of $n - 1 = 9 - 1 = 8$.

Since we want to estimate σ^2 with a confidence level of 95%,

$$\alpha = 1 - 0.95 = 0.05, \qquad \alpha/2 = 0.025, \qquad 1 - \alpha/2 = 0.975$$

So σ^2 will be within the following interval:

$$\frac{8(0.112)}{\chi^2_{0.025}} \leq \sigma^2 \leq \frac{8(0.112)}{\chi^2_{0.975}}$$

From Table 2.10, the values of $\chi^2_{0.025}$ and $\chi^2_{0.975}$ for a degree of freedom of 8 are, respectively, 17.53 and 2.18.

TABLE 2.10

Chi-square Distribution Table

d.f.	.995	.99	.975	.95	.9	.1	.05	.025	.01
1	0.00	0.00	0.00	0.00	0.02	2.71	3.84	5.02	6.63
2	0.01	0.02	0.05	0.10	0.21	4.61	5.99	7.38	9.21
3	0.07	0.11	0.22	0.35	0.58	6.25	7.81	9.35	11.34
4	0.21	0.30	0.48	0.71	1.06	7.78	9.49	11.14	13.28
5	0.41	0.55	0.83	1.15	1.61	9.24	11.07	12.83	15.09
6	0.68	0.87	1.24	1.64	2.20	10.64	12.59	14.45	16.81
7	0.99	1.24	1.69	2.17	2.83	12.02	14.07	16.01	18.48
8	1.34	1.65	2.18	2.73	3.49	13.36	15.51	17.53	20.09
9	1.73	2.09	2.70	3.33	4.17	14.68	16.92	19.02	21.67
10	2.16	2.56	3.25	3.94	4.87	15.99	18.31	20.48	23.21
11	2.60	3.05	3.82	4.57	5.58	17.28	19.68	21.92	24.72
12	3.07	3.57	4.40	5.23	6.30	18.55	21.03	23.34	26.22
13	3.57	4.11	5.01	5.89	7.04	19.81	22.36	24.74	27.69
14	4.07	4.66	5.63	6.57	7.79	21.06	23.68	26.12	29.14
15	4.60	5.23	6.26	7.26	8.55	22.31	25.00	27.49	30.58
16	5.14	5.81	6.91	7.96	9.31	23.54	26.30	28.85	32.00
17	5.70	6.41	7.56	8.67	10.09	24.77	27.59	30.19	33.41
18	6.26	7.01	8.23	9.39	10.86	25.99	28.87	31.53	34.81
19	6.84	7.63	8.91	10.12	11.65	27.20	30.14	32.85	36.19

So the confidence interval becomes

$$\frac{0.8976}{17.53} \leq \sigma^2 \leq \frac{0.8976}{2.18}$$

$$0.051 \leq \sigma^2 \leq 0.412$$

and $prob[0.051 \leq \sigma^2 \leq 0.412] = 0.95$

The probability for σ^2, to be between 0.051 and 0.412, is 0.95.

Now, what is the confidence interval for the standard deviation σ?

Since we already know the confidence interval for the variance is $0.051 \leq \sigma^2 \leq 0.412$, we can obtain the confidence interval for the standard deviation by finding the square root of the values for the confidence interval for the variance

$$\sqrt{0.051} \leq \sqrt{\sigma^2} \leq \sqrt{0.412}$$

Therefore, the confidence interval for the standard deviation will be

$$0.226 \leq \sigma \leq 0.642$$

SigmaXL provides a template that can help us find the confidence interval for the standard deviation (see Table 2.11).

We had found the variance to be $s^2 = 0.112$, therefore the standard deviation is

$$s = \sqrt{s^2} = \sqrt{0.112} = 0.335$$

TABLE 2.11

Calculate Confidence Interval for Sigma			
Enter Standard Deviation:		S	0.335
Enter Size of Sample:		n	9
Confidence level (enter as a percent):		100*(1-α)%	95.00%
		Lower Limit Sigma	0.226278107
		Upper Limit Sigma	0.641783245

Estimating sample sizes

In most cases, sampling is used in quality control to make an inference about a whole population because of the cost associated with actually studying every individual part of that population. Then again, the question of the sample size arises.

What size of a sample best reflects the condition of the whole population being estimated? Should we consider a sample of 150 products or a sample of 1000 products from a production line to determine the quality level of the output?

Sample size when estimating the mean

At the beginning of this chapter, we defined the sampling error E as being the difference between the sampling mean $\overline{X}$ and the population mean μ, $E = \overline{X} - \mu$.

We also have seen, when studying the sampling distribution of $\overline{X}$ that when μ is being determined, we can use the Z formula for sampling means.

$$Z_{\alpha/2} = \frac{\overline{X} - \mu}{\sigma / \sqrt{n}}$$

We can clearly see that the nominator is nothing but the sampling error E. We can therefore replace $\overline{X} - \mu$ by E in the Z formula and come up with

$$Z_{\alpha/2} = \frac{E}{\sigma/\sqrt{n}}$$

We can determine n from this equation

$$\sqrt{n} = \frac{Z_{\alpha/2}\sigma}{E}$$

$$n = \left(\frac{Z_{\alpha/2}\sigma}{E}\right)^2$$

Example 2.25 A production manager at a call center wants to know the average time an employee should spend on the phone with a customer. She wants to be within 2 min of the actual length of time and the standard deviation of the average time spent is known to be 3 min. What sample size of calls should she consider if she wants to be 95% confident of her result?

Solution:

$$Z_{\alpha/2} = 1.96, \quad E = 2, \quad \sigma = 3$$

$$n = \frac{(1.96 \times 3)^2}{2^2} = \frac{34.574}{4} = 8.643$$

Since we cannot have 8.643 calls, we can round up the result to 9 calls.

Using SigmaXL template From the menu bar, click on **SigmaXL**, then select **Templates and Calculators** and from the submenu, select **Sample Size-Continuous** (see Fig. 2.24).

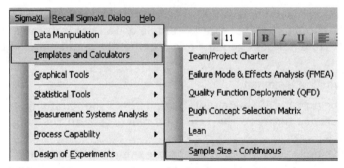

Figure 2.24

When the template appears, fill in the fields **S** and **delta** as shown in Table 2.12 and then click anywhere on the spreadsheet to populate the field **n.**

TABLE 2.12

Sample Size Calculator - Continuous Data			
Estimate of Standard Deviation:		S	3
Desired margin of error:		delta	2
Population Size (optional):		N	
Minimum Sample Size:		n	9
		n (adjusted for small N)	

The manager can be 95% confident that with a sample of 9 calls she can determine the average length of time an employee needs to spend on the phone with a customer.

Measurement Systems Analysis

- Jolof Medicals manufactures glucose meters to help patients with diabetes monitor their blood sugar. The Cayor-x275 model was released 3 months ago, but it has drawn a lot of criticism because it was found to be very unreliable. When testing their blood sugar using that model, the patients tend to obtain results that are far below the actual amount of sugar they have in their blood. A quality engineer was assigned the task to assess the reliability of model Cayor-x275 and determine the sources of its deficiencies.

- Galle-Basbe RFDM designs and manufactures radio frequency devices. It has received an important order from its customer Koussanar-Conveyors to manufacture laser optic barcode readers to control the traffic of products moved by conveyors. If the optical devices do not read the barcodes within the preset interval of time, products will pile up on the conveyors and jam them, and productivity would suffer. If they misread the barcodes, customers' shipments would end up being sent to wrong destinations. It is therefore extremely important for the devices to be precise and accurate so they can be used in distribution and manufacturing plants. After the design has been completed, a reliability engineer is charged with testing the devices for fitness for use.

- Reynolds manufactures and sells diesel engines and the replacement parts for the engines. The distribution center located in Memphis, Tennessee has been having problems with inventory accuracy for the parts that weigh less than 1 oz. The warehouse is always either overstocking those parts or back ordering them because it ran out of stock. The inventory manager believes that this is because when more than 50 pieces of those parts are ordered, the employees do not have to count them, they can just put them on scales to measure their weight and determine the quantities. The quality engineer has been assigned with the task of ensuring that the measurement process is reliable and that it is not causing the uncontrolled variability in the inventory.

Before creating control charts and analyzing a process' capabilities, it is necessary to ensure that the methods and tools used to inspect, test, measure, analyze, or audit the process or the products generated by that process provide accurate, precise, and reliable information. If a faulty gauge is used to measure the conformance of a CTQ to its preset standards, the results obtained can only be flawed. The performance of the gauges used to test and generate measurements on a process yield is a quintessential component of process engineering. The determination of the tools and methods used to assess the quality of the data generated through a measurement system should be the first consideration when appraising the performance of a production process.

The data generated by a measurement process always exhibits some variability and the variations in measurements can only come from two sources: variations due to actual differences between the measured parts and variations due to the measurement process (how the parts were measured). The variance is used to measure those variations.

Suppose that a quality controller wants to determine the sources of variations between motor mounts. The variations can come from either the actual differences between the parts themselves, the measurement process, or the interaction between the parts and the measurement process.

$$\sigma^2_{\text{Total}} = \sigma^2_{\text{Motor Mount}} + \sigma^2_{\text{Measurement Process}} + (\sigma^2_{\text{Motor Mount}} * \sigma^2_{\text{Measurement Process}})$$

σ^2_{Total} measures the total variations.

$\sigma^2_{\text{Motor Mount}}$ measures the variations due to the actual differences between the motor mounts.

$\sigma^2_{\text{Measurement Process}}$ measures the variations due to how the measurements were taken.

$\sigma^2_{\text{Motor Mount}} * \sigma^2_{\text{Measurement Process}}$ measures the interaction between the parts and the measurement process.

The variations due to the measurement process $\sigma^2_{\text{Measurement Process}}$ can be subdivided further between the variations due the operator (how he took the measurements) and the variations due to the gauge (the instrument used to collect the measurements).

$$\sigma^2_{\text{Measurent Process}} = \sigma^2_{\text{Operator}} + \sigma^2_{\text{Gauge}}$$

Therefore, when assessing the sources of variations, three components and their interactions are considered: part-to-part variations, variations due to the operators, and variations due to the gauge and the interactions motor mount/operator, operator/gauge, motor mount/gauge, and motor mount/gauge/operator:

$$\sigma^2_{\text{Total}} = \sigma^2_{\text{Motor Mount}} + \sigma^2_{\text{Operator}} + \sigma^2_{\text{Gauge}} + (\sigma^2_{\text{Motor Mount}} {}^* \sigma^2_{\text{Operator}})$$

$$+ (\sigma^2_{\text{Operator}} {}^* \sigma^2_{\text{Gauge}}) + (\sigma^2_{\text{Motor Mount}} {}^* \sigma^2_{\text{Gauge}}) + (\sigma^2_{\text{Motor Mount}} {}^* \sigma^2_{\text{Operator}} {}^* \sigma^2_{\text{Gauge}})$$

If significant variations are present in a production process, measurement systems analysis (MSA) can be used to determine their sources.

The control charts and the analysis of variance (ANOVA) are among the tools used by MSA to pinpoint the sources of variations. The control charts visualize the patterns of the measurements while the ANOVA determines the significance of the factors contributing to the variations.

Example 2.26 A car manufacturer wants to reduce the noise generated by one of its models and has determined that most of the rattling noise is coming from the loose connection between the motor mount (a motor mount is a device that connects different pieces of a car engine to their framework) and the chassis. The thickness of the motor mount is deemed critical to quality. The quality engineer assigned to the task uses an electronic caliper to assess the conformity of a random sample of 20 motor mounts. He measures each of the 20 motor mounts three times and tabulates the results as shown in Table 2.13.

TABLE 2.13

Motor mount	M1	M2	M3	Motor mount	M1	M2	M3
1	10.0264	10.0391	10.0348	11	11.8616	11.8781	11.8212
2	10.9938	10.98276	10.9325	12	10.0264	10.0391	10.0348
3	10.0332	10.0299	10.0354	13	10.0016	10.0446	10.0036
4	10.0008	9.9523	9.9589	14	10.9879	10.9569	10.9655
5	10.0032	10.0035	10.0094	15	10.0032	10.0035	10.0094
6	11.0204	10.9823	10.9956	16	10.0008	9.9523	9.9589
7	10.0032	10.0035	10.0094	17	10.0264	10.0391	10.0348
8	10.0032	10.0035	10.0094	18	10.9938	10.98276	10.9325
9	10.9938	10.98276	10.9325	19	10.9938	10.98276	10.9325
10	10.0264	10.0391	10.0348	20	10.0032	10.0035	10.0094

The differences across columns M1, M2, and M3 for each line show variation in measurements for the same part while the differences within each column show the variations between parts.

Figure 2.25 shows SigmaXL generated $\overline{X} - R$ control charts. The interpretation that is made for these control charts will be different from the one created when monitoring a production process with statistical process control. Since each motor mount is a sample and the measurements M1, M2, M3 are the pieces of the sample, each dot on the charts represents a mean measurement for one motor mount. The $\overline{X}$ chart shows the significance of the part-to-part variations, while the R chart shows how consistent the measurement process has been. If the gauge used to measure the parts is good and the same operator measured every part in a consistent manner, the R chart is expected to be in control and stable; otherwise, it will show out of control patterns.

The $\overline{X}$ chart in Fig. 2.25 shows just about every dot is out of control. These extreme disparities between the parts indicate that the part-to-part variations are very significant. The R chart is stable and in control. This means that the operator did not

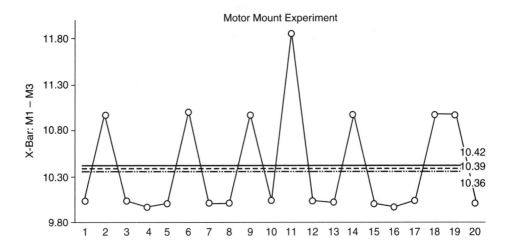

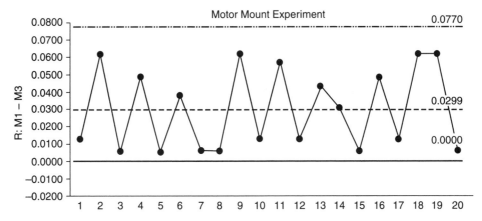

Figure 2.25

have any problem taking measurements, there is consistency in the way he measured the motor mounts, and both the gauge and the operator performed well. Consequently, the sources of the variations in the measurements are traceable to the differences in size between the pieces tested.

SigmaXL output for motor mount experiment (obtained from Motor mount.xls) is shown in Fig. 2.25.

While the control charts visualize the sources of variations, ANOVA quantifies the significance level of the sources of variation. When using the ANOVA single factor completely randomized experiment, each part is considered as a treatment factor. Since 20 pieces of motor mounts were used, the degree of freedom for the factors would be 19 and since each test was replicated three times, we ended up with 60 measurements. Therefore, a degree of freedom for the total was 59 and 40 degrees of freedom would be left for the error term.

The null hypothesis for the ANOVA experiment suggests that there is no difference between the motor mounts and the alternate hypothesis suggests that at least one motor mount is different.

The Minitab output in Table 2.14 shows that the p-value for the factors is equal to 0.000, which means that the null hypothesis should be rejected and the conclusion is that there are significant differences between the motor mounts.

TABLE 2.14   Minitab Output

Source	DF	SS	MS	F	P
Factor	19	18.18331	0.95702	2402.77	0.000
Error	40	0.01593	0.00040		
Total	59	18.19924			

$S = 0.01996$ R–Sq = 99.91%   R–Sq(adj) = 99.87%

Precision and Accuracy

Two kinds of errors can result from using a measurement process: accuracy and precision. Accuracy refers to how close the measurements taken reflect the actual true value of the CTQ being measured and precision refers to how consistent the gauge is at getting the same results every time it measured the same object. It is therefore possible to be precise and not be accurate and vice versa.

Suppose that a sample of 100 brake pads that are known to weigh 1.5 lb each are tested using an electronic scale and the scale shows that each pad weighs 1.45 lb. The scale would not be considered accurate because it fails to give the exact true measurement of 1.5 lb, but it should be considered precise because it obtains the same results at every test.

Measurement errors due to precision

Precision refers to the consistency of the measurements obtained from repeated tests. The two important aspects of precision that are considered are repeatability and reproducibility. Repeatability addresses the variations obtained when a single instrument is used by an operator who is measuring a CTQ characteristic of the same part. When the same instrument is used by the same operator testing the same part several times, any variation in the results of the tests can only be traced to the instrument used in the test. Suppose that an appraiser uses the same electronic caliper to test 10 times the diameter of the same exhaust system under the same conditions. If the results that he obtains after the tenth time are not all consistent, the variations can only be traced to the caliper. Therefore, repeatability is concerned with the gauge used to conduct an experiment.

Example 2.27   If the weights of the shipments were not correctly reported by the MDC warehouse, the shipping costs incurred by the company would be exorbitant. A scale has been acquired for the shipping department to weigh the customers' shipments. Bill has been assigned the task to test the scale. He weighs the same pallet 10 times on the new scale and obtains the results shown in Table 2.15. The table is in the file *MDC.MTW*

TABLE 2.15

Part #	Operator	Measurements
1	Bill	14.9988
1	Bill	15.0005
1	Bill	15.0017
1	Bill	14.9999
1	Bill	14.9996
1	Bill	15.0006
1	Bill	14.9997
1	Bill	14.9997
1	Bill	14.998
1	Bill	14.9988

With Minitab, we can use the gauge run chart to visualize the distribution of the measurements.

Open *MDC.MWT* and from the menu bar, click on **Stat**, then click on **Quality Tools**. From the submenu, select **Gauge Study** and then **Gauge Run Chart**.

Enter "Part Number" in the **Part Number** field, "Operator" in the **Operator** field, and "Measurements" in the **Measurement Data** field. Then click on the **OK** button to obtain Fig. 2.26.

The graph shows that the data are randomly distributed between 14.9980 and 15.0020. If this range (0.004) is acceptable, the scale can be considered fit for use.

Reproducibility addresses the average measurements obtained when different appraisers use the same measuring device to test the same part. Reproducibility is about the measurement system, the variations that result from inconsistencies in the way the operators conduct the tests, or the interactions between the operators and the gauges used in the testing.

Example 2.28   Mourid Trucking is a company that ships the customers' orders. It has been complaining about the actual weights of the shipments being different from what is reported by the warehouse, causing it to fail at weigh stations. Jenny, Marquel, and

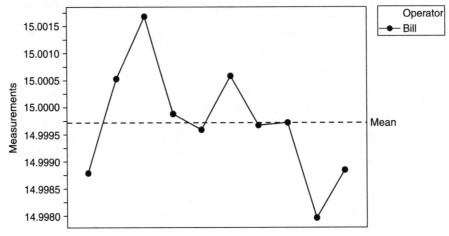

Gage Run Chart of Measurements by Part Number, Operator

Figure 2.26

Joe are the employees at the shipping department responsible for determining the weight of the shipments before they are loaded on the trailers. Their supervisor asks them to take the same pallet and weigh it three times. The results that they obtained are summarized in Table 2.16. The data are contained in the file *Pallet.MTW*.

TABLE 2.16

Part	Operators	Measurements
Pallet	Jenny	70.8894
Pallet	Jenny	71.9898
Pallet	Jenny	70.9876
Pallet	Marquel	70.5440
Pallet	Marquel	70.5414
Pallet	Marquel	69.0260
Pallet	Joe	70.5895
Pallet	Joe	70.5896
Pallet	Joe	70.5710

Using Minitab, we obtain the following gauge run chart shown in Fig. 2.27.

The graph shows that while Joe has been consistent with his measurements, Marquel and Jenny have been having difficulties measuring the parts.

Example 2.29   Reynolds manufactures and sells diesel engines and the replacement parts for the engines. The distribution center located in Memphis, Tennessee has been having problems with inventory accuracy for the parts that weigh less than 1 oz. The warehouse is always either overstocking those parts or back ordering them because it ran out of stock.

Gage Run Chart of Measurement by Part, Operator

Gage name:
Date of study:

Reported by:
Tolerance:
Misc:

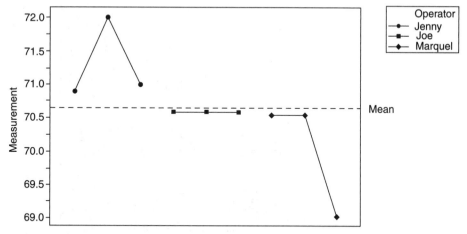

Figure 2.27

The inventory manager believes that this is because when more than 50 pieces of those parts are ordered, the employees do not have to count them. For the sake of saving time, they can just put them on a scale to measure their weight and determine the quantities.

The quality engineer has been assigned with the task of ensuring that the measurement process is reliable and that it is not causing the uncontrolled variability in the inventory.

He selects three employees and has them each measure 15 parts using the scale. The measurements obtained are summarized in Table 2.17.

Using SigmaXL   SigmaXL offer a very easy and practical way to prepare and organize data for MSA. Open SigmaXL and then open the file *Inventory Accuracy.xls*. From the menu bar, click on **SigmaXL** and then click on **Measurement Systems Analysis** on the menu list. Then select **Create Gage R&R (Crossed) Work Sheet** as shown in Fig. 2.28.

The **Create Gage R&R (Crossed) Work Sheet** dialog box appears. Fill it out as shown in Fig. 2.29.

Press the **OK>>** button.

When the **Create Gage R&R (Crossed) Worksheet** appears, enter the measurements in the **Measurement** column as shown in Table 2.18.

Click on **SigmaXL** on the menu bar and then click on **Measurement Systems Analysis** on the menu list and then on **Analyze Gage R&R Crossed**. The area should be selected when the **Analyze Gage R&R Crossed** box appears. Press the **Next>>** button. Enter **Part, Operator,** and **Measurement** in their respective fields and the press **OK>>** to get the results in Table 2.19.

TABLE 2.17

Bill					Ron					Tim				
M1	M2	M3	$\overline{X}_{Bill}$	Range	M1	M2	M3	$\overline{X}_{Ron}$	Range	M1	M2	M3	$\overline{X}_{Tim}$	Range
151.023	154.409	151.419	152.284	3.386	151.182	148.715	144.955	148.284	6.227	150.439	149.767	145.786	148.664	4.653
149.976	149.234	147.413	148.874	2.563	151.115	149.711	150.429	150.418	1.404	150.283	151.192	146.958	149.478	4.234
151.448	148.027	146.285	148.587	5.162	149.979	151.372	151.77	151.040	1.791	150.099	149.078	156.068	151.748	6.991
150.227	150.34	151.15	150.572	0.923	149.597	146.323	150.041	148.654	3.718	149.877	154.344	146.234	150.152	8.111
150.066	149.041	153.739	150.949	4.697	151.41	150.831	148.433	150.225	2.977	150.914	149.899	155.714	152.176	5.815
149.77	149.319	149.724	149.604	0.451	152.308	146.741	144.806	147.952	7.502	149.598	150.271	153.622	151.164	4.023
149.361	153.819	153.568	152.249	4.458	150.845	146.731	146.484	148.020	4.362	150.382	150.978	148.577	149.979	2.401
149.961	153.763	150.517	151.414	3.802	150.883	149.999	151.019	150.634	1.020	149.665	153.676	150.427	151.256	4.011
149.912	148.745	150.971	149.876	2.227	150.221	154.039	153.656	152.639	3.819	150.949	146.551	146.476	147.992	4.473
149.425	146.14	156.123	150.563	9.983	149.413	152.846	156.92	153.060	7.507	152.133	150.722	150.259	151.038	1.874
149.225	150.159	145.507	148.297	4.652	150.887	149.631	150.774	150.431	1.257	150.089	152.494	155.869	152.817	5.780
151.443	150.192	150.056	150.564	1.387	150.691	150.765	146.262	149.239	4.503	149.913	149.516	147.537	148.989	2.376
148.648	148.534	144.9	147.361	3.748	150.035	146.072	150.153	148.753	4.081	148.497	149.827	145.364	147.896	4.463
150.952	151.054	148.278	150.095	2.777	147.893	150.028	149.352	149.091	2.135	150.259	151.031	150.434	150.575	0.772
149.726	149.148	152.943	150.606	3.794	149.413	148.24	156.074	151.242	7.833	149.643	143.215	148.371	147.076	6.428
		Means	150.126	3.601			Means	149.979	4.009			Means	150.067	4.427

SigmaXL
Recall SigmaXL Dialog
Help
Data Manipulation
Templates and Calculators
Graphical Tools
Statistical Tools
Measurement Systems Analysis
Process Capability
Basic MSA Templates
Create Gage R&R (Crossed) Worksheet
10
100%
I
J
K
L

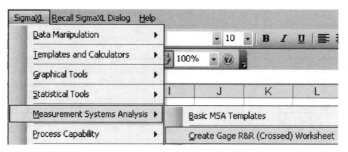

Figure 2.28

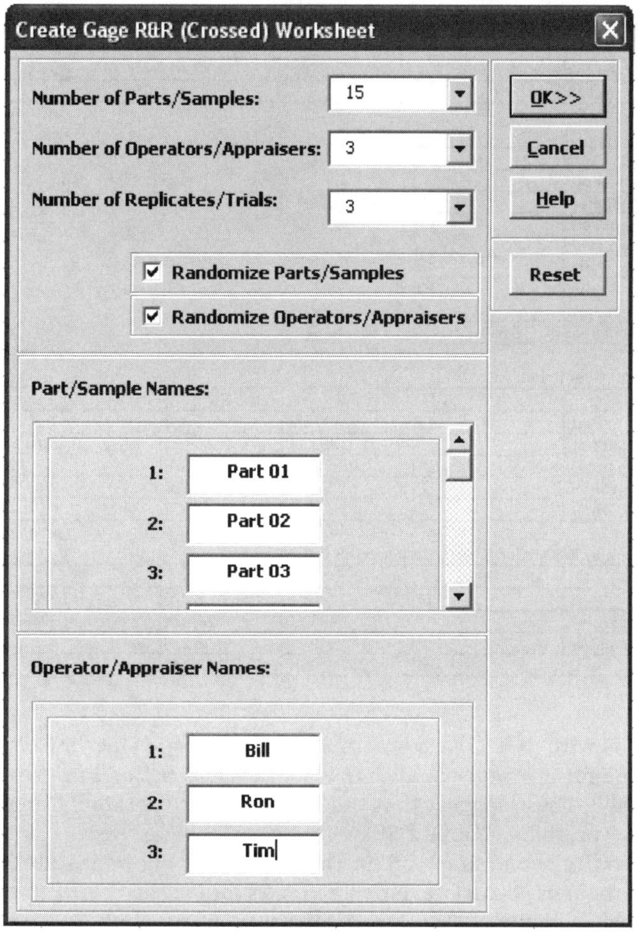

Create Gage R&R (Crossed) Worksheet
Number of Parts/Samples:
15
Number of Operators/Appraisers:
3
Number of Replicates/Trials:
3
OK>>
Cancel
Help
Reset
Randomize Parts/Samples
Randomize Operators/Appraisers
Part/Sample Names:
1: Part 01
2: Part 02
3: Part 03
Operator/Appraiser Names:
1: Bill
2: Ron
3: Tim

Figure 2.29

TABLE 2.18

Gage R&R Study (Crossed) Worksheet

Gage Name:	Inventory Accuracy
Date of Study:	1/5/2007
Performed By:	Raymond Tate
Notes:	Parts weight

Enter the measurements here

Run Order	Std. Order	Part	Operator	Measurement
1	123	Part 11	Tim	
2	130	Part 14	Tim	

TABLE 2.19

Gage R&R Study (Crossed) Report

Gage Name:	Inventory Accuaracy
Date of Study:	1/5/2007
Performed By:	Raymond Tate
Notes:	Parts weight

Process Tolerance (USL - LSL):	
Historical Process Standard Deviation:	
Standard Deviation Multiplier:	6
Alpha to Remove Interaction:	0.1
Confidence Level:	90.0
Number of Parts:	15
Number of Operators:	3
Number of Replicates:	3
Design Type:	Balanced

Analysis of Variance with Part * Operator Interaction:

Source	DF	SS	MS	F	P
Part:	14	90.686	6.478	0.827667	0.6358
Operator:	2	0.495268	0.247634	0.031641437	0.9689
Part * Operator:	28	219.14	7.826	1.353	0.1443
Repeatability:	90	520.73	5.786		
Total:	134	831.04	6.202		

Analysis of Variance without Part * Operator Interaction (P for Interaction >= 0.1):

Source	DF	SS	MS	F	P
Part:	14	90.686	6.478	1.033096339	0.4259
Operator:	2	0.495268	0.247634	0.039494924	0.9613
Repeatability:	118	739.86	6.270		
Total:	134	831.043417	6.201816545		

The ANOVA with Part*Operator interaction shows a p-value for part equal to 0.6358 and a p-value for operator equal to 0.9689. The interaction Part*Operator has a p-value of 0.1443. These suggest that the parts and the operators are not significantly affecting the variability (Table 2.20).

Reproducibility accounts for 0% in the variance, while repeatability accounts for 99.63% and the part-to-part variation accounts for 0.37%.

Reproducibility measures the variability between operators. Because it accounts for 0% of the variability, the problem is not with the employees, nor is it with the part-to-part variations since 0.37% is negligible and can be the result of common causes.

TABLE 2.20

Gage R&R Metrics	StDev	StDev Lower 90% CI	StDev Upper 90% CI	6 * StDev	% Total Variation (TV)
Gage R&R:	2.504	2.240	2.777	15.024	99.82
Operator (AV Appraiser Variation):	0	0	0	0	0.00
Part * Operator (INT Interaction):	0	0	0	0	0.00
Reproducibility (SQRT(AV^2 + INT^2)):	0	0	0	0	0.00
Repeatability (EV Equipment Variation):	2.504	2.264	2.807	15.024	99.82
Part Variation (PV):	0.151846	0	0.916543	0.911076	6.05
Total Variation (TV):	2.509	2.264	2.789	15.052	100.00

Gage R&R Metrics	Variance Component	Variance Component
Gage R&R:	6.270	99.63
Operator:	0	0.00
Part * Operator:	0	0.00
Reproducibility:	0	0.00
Repeatability:	6.270	99.63
Part Variation:	0.023057198	0.37
Total Variation:	6.293	100.00

Gage R&R Metrics	NDC	NDC Lower 90% CI	NDC Upper 90% CI
Number of Distinct Categories (Signal-to-Noise Ratio: 1.41 * PV/R&R):	0.1	0.0	0.5

Repeatability measures the variability due to the gauge, the instrument used in the testing. In this case, repeatability contributes to 99.63% of the variation, which means that the sources of the variations are essentially found in the scale used to weigh the parts.

Variations due to accuracy

Accuracy is a measure of deviation; it addresses the deviation of the measurement process from the actual true values of the CTQs being measured. If a known actual weight of an object were 15 lb and the measurement taken shows that the object weighs 14 lb, we would conclude that the measurement system is inaccurate and biased and that it deviates from the true value by 1 lb.

If only one measurement is taken to determine the CTQ characteristic of an object, one error can make the result obtained misleading. Therefore, it is necessary to take a representative sample of the objects being measured.

If a sample of objects is being tested, for a measurement process to be accurate, the data gathered should only exhibit common cause variations. Such a condition implies lack of bias and linearity. Bias is defined as the deviation of the measurement results from the true values of the CTQs, while linearity refers to gradual proportional variations in the results of the measurements when the object being measured is incrementally increased.

Gauge bias

Bias assesses the accuracy of the measurement system by determining the difference between the results of the measurement system and the actual value of the part being measured. If the reference, the true length of a part, is 25 in. and after measuring it, the result we obtain is 23 in., we would conclude that the measurement system is biased by 2 in.

If a sample of measurements is used to estimate the CTQ of a part, the following formula can be used to estimate the gage bias.

$$\text{Bias} = \frac{\sum_{i-1}^{n} x_i}{n} - q$$

with $\frac{\sum_{i-1}^{n} x_i}{n}$ being the average measurement and q the true value of the part being measured. The equation is read as the difference between the average measurement result and the true value of the part (see Fig. 2.30).

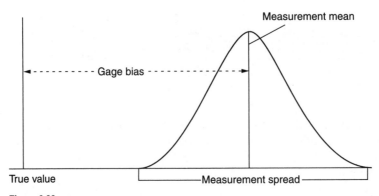

Figure 2.30

A gauge bias assesses the extents to which the mean measurement deviates from the actual value of the product being measured. Since all the measurements taken are just a sample of the infinite number of the possible measurements, one way of estimating bias would be to measure the statistical significance of the difference between the true known value and the sample mean. This can be done by hypothesis testing. The null hypothesis would consist of stating that there is no difference between the measurements' mean and the actual value of the part being measured:

$$H_0 : \overline{X} = q$$

And the alternate hypothesis would state the opposite

$$H_1 : \overline{X} \neq q$$

where $\overline{X}$ is the sample's mean and q is the true value of the part.

If the number of measurements taken is relatively small, we can use the t-test to test the hypothesis. In that case,

$$t_{\alpha/2, n-1} = \frac{\overline{X} - q}{s/\sqrt{n}}$$

$$df = n - 1$$

Example 2.30 The measurement process for weight of raw materials received at a depot is being audited. A bag of fertilizer known to weigh 10 lb is being tested. The bag was weighed 20 times and the measurements obtained are summarized in Table 2.21.

TABLE 2.21

10.0526	9.9599	9.9156	9.9356
9.9842	10.0279	9.9707	9.9848
9.9151	10.0282	9.9898	10.0169
9.9983	10.1008	10.0365	9.997
10.0245	9.9977	10.1136	10.0793

1 Sample t-test	
HO: Mean (Mu) = 10	
Ha: Mean (Mu) Not Equal To 10	
	test
Count	20
Mean	10.00645
StDev	0.054593083
SE Mean	0.012207385
t	0.528369
P-value (2-sided)	0.6034
UC (2-sided, 95%)	10.03200035
LC (2-sided, 95%)	9.981

Determine if the measurement process is accurate for an alpha level of 0.05.

Solution: The sample mean is equal to 10.00645; therefore,

$$\text{Gauge bias} = 10.00645 - 10 = 0.00645$$

To determine if the process is biased or not, we can run a hypothesis testing. The null hypothesis for the t-test would be

$$H_0 : \overline{X} = 10$$

Table 2.21 is the SigmaXL output using 1 Sample t-Test for *Raw material.xls*.

The p-value is equal to 0.6034; therefore, we fail to reject the null hypothesis and have to conclude that there is not a statistical significant difference between the true value and the measurements' mean.

Gauge linearity

In Example 2.30, the measurements taken to assess the accuracy of the process were from only one part. For a measurement system to be accurate not only should it be unbiased when measuring one object, but it should be consistent when the dimensions of the object vary. Since the scale used in Example 2.30 is very much likely to be used to weigh loads that are more or less than 10 lb, we expect it to be unbiased for loads of all sizes. If the loads that are being weighed are incrementally increased, the scale should display measurements that are proportional to the incremental changes for the process to be deemed accurate.

Even if the gauge is biased, if it exhibits linearity, we should expect the same proportional variations. Suppose that we are using a voltmeter to measure the voltage of the current that flows through an electrical line. If the actual voltage applied is 120 V and that voltage is doubled and then tripled, we should expect it to read 120 V, then 240 V, and finally 360 V if the voltmeter is accurate.

If the first reading of the voltmeter was not exact and was off by 5 V, we should expect the readings to be 125 V for the first reading, 250 V for the second reading, and 375 V for the third reading. If these results are obtained, we can conclude that the gauge exhibits linearity.

To run a gauge linearity test, we can use a regression analysis to determine the regression line and observe the spread of the data plots of the gauge measurements about the line.

The regression analysis would be a simple linear one with the independent variables being the known actual values and the dependent variables being the gauge bias.

The equation of the regression line is under the form of

$$Y = aX + b$$

If a scatter plot with a regression line is used to visualize the relationship between the known actual CTQ values and the gauge is accurate and precise, all the measurements obtained should be on the regression line, and the equation for the regression line would be

$$Y = X$$

in other words, if in the equation $Y = aX + b$, $a = 1$ and $b = 0$, we would conclude that the gauge is a perfect instrument to measure the parts because every gauge measurement would be equal to the true value of the part being measured and therefore the bias would be equal to 0 and the regression plot would look like Fig. 2.31.

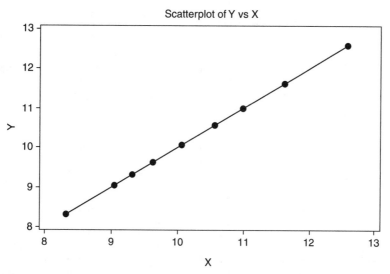

Figure 2.31

To have a good estimate of the measurements, each part should be measured several times (at least four) and the bias would be the difference between the known actual value and the mean measurement for each part. This is if the regression equation relates the actual known true values, the reference values, with the measurements taken.

Linearity can be assessed by creating a regression equation for the bias using the reference values.

$$\text{Bias} = ax + b$$

$$\text{Linearity} = |a| \times \text{process variation}$$

The process variation is usually within the Six Sigma tolerance range.

$$\%\text{Linearity} = (\text{linearity/process variation}) \times 100$$

The slope and the intercept are used to make an inference about the correlation between the bias and the reference value. Here again, linearity is estimated based on the slope of the equation but because bias is used instead of the mean measurements, a lower slope would indicate better linearity, that is, the lower the slope, the better the linearity.

Example 2.31   A scale is used to measure the weight of pistons. The true values of the pistons are known and five measurements for each piston are taken using the same scale. The results of the measurements are summarized in Table 2.22. We want to find the equation of the regression line to estimate the bias at any value and determine if the scale is a good gauge to measure the weight of the parts.

Solution: SigmaXL output for the equation of the regression line (*Pistons. xls*) is given as shown in Table 2.23.

TABLE 2.22

True value	M1	M2	M3	M4	M5	Mean	Bias
5	5.000	4.980	4.990	5.001	5.001	4.994	−0.006
10	9.980	10.000	10.003	10.009	10.007	10.000	0.000
15	14.990	14.980	14.980	15.000	14.980	14.986	−0.014
20	20.004	19.990	19.984	19.990	19.980	19.990	−0.010
25	25.008	24.920	24.976	25.000	25.000	24.981	−0.019
30	30.003	29.990	29.980	29.990	29.990	29.991	−0.009
35	34.980	34.990	34.980	35.000	35.000	34.990	−0.010
40	40.000	40.000	40.000	39.970	40.001	39.994	−0.006
45	44.998	45.000	45.000	45.000	45.000	45.000	0.000
50	50.009	50.000	50.098	50.005	50.008	50.024	0.024
55	55.002	54.998	55.009	54.998	55.008	55.003	0.003
60	60.000	60.000	60.000	60.005	60.030	60.007	0.007
65	65.040	64.960	65.000	65.008	65.000	65.002	0.002
70	70.001	69.990	69.967	70.000	70.098	70.011	0.011
75	75.000	75.009	74.990	75.000	75.000	75.000	0.000
80	80.002	80.001	80.002	79.967	80.034	80.001	0.001
85	85.040	84.980	84.920	85.000	85.090	85.006	0.006
90	90.020	89.986	90.005	89.180	90.840	90.006	0.006
95	95.020	95.074	94.960	95.000	94.980	95.007	0.007
100	100.030	100.000	100.000	99.990	100.008	100.006	0.006

TABLE 2.23

Multiple Regression Model: Bias = (-0.010724) + (2.017E-04) * True Value

Model Summary:	
R-Square	36.77%
R-Square Adjusted	33.26%
S (Root Mean Square Error)	0.008038308

The slope of the regression line, which represents the percent linearity, is 0.000202 and the y-intercept is equal to −0.010724.

The gauge linearity is estimated based on the slope of the regression equation for the bias:

$$Bias = 0.000202 \times true \ value - 0.010724$$

$$Linearity = |\,slope\,| \times process \ variation = |\,0.000202\,| \times 6 = 0.00121$$

Linearity is generally measured in terms of its percentage of the process variation:

$$\%Linearity = [linearity/process \ variation] \times 100\% = [0.00121/6] \times 100\% = 0.02\%$$

Let us use Minitab to verify our results.

Open the file *Pistons.MTW* and, from the menu bar, click on **Stat** and then on **Quality Tools.** From the submenu, select **Gauge Study** and then **Gauge Linearity and Bias Study**.

The **Gauge Linearity and Bias Study** box appears; fill it out as indicated in Fig. 2.32. Press the **OK** button to obtain Fig. 2.33.

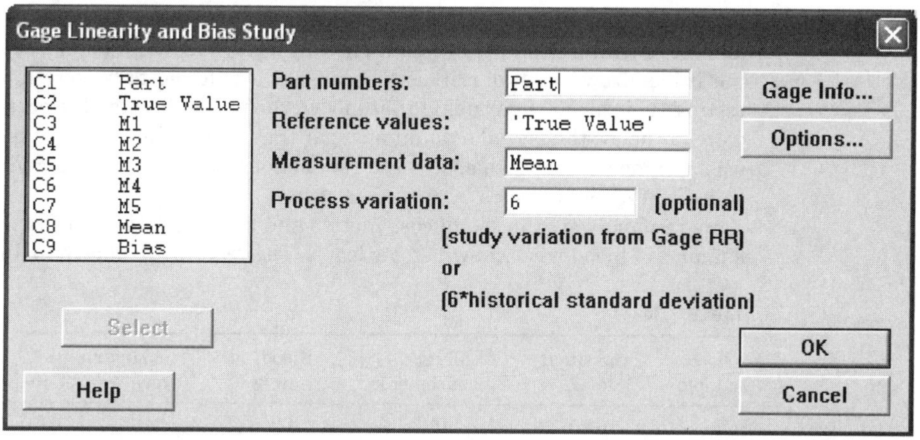

Figure 2.32

The results that we have obtained with Minitab perfectly match our calculations. The slope for the bias is equal to 0.00020195, the measure of linearity is equal to 0.00121, and the percentage linearity is equal to 0.0%.

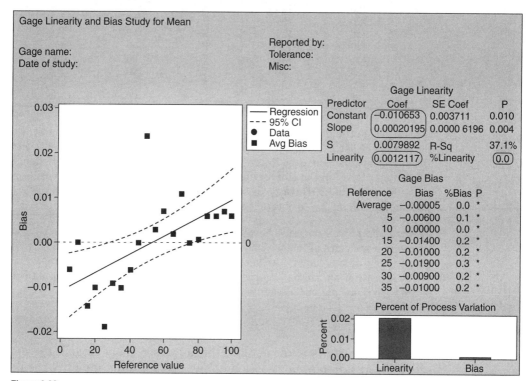

Figure 2.33

The slope is very low (0.00020195), which suggests that the gauge linearity is good because the lower the slope, the better the gauge linearity.

Example 2.32 Jolof Medicals manufactures glucose meters to help patients with diabetes monitor their blood sugar. The Cayor-x275 model was released 3 months ago but it has drawn a lot of criticism because it was found to be very unreliable. When testing their blood sugar using that model, the patients tended to obtain results that were far below the actual amount of sugar they had in their blood. A quality engineer was assigned the task to assess the reliably of model Cayor-x275 and determine the sources of its deficiencies.

He randomly selects six glucose meters and tests each of them five times using five samples of blood with known sugar levels. The results are shown in Table 2.24.

TABLE 2.24

Blood sample	True sugar level	Meter measurement	Blood sample	True sugar level	Meter measurement
A	105	110	D	116	121
A	105	110	D	116	121
A	105	110	D	116	121
A	105	110	D	116	121
A	105	111	D	116	122
B	120	125	E	110	115
B	120	125	E	110	115
B	120	125	E	110	115
B	120	125	E	110	115
B	120	126	E	110	116
C	125	130	F	100	105
C	125	130	F	100	105
C	125	130	F	100	105
C	125	130	F	100	105
C	125	131	F	100	106

Solution: Open the file *Bloodsugar.MTW* and from the menu bar, click on **Stat**. From the drop down list, select **Quality Tools** and then **Gage Study** and then click on **Gage Linearity and Bias Study.** Enter "Blood samples" in the **Part Numbers** field, enter "True sugar level" in the **Reference Value** field, and enter "Meter measurements" in the **Measurement Data** field. In the **Process Variation** field, type in "**6**". Then click on the **OK** button to obtain the output in Fig. 2.34.

Interpretation The constant for the equation is equal to 5.2 with a slope of 0, and therefore a measure of linearity equal to 0 and percentage linearity equal to 0. The gauge is biased but linear. The variations in the measurements are proportional to the actual variations in the sugar levels in the blood pools. Linearity accounts for 0% of the variations and bias accounts for 86.7% of the variations.

Attribute Gauge Study

Some CTQ characteristics do not give the appraiser many options; the product meets a given standard or it does not. The gauge cannot quantify the degree to which the standards are met. The attribute gauge studies are like binary

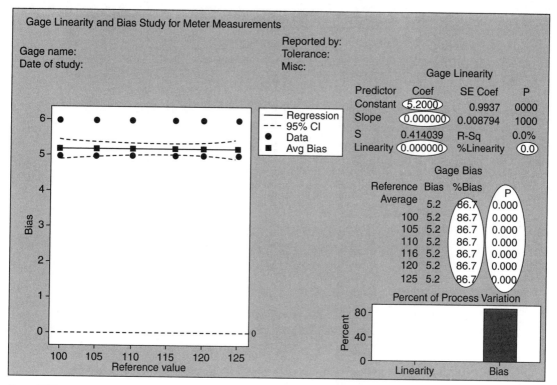

Figure 2.34

studies—when they compare the parts to a unique standard only two outcomes can occur, a pass or a fail, and there is no in-between. Attribute gauge studies estimate the amount of bias and repeatability of a measurement system when the response is a binary attribute variable.

Example 2.33 Galle-Basbe RFDM designs and manufactures radio frequency devices. It has received an important order from its customer Koussanar-Conveyors to manufacture laser optic barcode readers to control the traffic of products moved by conveyors. If the optical devices do not read the barcodes within the preset interval of time, products will pile up on the conveyors and jam them, and productivity would suffer. If they misread the barcodes, customers' shipments would be sent to the wrong destinations. It is therefore extremely important for the devices to be precise and accurate in order for them to be used in distribution and manufacturing plants.

After the design has been completed, a reliability engineer is charged with testing the devices for fitness for use. He collects a sample of 20 randomly chosen barcode readers, spreads them along a conveyor, places a box with barcodes on it in front of the readers, and lets the box pass through all the readers. When the box reaches the end of the conveyor, he repeats the test again with another box. When he is finished, he has another tester repeat the test exactly as he did with the same boxes and the same conveyors and barcode readers. He then summarizes the "pass" and "fail" data as shown in Table 2.25.

TABLE 2.25

Barcode Readers	Operator 1		Operator 2	
	Test 1	Test 2	Test 1	Test 2
1	P	P	F	P
2	P	P	P	P
3	P	P	P	P
4	P	P	P	P
5	P	P	P	P
6	P	P	P	P
7	F	P	P	P
8	P	P	P	P
9	P	P	P	P
10	P	P	P	P
11	P	P	P	P
12	F	F	F	F
13	P	P	P	P
14	P	P	P	P
15	P	P	P	P
16	P	P	F	P
17	F	F	P	F
18	P	P	P	P
19	P	P	P	P
20	P	P	P	P

A reader is only acceptable if it passes all tests; otherwise, it has to be rejected and the reason for failure investigated so that the production process to generate the reader can be improved.

SigmaXL output is shown in Table 2.26.

TABLE 2.26

Within Appraiser Agreement:		# Inspected	# Matched	Percent	95% LC	95% UC	Fleiss' Kappa	Fleiss' P-Value
	1	20	19	95.00	75.13	99.87	0.7714	0.0003
	2	20	17	85.00	62.11	96.79	0.3143	0.0799

Between Appraiser Agreement:	# Inspected	# Matched	Percent	95% LC	95% UC	Fleiss' Kappa	Fleiss' P-Value
	20	16	80.00	56.34	94.27	0.5429	0.0000

SigmaXL output provides two types of agreements, the within appraiser and the between appraiser. The statistics of interest here are the Fleiss Kappa coefficient of correlation, which is a chance corrected index of agreement for discrete data. It is a ratio of the observed excess over chance agreement to the maximum possible excess over chance. The coefficient is a within the range of -1 to $+1$. If it is equal to $+1$, there is a perfect agreement, if it is equal to 0, the observed agreement is equal to the chance agreement, and if it is equal to -1, then there is perfect disagreement. The Kappa value of 0.7714 indicates an acceptable measurement system. Operator 2 needs to improve. The between appraiser agreement shows grounds for improvement.

Assessing a Processes Ability to Meet Customers'
Expectations—Process Capability Analysis

Statistical process control (SPC) enables the producer to determine if his production process is stable and in control; in other words, if the process is yielding products that are consistent and in a manner that makes it possible for the producer to make predictions on future trends. What the SPC does not tell the producer is whether the products generated by his process meet the customers' expectations.

To determine if the process is generating products that meet customers' expectations, process capabilities indices are used. Process capabilities indices determine how effective a production process is at meeting customers' needs. The design engineers assess and determine the customers' needs (through techniques developed using surveys, quality function deployment (QFD), Kano analysis, etc.), which are then translated into CTQ characteristics and integrated in the designs of the products. The CTQs are those characteristics whose absence or lack of conformity would reduce quality of the products.

When we sign a contract with a cellular phone company, we have clear expectations that may have been explicitly written in the contract or implicitly agreed upon. We expect to be able to make and receive phone calls and be able to send and receive text messages, for instance. Therefore, clear reception of calls is critical to the quality of the service provided by the cellular phone company. If we try to make a call and it fails, the process used to provide us with the service we paid for would have failed to meet our expectations.

The engineers define the level of each CTQ that would optimize the customers' satisfaction and they specify that level as being the target that needs to be met when the production is in progress.

When we buy a bottle of fruit juice, we can see that some of the specified CTQs are already explicitly labeled on the bottle. For instance, we can see that the amount of carbohydrate is 45 g, the amount of sugar is 43 g, the amount of fat is 0 g, the total volume of juice in the bottle is 0.5 L, etc. Each one of these is critical to quality and is a specified target that the design engineers have determined to be the level of CTQ that would optimize the customers' satisfaction. However, the engineers know that no matter how well the production processes are designed, when they are in progress it will be impossible for them to only generate products that exactly meet the engineered specified targets. This is because variations (as explained in the section on statistical process control) are a constant, an inherent part of every production process. Therefore, it is impossible to produce identical products that all match the targets. Consequently, the engineers have to define a tolerance within which variations are deemed acceptable.

The target volume of juice in the bottles is 0.5 L, but if a customer buys a bottle that only contains 0.498 L or 0.502 L, he may not even notice the difference. Therefore, he would consider these two volumes as being acceptable. However, if he buys a bottle of juice and finds out that it only contains 0.40 L, he would be unlikely to buy from that manufacturer again because his expectations would not have been met. If the bottles happen to contain more than 0.60 L of juice, this would end up being prejudicial to the manufacturer because he would have been selling more for less but if the content in the bottles is 0.502 L, the manufacturer could consider it as acceptable.

So when determining the tolerance around the target, the design engineers need to consider both the customers' expectations and the cost of production. Based on customers' expectations and production cost, the engineers can determine that any bottle that contains between 0.498 and 0.502 L is good enough to be sold. In that case, the target for the volume of juice would be 0.5, and 0.498 and 0.502 would be called the upper specified limit (USL) and lower specified limit (LSL), respectively.

If the production process is set in such a way that every bottle that comes from the production lines contains a volume of juice that is within the specified interval, then the production process is said to be capable. If it generates bottles that are outside that interval, the process is then said to be incapable.

The specification of the target for the CTQ and the tolerance around the target is determined prior to starting the production process. Once production is in progress, the producer uses control charts to determine if the process is stable and in control using the control limits as indicators of control and stability. If the variations are contained within the control chart in a random manner, the process is deemed stable and in control. However, a stable and in control process does not necessarily mean that the entire production yield is within the engineered specified limits and therefore meets customers' expectations. This is because the specified limits and the control limits are two distinct and unrelated indicators. The design engineers determine the specified limits prior to the beginning of the production process, while the production process in progress generates the control limits. There is no statistical relationship between the two.

Example 2.34   In the case of the fruit juice producer, let us suppose that the target volume is still 0.5 L and that the engineered specified limits are USL = 0.502 and LSL = 0.498. The sample means in Table 2.27 are plotted on a control chart. We want to determine if the process has been generating only good quality products.

TABLE 2.27

Volume	0.5	0.509	0.506	0.505	0.503	0.506	0.493	0.507	0.502	0.5	0.499	0.508	0.5
Volume	0.5	0.509	0.5	0.492	0.504	0.497	0.496	0.504	0.5		0.5	0.498	0.504

Using SigmaXL, we obtain the control chart in Fig. 2.35. Figure 2.35 shows that the process is stable and in control with UCL = 0.51816, CL = 0.50, and LCL = 0.48448.

Figure 2.36 shows a probability plot that suggests normality; the samples are normally distributed within the control limits.

But Fig. 2.37 shows that even though the process is in control and stable, the spread of the data as shown by the histogram extends far outside the specified limits which are USL = 0.502 and LSL = 0.498. The part of the histogram on the left of the LSL and on the right of the USL represents the products that are of poor quality. So even though the process is stable and in control, it is still generating poor quality products.

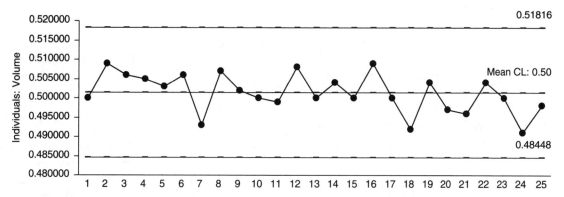

Figure 2.35

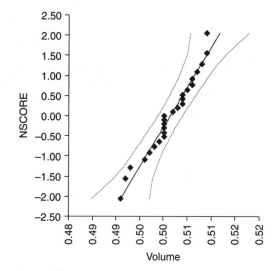

Figure 2.36

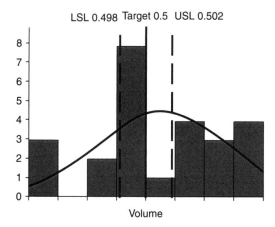

Figure 2.37

To determine if the production process is yielding good quality products, it is necessary to combine the product specifications with the control charts to generate capability indices used to assess precisely how capable the processes are. Several indices are used to measure process capabilities.

Process Capabilities with Normal Data

Process capability indices can be divided into two groups: the indices that measure the processes *potential* capabilities and the ones that measure their *actual* capabilities. The potential capability indices determine how capable a process would be if certain conditions were met, essentially if the mean of the processes' natural variability were centered to the target of the engineered specifications. The actual capability indices determine how the process is actually doing. The process capability indices are unitless; they are not expressed in terms of a predetermined unit of measurement. The characterization of the process capabilities should take into account the production process variations that occur while it is in progress. Those variations are measured in terms of how the process shifts from its original position due to a combination of common causes of variations and adjustments made to correct special causes. The processes' shifts have an impact on the spreads of the control charts and consequently the processes' ability to meet customers' expectations in the long run. When a process' ability to meet customers' expectations is assessed in the short term, process capability indices such as C_p, C_{pk}, and C_r are used and when the long-term performance is assessed, process performance indices such as P_p, P_{pk}, and P_r are used. The formulae used to compute the capability and the performance indices look very much alike with the exception that the main variable in the formulae that remains the value of sigma is different based on whether what is being addressed is the long-term or short-term variations.

Estimating Sigma

Since the process spread is always equal to six sigma for Shewhart control charts, the variable that explains the range between the upper control limit (UCL) and the lower control limit (LCL) is sigma, the standard deviation. The value of sigma depends on the short-term and the long-term variations. The short-term sigma accounts only for the natural variability of the process, while the long-term sigma takes into account the process shifts due to a combination of variations due to both common and assignable causes.

Short-term sigma

Short-term sigma as seen in Chap. 5 can be calculated in several ways.

- Standard deviation estimated based on $\overline{R}$

$$\bar{\sigma}_{ST} = \frac{\overline{R}}{d_2}$$

- Standard deviation estimated based on samples

$$\tilde{\sigma}_{ST} = \frac{\bar{s}}{c_4}$$

- Standard deviation estimated for moving range

$$\tilde{\sigma}_{ST} = \frac{\overline{MR}}{d_2} = \frac{\overline{MR}}{1.128}$$

Long-term sigma

Long-term sigma is used to calculate the measure of the expected process performance over a period of time for the process to generate the conforming 99.73% of its output.

$$\tilde{\sigma}_{LT} = \frac{\sqrt{\dfrac{\sum (x_i - \bar{x})^2}{kn - 1}}}{c_{4kn}}$$

where the numerator is the standard deviation based on all the k samples of n measurements in the control charts. The c_4 term is used as a bias factor.

Potential Capabilities

The potential capability indices indicate how capable the process would be at meeting customers' expectations if certain conditions were met. The potential indices that are more frequently used are C_p and C_r. The output for most production processes is normally distributed. For a sigma scaled normal graph, 99.73% of the observations would be concentrated between $\pm 3\sigma$ from the mean, within a 6σ range.

Short-term potential capabilities, C_p and C_r

Process capabilities are generated by comparing the spread of the control charts to the one of the product specification. A process is said to be potentially capable if the spread of the natural variations is smaller than the spread of the specified limits. This is so when the ratio of the specified range to the one of the control limits is greater than 1. In other words, the following ratio should be greater than 1.

$$C_p = \frac{USL - LSL}{UCL - LCL}$$

The range of the control charts is obtained by subtracting the LCL from the UCL and since $UCL = \mu + 3\sigma_{ST}$ and $LCL = \mu - 3\sigma_{ST}$,

$$\text{Range} = UCL - LCL = (\mu + 3\sigma_{ST}) - (\mu - 3\sigma_{ST}) = \mu + 3\sigma_{ST} - \mu + 3\sigma_{ST} = 6\sigma_{ST}$$

Therefore, the range of the control chart is always equal to $6\sigma_{ST}$.
  Therefore,

$$C_p = \frac{USL - LSL}{UCL - LCL} = \frac{USL - LSL}{6\sigma_{ST}}$$

$C_p = 1$ if the specified range equals the range of the natural variations of the process, in which case the process is said to be barely capable. It has the *potential* to produce only nondefective products if the process mean is centered to the specified target. Approximately 0.27% or 2700 parts per million are defective.
  $C_p > 1$ if the specified range is greater than the range of the control limits. The process is *potentially* capable if the process mean is centered to the engineered specified target and is (probably) producing products that meet or exceed the customers' requirements.
  $C_p < 1$ if the specified range is smaller than the range of the control limits and the process is said to be incapable; in other words, the company is producing junk.
  Another way of expressing the short-term potential capability would be through the use of the capability ratio, C_r. It indicates the proportion or percentage of the specified spread that is needed to contain the process range for the production process to be capable. Let us note that it is not the proportion that is necessarily actually occupied.

$$C_r = \frac{1}{C_p} = \frac{UCL - LCL}{USL - LSL} = \frac{6\sigma_{ST}}{USL - LSL}$$

Example 2.35   The specified limits for the diameter of car tires are 15.6 for the upper limit and 15 for the lower limit with a process mean of 15.3 and a standard deviation of 0.09. Find C_p and C_r. What can we say about the process' capabilities?

Solution:

$$C_p = \frac{USL - LSL}{6\sigma} = \frac{15.6 - 15}{6(0.09)} = \frac{0.6}{0.54} = 1.111$$

$$C_p = 1.111$$

$$C_r = \frac{1}{1.111} = 0.9$$

Since C_p is greater than 1 and therefore C_r is less than 1, we can conclude that the process is potentially capable if the process mean is centered to the specified target. SigmaXL template gives the output shown in Table 2.28.

TABLE 2.28

Calculate Process Capability Indices: Cp, Cpk; Pp, Ppk				
Enter Mean:			X-Bar	15.3
Enter Standard Deviation:			S	0.09
Enter USL:				15.6
Enter LSL:				15
			Cp, Pp	1.11
			Cpk, Ppk	1.11
			Cpu, Ppu	1.11
			Cpl, Ppl	1.11

Long-term potential performance

The long-term process performance is calculated using similar methods as C_p and C_{pk} with the difference being that the standard deviation used in this case applies to the long-term variations

$$P_p = \frac{USL - LSL}{6\sigma_{LT}}$$

and the performance ratio is

$$P_r = \frac{6\sigma_{LT}}{USL - LSL}$$

Actual capabilities

The potential capability indices only consider the range of the specified limits and the spread of the process. Those indices only show potentialities because the spread of the process natural variability can be smaller than the specified range while the process is still generating defects. The right side of Fig. 2.38 shows a control chart with a spread that is smaller than the specified range but because the process is not centered, the process is generating defects.

If the process mean is not centered to the specified target, C_p would only tell which of the two ranges (process control limits and engineered specified limits) is wider but it would not be able to inform on whether the process is generating defects. In that case, another capability index is used to determine a process'

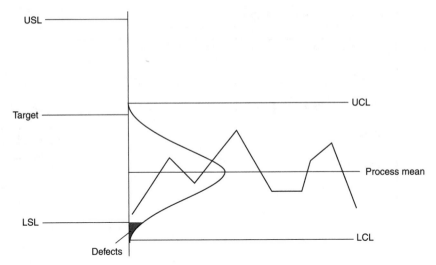

Figure 2.38

ability to respond to customer requirements. The C_{pk} measures how much of the production process really conforms to the engineered specifications. The k in C_{pk} is called the k-factor; it measures the level of variation of the process mean from the specified target.

$$C_{pk} = (1-k)C_p$$

where

$$k = \frac{\left| (USL + LSL)/2 - \overline{\overline{X}} \right|}{(USL - LSL)/2}$$

$k = 0$ means that the process is perfectly centered and therefore $C_{pk} = C_p$.

If $k \neq 0$, then $C_{pk} = \min\{C_U, C_L\}$, C_{pk} would be the smaller of the two. where

$$C_{UL} = \frac{1}{3} Z_{UL}$$

$$C_{LL} = \frac{1}{3} Z_{LL}$$

and

$$Z_{UL} = \frac{USL - \overline{\overline{X}}}{\sigma_{ST}}$$

and

$$Z_{LL} = \frac{\overline{\overline{X}} - LSL}{\sigma_{ST}}$$

The long-term process capability indices are obtained similarly

$$P_{pk} = \min \left\{ \frac{1}{3} Z_{UL_{LT}}, \frac{1}{3} Z_{LL_{LT}} \right\}$$

where

$$Z_{UL_{LT}} = \frac{USL - \overline{\overline{X}}}{\sigma_{LT}}$$

and

$$Z_{LL_{LT}} = \frac{\overline{\overline{X}} - LSL}{\sigma_{LT}}$$

Capability indices and parts per million

The process capability indices are unitless; in other words, they are not expressed in terms of meters, pounds, or grams. For that reason, it is not always easy to explain their meaning to a person who is not statistics savvy.

For instance, if at the measure phase of a Six Sigma project, the Cpk is found to be 0.78 and at the end of the project, after improvement, it becomes 1.26, all that can be said is that there has been improvement in the production process. However, based only on these two numbers, one cannot easily explain to a person who is not statistics savvy the amount of reduction of defects from the process. The quantification of the parts per million (PPM) that fall outside the specified limits can help alleviate that shortcoming. PPM measures how many parts out of every million produced are defective. Estimating the number of defective parts out of every million produced makes it easier for anyone to visualize and understand the quality level of a production process.

Process capability and Z transformation

For normally distributed data, it is easy to show the relationship between C_{pk} and the Z distribution and, from the Z transformation, the number of defective PPM can be estimated.

Remember the Z formula from the normal distribution

$$Z = \frac{X - \mu}{\sigma}$$

If

$$C_{UL} = \frac{1}{3} Z_{UL} = \frac{USL - \overline{\overline{X}}}{3\sigma_{ST}}$$

and

$$C_{LL} = \frac{1}{3} Z_{LL} = \frac{\overline{\overline{X}} - LSL}{3\sigma_{ST}}$$

In other words,

$$C_{pk} = \frac{Z_{min}}{3}$$

or

$$Z_{min} = 3C_{pk}$$

That formula enabled us to calculate the probability for an event to happen and the cumulative probability for the event to take place if the data being considered are normally distributed. The same formulae can be used to calculate the PPM. The total PPM is obtained by adding the PPM on each side of the specified limits.

$$PPM_{UL} = 10^6 \Omega(Z_{UL}) = 10^6 \Omega\left(\frac{USL - \mu}{\sigma}\right)$$

$$PPM_{LL} = 10^6 \Omega(Z_{LL}) = 10^6 \Omega\left(\frac{\mu - LSL}{\sigma}\right)$$

$\Omega(Z_{UL})$ and $\Omega(Z_{LL})$ represent the values of Z_{LL} and Z_{UL} obtained from the normal probability table.

$$PPM = PPM_{LL} + PPM_{UL}$$

There is a constant relationship between C_{pk}, Z_{min}, and PPM when the process is centered (see Table 2.29).

Example 2.36   The weight of a brake drum is critical to quality. The specifications for the weight are 9.96 and 10.04 lb for the USL and the LSL, respectively, with a target of 10 lb. The data in Table 2.30 represent samples that were used to construct the control charts used to monitor the production process. What can we say about the process capability?

Solution: The process mean is once again obtained while the production is in progress. It is equal to 9.997202 and the process standard deviation is 0.019966. Therefore,

$$C_p = \frac{10.04 - 9.96}{6(0.019966)} = 0.668$$

$$C_{UL} = \frac{10.04 - 9.997202}{3(0.01996)} = 0.715$$

$$C_{LL} = \frac{9.997202 - 9.96}{3(0.019966)} = 0.621$$

Since C_{pk} is equal to the lowest value between C_U and C_L, it is therefore equal to 0.62.

TABLE 2.29

C_{pk}	Z_{min}	PPM
0.50	1.50	133,600
0.52	1.56	118,760
0.55	1.64	100,000
0.78	2.33	20,000
0.83	2.50	12,400
1.00	3.00	2,700
1.10	3.29	1,000
1.20	3.60	318
1.24	3.72	200
1.27	3.80	145
1.30	3.89	100
1.33	4.00	63
1.40	4.20	27
1.47	4.42	10
1.50	4.40	7.00
1.58	4.75	2.00
1.63	4.89	1.00
1.67	5.00	0.600
1.73	5.20	0.200
2.00	6.00	0.002

TABLE 2.30

Sample ID	M1	M2	M3	M4	M5	$\overline{X}$	R
1	9.9668	10.0025	10.0208	9.991	9.9905	9.99432	0.054
2	9.9922	9.9838	10.0006	9.9966	10.0073	9.9961	0.0235
3	10.0041	9.9365	9.9718	10.006	9.968	9.97728	0.0695
4	10.0113	9.9771	10.0159	10.0345	9.9914	10.00604	0.0574
5	10.0035	9.995	10.0042	10.0087	10.0073	10.00374	0.0137
6	10.0501	9.9736	10.0045	9.9924	9.9918	10.00248	0.0765
7	10.0044	10.0178	10.0152	9.9769	10.0285	10.00856	0.0516
8	9.9791	10.0225	9.9632	9.9981	9.9897	9.99052	0.0593
9	9.9967	9.9964	9.9924	10.0117	9.9953	9.9985	0.0193
10	10.0359	9.9874	10.0377	10.0326	9.9819	10.0151	0.0558
11	9.9722	10.0142	9.9755	9.9906	10.0228	9.99506	0.0506
12	9.9894	9.9608	10.0243	10.0098	9.9882	9.9945	0.0635
13	10.0071	9.9956	9.9946	10.0148	10.015	10.00542	0.0204
14	9.9998	9.9849	9.9978	9.9773	10.0168	9.99532	0.0395
15	10.0274	9.9941	10.0037	10.0048	9.9852	10.00304	0.0422
16	9.9331	9.9751	10.0002	9.978	9.9891	9.9751	0.0671
17	9.9966	10.0259	9.9685	9.9809	9.997	9.99378	0.0574
18	9.9962	10.0133	9.99	10.0208	10.0104	10.00614	0.0308
19	9.98	9.9763	9.9829	10.01	9.9826	9.98636	0.0337
20	9.9855	9.9892	10.0018	10.0212	9.9807	9.99568	0.0405
21	9.9746	9.9986	10.0357	10.0484	10.0147	10.0144	0.0738
22	9.994	9.9767	9.9772	9.9897	10	9.98752	0.0233
23	10.0058	9.9851	10.0145	9.9947	9.9735	9.99472	0.041
24	10.0076	9.9872	9.9705	10.0047	9.9825	9.9905	0.0371
25	9.9794	9.9923	10.0202	10.0041	10.0033	9.99986	0.0408
					Mean	9.997202	0.045692
					Standard Dev	0.019966	

$$PPM_{UL} = 10^6 \Omega(Z_{UL}) = 10^6 \Omega\left(\frac{USL-\mu}{\sigma}\right) = 10^6 \Omega\left(\frac{10.04-9.997202}{0.01996}\right) = 10^6 \Omega(2.144)$$

On the Z score table, 2.14 corresponds to 0.4838

Thus, alpha (defects) = $0.5 - 0.4838 = 0.0162$

$$PPM_{UL} = 10^6 (0.0162) = 16200$$

$$PPM_{LL} = 10^6 \Omega(Z_{LL}) = 10^6 \Omega\left(\frac{\mu-LSL}{\sigma}\right) = 10^6 \Omega\left(\frac{9.997202-9.96}{0.01996}\right) = 10^6 \Omega(1.864)$$

On the Z score table, 1.86 corresponds to 0.4686

Thus, alpha (defects) = $0.5 - 0.4686 = 0.0314$

$$PPM_{LL} = 10^6 (0.0314) = 31400$$

$$\text{Total } PPM = PPM_{LL} + PPM_{UL} = 16200 + 31400 = 47600$$

Potential (within) capability

C_p	0.668
CPL	0.62
CPU	0.715
C_{pk}	0.62

Expected (overall) performance

PPM < LSL	31400
PPM > USL	16200
PPM Total	47600

The C_p is equal to 0.668, which means that the process is not even potentially capable. The C_{pk} is equal to 0.62; therefore the process is generating defects that the total PPM estimates to 47,6000 PPM.

Figure 2.39 summarizes how the defects are spread about the specified limits.

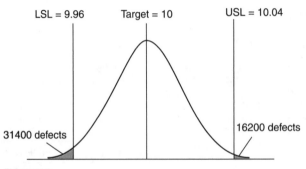

Figure 2.39

Using SigmaXL to verify the results, open the file *Brakedrum.xls* and select the area containing the data. From the menu bar, click on **SigmaXL**, then select **Process Capability.** From the submenu, click on **Capability Combination Report (Subgroup)** (see Fig. 2.40).

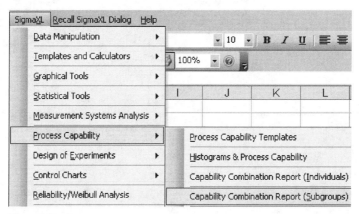

Figure 2.40

The **Capability Combination** dialog box should appear with the field **Please select your data** already filled (Fig. 2.41).

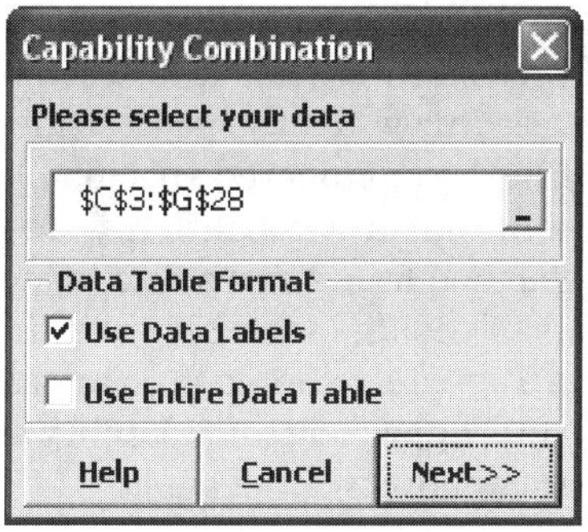

Figure 2.41

Press the **Next>>** button. The **Capability Combination Report (subgroup)** box appears. Make sure that the option **Subgroup across rows** (2 or more numeric data columns) is selected and fill out the dialog box as indicated in Fig. 2.42.

Press the **OK>>** button to get the results shown in Table 2.31. The results are used to create an X-bar chart shown in Fig. 2.43.

Minitab output

The minor differences in the results of SigmaXL, Minitab, and the computations that we performed are due to the rounding effects.

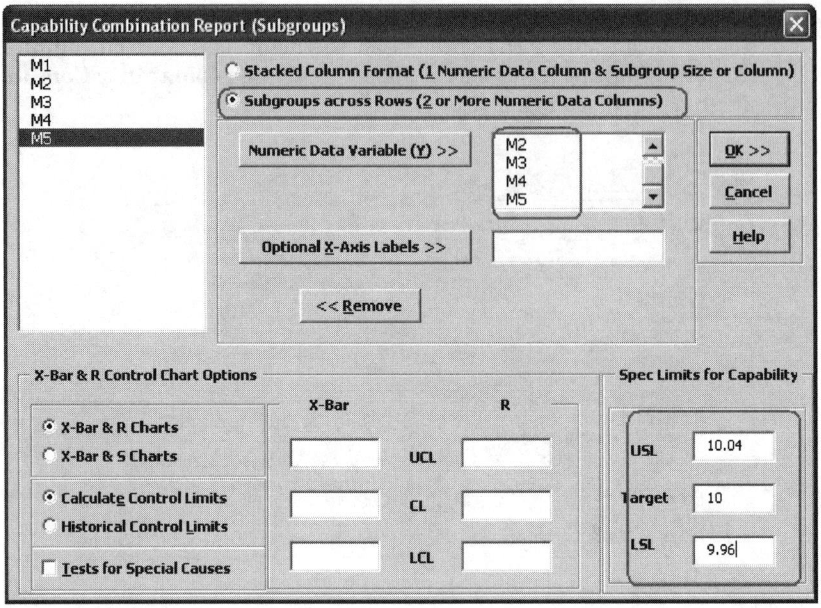

Figure 2.42

Taguchi's capability indices C_{PM} and P_{PM}

Thus far, all the indices that were used (C_p, C_{pk}, P_p, P_{pk}, and C_r) only considered the specified limits, the standard deviation, and, in the cases of C_{pk} and P_{pk}, the production process mean. None of these indices take into account the variations within tolerance, the variations that occur while the process mean fails to meet the specified target but is still within the engineered specified limits. Taguchi's approach to process control suggests that any variation from the engineered target, be it within or outside the specified limits, is a source of defects and a loss to society. That loss is proportional to the distance between the process mean and the specified target. Because of Taguchi's approach to tolerance around the engineered target, the definition and approach to capability measures differ from that of the traditional process capability analysis.

Figure 2.44 shows a process that is perfectly with the specified limits with a C_{pk} of 1.05, but because all the observations do not match the target, C_{pm} is less than 1. According to Taguchi, this process should be considered incapable. Figure 2.45 shows a process depicted by a histogram that is fully within the specified limits. For that reason, C_p, C_{pk}, P_p, and P_{pk} are all greater than 1. Therefore, the process is deemed capable but the process mean fails to meet the specified target of 14.5. Consequently, according to Taguchi, the process should not be considered capable.

The index used to measure Taguchi's process capability is C_{pm} and in this case, it is equal to 0.62.

$$C_{pm} = \frac{USL - LSL}{6\sqrt{\sigma_{ST}^2 + (\mu - T)^2}}$$

TABLE 2.31

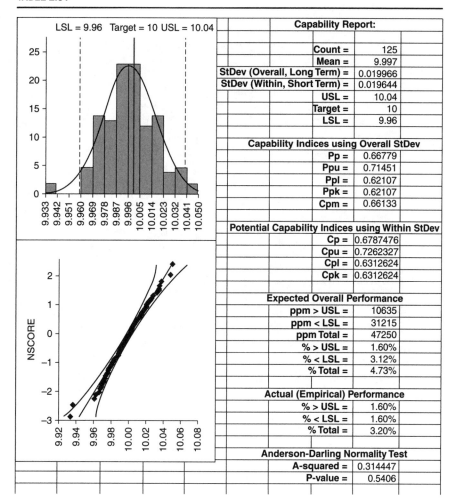

Capability Report:	
Count =	125
Mean =	9.997
StDev (Overall, Long Term) =	0.019966
StDev (Within, Short Term) =	0.019644
USL =	10.04
Target =	10
LSL =	9.96
Capability Indices using Overall StDev	
Pp =	0.66779
Ppu =	0.71451
Ppl =	0.62107
Ppk =	0.62107
Cpm =	0.66133
Potential Capability Indices using Within StDev	
Cp =	0.6787476
Cpu =	0.7262327
Cpl =	0.6312624
Cpk =	0.6312624
Expected Overall Performance	
ppm > USL =	10635
ppm < LSL =	31215
ppm Total =	47250
% > USL =	1.60%
% < LSL =	3.12%
% Total =	4.73%
Actual (Empirical) Performance	
% > USL =	1.60%
% < LSL =	1.60%
% Total =	3.20%
Anderson-Darling Normality Test	
A-squared =	0.314447
P-value =	0.5406

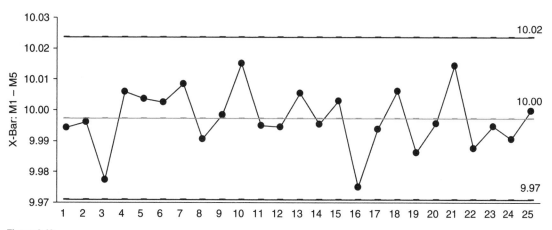

Figure 2.43

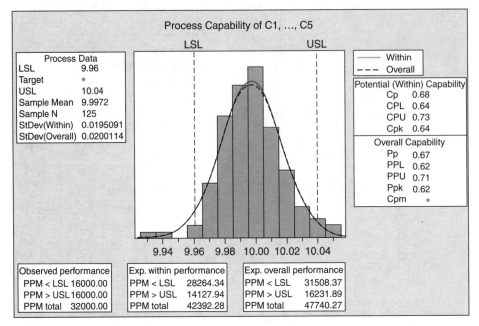

Figure 2.44

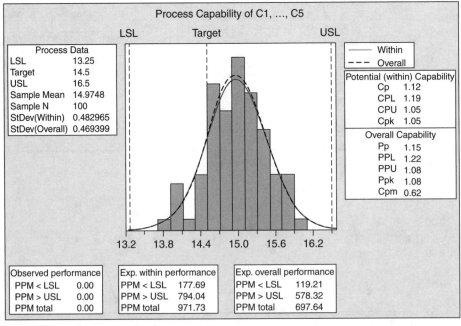

Figure 2.45

Therefore, C_{pm} is a function of both the process short-term variance and the difference between the process mean μ and the specified target T. If the process mean meets the specified target, $\mu - T = 0$, and therefore

$$C_{pm} = \frac{USL - LSL}{6\sqrt{\sigma_{ST}^2 + (\mu - T)^2}} = \frac{USL - LSL}{6\sqrt{\sigma_{ST}^2 + 0}} = \frac{USL - LSL}{6\sigma_{ST}} = C_p$$

Example 2.37   A machine produces parts with the following specified limits:

$$USL = 30$$

$$LSL = 28$$

$$\text{Specified target} = 29$$

The standard deviation is determined to be 0.50 and the process mean 29.5. Find the value of C_{pm}. Compare the C_{pm} with C_{pk}.

If the process mean were equal to 29, what could we say about C_{pm} and C_p?

Solution: The value of C_{pm} is

$$C_{pm} = \frac{USL - LSL}{6\sqrt{\sigma_{ST}^2 + (\mu - T)^2}} = \frac{30 - 28}{6\sqrt{(0.50)^2 + (29.5 - 29)^2}} = \frac{2}{6\sqrt{0.25 + 0.25}} = \frac{2}{4.25} = 0.471$$

Comparing C_{pm} and C_{pk}

$$C_{UL} = \frac{30 - 29.5}{3(0.50)} = 0.333$$

$$C_{LL} = \frac{29.5 - 29}{3(0.5)} = 0.333$$

Therefore,

$$C_{pk} = 0.333$$

C_{pk} is significantly lower than C_{pm}.

If the process mean met the target of 29,

$$C_{pm} = \frac{USL - LSL}{6\sqrt{\sigma_{ST}^2 + (\mu - T)^2}} = \frac{30 - 28}{6\sqrt{(0.50)^2 + (29 - 29)}} = \frac{2}{6(0.50)} = 0.667$$

and

$$C_p = \frac{USL - LSL}{6\sigma} = \frac{2}{6(0.50)} = 0.667$$

C_p would have been equal to C_{pm}.

Example 2.38   The amount of inventory kept at Touba Warehouse is critical to the performance of that plant. The objective is to have an average of 28-day supply of inventory with a tolerance of USL of 33 and an LSL of 24.

The data on the file *Touba-Warehouse.MTW* represent a sample of DSI.

- Run a capability analysis to determine if the production process used thus far has been capable.
- Is there a difference between C_{pm} and C_p? Why?
- The tolerance limits and the target have been kept as they are, but the process mean has been improved to where it meets the target of 28. What effect would that improvement have on C_p as compared to C_{pm}?

Solution: Open the file *Touba warehouseI.MPJ*. On the tool bar, click on **Stat**, then on **Quality Tools,** select **Capability Analysis,** and click on **Normal.** Fill out the dialog box as indicated in Fig. 2.46.

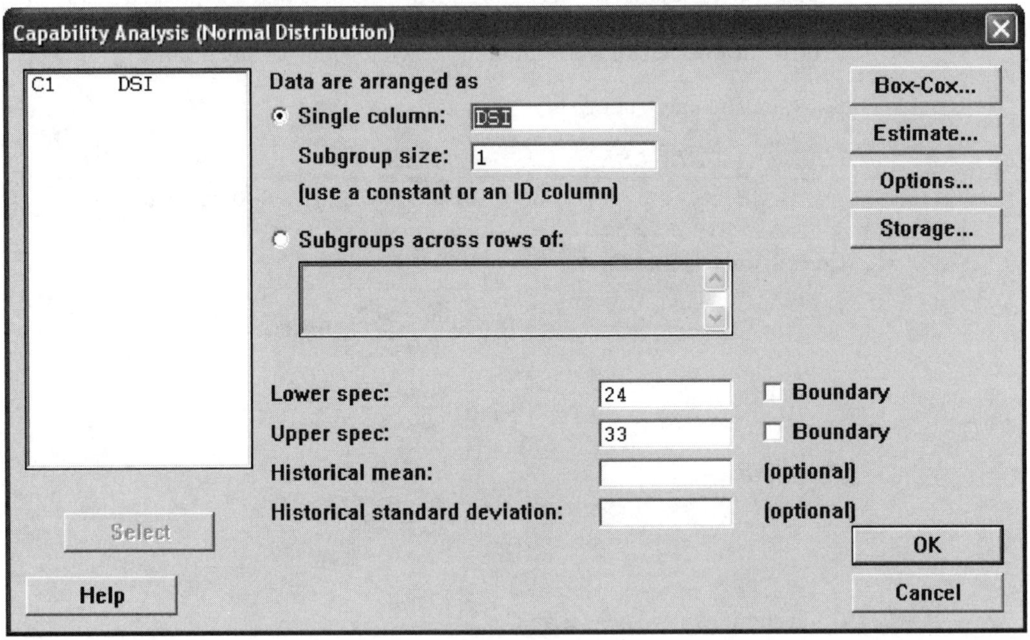

Figure 2.46

Click on **Options...** and on the **Capability Analysis Options** dialog box, type in 28 in the **Target (add C_{pm} to table)** field. Leave 6 in the **K** field. Click on **OK** and then on **OK** again to get the output in Fig. 2.47.

Interpretation   The data plot shows that all the observations are well within the specified limits and not a single one comes anywhere close to any one of the limits, yet all of them are concentrated in-between the LSL and the target. The fact that not a single observation is outside the specified limits generated a PPM equal to 0 for the observed performance. $C_{pk} = 1.06$ suggests that the process is capable. However,

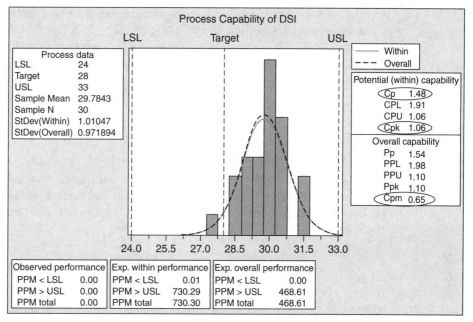

Figure 2.47

from Taguchi's approach, the process with a $C_{pm} = 0.65$ is not capable because even though all the observations are within the specified limits, the process is not centered to the target.

Process Capability Analysis with Nonnormal Data

Thus far, one of the assumptions for a process capability analysis has been the normality of the data. C_{pk}, P_{pk}, and parts per million (PPM) were calculated using the Z transformation, therefore assuming that the data being analyzed were normally distributed. If we elect to use the normal option in Minitab for process capability analysis and the normality assumption is violated because the data are skewed in one way or another, the resulting values of C_{pk}, C_p, P_p, P_{pk}, and PPM would not reflect the actual process capability.

Not all process outputs are normally distributed. The daily numbers of calls or the call times at a call center, for instance, are not in general normally distributed unless a special event makes it so. In a distribution center where tens of employees pick, pack, and ship products, the overall error rate at inspection is not normally distributed because it depends on many factors, such as training, the mood of the pickers, the Standard Operating Procedures (SOPs), etc.

It is advised to test the normality of the data being assessed before conducting a capability analysis.

There are several ways process capabilities can be assessed when the data are not normal:

- If the subsets that compose the data are normal, the capabilities of the subsets can be assessed and their PPMs aggregated.

- If the subsets are not normal and the data can be transformed using the Box-Cox or natural log for parametric data or Logit transformation for binary data, transform the data before conducting the analysis.

- Use other distributions to calculate the PPM.

Normality Assumption and Box-Cox Transformation

One way to overcome the nonnormality of the data is to use the Box-Cox transformation. The Box-Cox transformation converts the observations into an approximately normal set of data. The formula for the transformation is given as

$$y = \frac{x^\lambda - 1}{\lambda}$$

where y is the response factor and λ is the transformation parameter.

If $\lambda = 0$, the denominator would equal 0. To avoid that hurdle, the natural log would be used instead.

$$y = \ln x$$

Example 2.39   The data in the file *Normality.xls* is known to follow Poisson distribution. Using the Box-Cox transformation in SigmaXL, normalize the data.

Solution: Open SigmaXL, then open the file *Normality.xls*. Select the area that contains the data before clicking on the **SigmaXL** from the menu bar. Click on **Data Manipulation** on the menu and from the submenu, select **Box-Cox transformation**. When the **Box-Cox transformation** dialog box appears, press the **Next>>** button, then click on **Numerical Data Variable (Y)>>**, and then click on the **OK>>** button to get the normalized data (Table 2.32).

SigmaXL displays the initial data and the normalized data along with the p-value for the Anderson-Darling test.

Example 2.40   Open the file *boxcoxtrans.MPJ* and transform the data.

Solution: Open the file *boxcoxtrans.MPJ*. Click on **Stat**, select **Control Charts,** and then click on **Box-Cox transformation.** In the **Box-Cox transformation** dialog box, leave "All observations for the chart are in one column" in the first field. Select **C1** for the second textbox. Type 1 in **subgroup size** field. Click on the **Options** button and type C2 in "store transformed data in:" Click **OK** twice.

The system should generate a second column that contains the data yielded by the transformation process. The normality of the data in column C2 can be tested using the probability plot. The graph in Fig. 2.48 plots the data before and after transformation.

TABLE 2.32

Box-Cox Power Transformation:Normality		Normality	Transformed Data (Y**0.50)
		17	4.123
Optimal Lambda	0.340000	14	3.742
Final Lambda	0.500000	14	3.742
UC Lambda (95%)	2.196	8	2.828
LC Lambda (95%)	−1.431	15	3.873
Anderson darling normality test for transformed data:		14	3.742
A-squared	0.306527	22	4.690
AD P-value	0.5438	12	3.464
		10	3.162
		24	4.899
		13	3.606
		23	4.796
		18	4.243
		17	4.123
		12	3.464
		14	3.742
		18	4.243
		10	3.162
		16	4
		13	3.606
		9	3
		14	3.742
		12	3.464
		11	3.317
		6	2.449
		15	3.873
		12	3.464
		10	3.162
		17	4.123
		12	3.464

Box-Cox Power Transformation: Normality

The Anderson-Darling hypothesis testing for normality shows an infinitesimal p-value of less than 0.005 for C1 (before transformation), which indicates that the data are not normally distributed. The same hypothesis testing for C2 (after transformation) shows a p-value of 0.819 and the graph clearly shows normality.

Process Capability Using Box-Cox Transformation

Example 2.41   The data in the file *downtime.MPJ* measure the time between machine breakdowns. A normality test has revealed that the data are far from being normally distributed; in fact, they follow an exponential distribution. Yet, if the engineered specified limits for the downtimes were set at 0 for the LSL and 25 for the USL, and if we run a capability test assuming normality, we would end up with the results in Fig. 2.49.

It is clear that no matter what type of unit of measurement is being used, the time between machine breakdowns cannot be negative. We set the LSL at 0, but the PPM

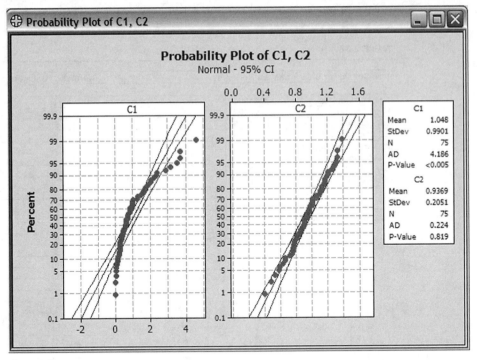

Figure 2.48

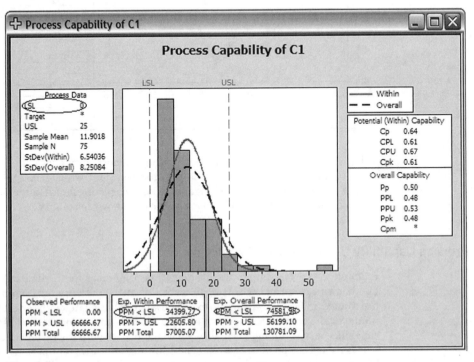

Figure 2.49

for the lower specification is 34399.27 for the within performance and 74581.98 for the overall performance. This suggests that some machines might break down at negative units of time measurements. This is because the normal probability Z transformation was used to calculate the probability for the machine breakdowns to occur even though the distribution is exponential.

One way to correct that problem is through the transformation, the normalization of the data. For this example, we will use the Box-Cox transformation and instead of setting the lower limit at 0, we increase it to 1 unit of time measurement. The process of estimating the process capabilities using Minitab is the same as the one we previously did with the exception that

- We have to click on the **Box-Cox...** button.

- Put a check mark by Box-Cox Power Transformation (W = Y**Lambda).

- We leave the option at **Use optimal Lambda** and click on the **OK** button to obtain the output in Fig. 2.50.

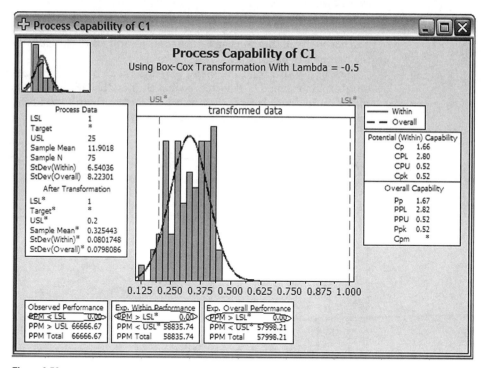

Figure 2.50

The process is still incapable but in this case, the transformation has yielded a PPM equal to 0 for the lower specification. In other words, the probability for the process to generate machine breakdowns at less than 0 units of measurements is 0.

Example 2.42 WuroSogui-Stream is a call center that processes customer complaints over the phone. The longer the customer services associates stay on the phone with the customers, the more associates will be needed to cater to the customers' needs, which would result in extra operating cost for the center. The quality control department set the specifications for the time that the associates are required to stay on the phone with the customers. They are expected to expedite the customer's concerns in 10 min or less. Therefore, in this case, there is no LSL and the USL is 10 with a target of 5 min. The file *wurossogui.MPJ* contains data used to create a control chart to monitor production process at the call center.

1. What can be said about the normality of the data?
2. What happens if a normal process capability analysis is conducted?
3. If the data are not normally distributed, run a process capability analysis with a Box-Cox transformation
4. Is the process capable?
5. If the organization operates under Taguchi's principles, what could we say about the process capabilities?
6. Compare C_{pk} with C_{pm}.
7. What percentage (not PPM) of the parts produced is likely to be defective for the overall performance?

Solution:

1. The normality of the data can be tested in several ways; the easiest way is through the probability plot.

   Click on **Graph** from the menu bar, and then click on **probability plot.** The **single** option should be selected, so just click on the **OK** button. The **Probability plot-Single** dialog box pops up, select **C1** for the **Graph variable** textbox before clicking on the **OK** button. The graph in Fig. 2.51 pops up.

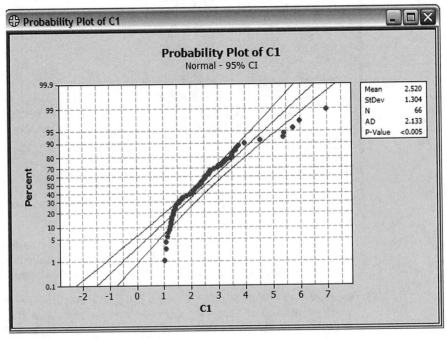

Figure 2.51

The graph itself shows that the data are not normally distributed for a confidence interval of 95%. Many of the dots are scattered outside confidence limits and the Anderson-Darling hypothesis test for normality yielded an infinitesimal p-value of less than 0.005. Therefore, we have to conclude that the data are not normally distributed.

2. If we had conducted a process normal capability analysis, we would have obtained a C_{pk} and PPM that were calculated based on the normal Z transformation. Since the Z transformation cannot be used to calculate a process capability for non-normal data unless the data have been normalized, the results obtained would have been misleading.

3. Open the file *wurossogui.MPJ*. From the menu bar, click on **Stat,** then select **Quality Tools**, then select **Capability Analysis** from the drop down list and click on **Normal.** Select the single column option and select **C1** for that field. For **Subgroup size**, type 1. Leave the **Lower Spec** field empty and type 10 in the **Upper Spec** field. Click on the **Box-Cox** button. Put a check mark by **Box-Cox power transformation (W = Y**Lambda)** and click on the **OK** button. Click on the **Options** button and type 5 in the **Target (adds CPM to table)** field. Put a check mark by **Include confidence interval** and click on the **OK** button. Then click on **OK** again to obtain the graph in Fig. 2.52.

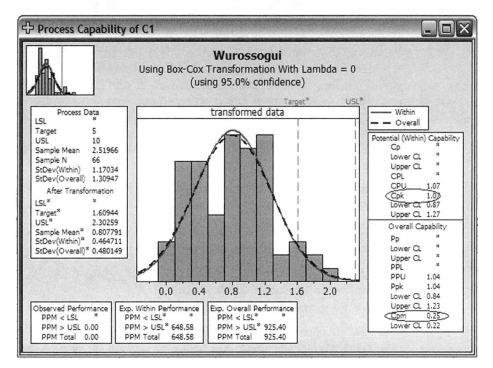

Figure 2.52

4. Based on the $C_{pk} = 1.07$, we can conclude that the process is barely capable and that the results show opportunities for improvement.

5. If the organization operates under Taguchi's principles, we would have to conclude that the process is incapable because $C_{pm} = 0.25$ and this is because while all the

observations are within the specified limits, the process is not centered to the target and most of the observations do not match the target 1.6.

6. $C_{pk} = 1.07$ and $C_{pm} = 0.25$. The difference is explained by the fact that Taguchi's approach is very restrictive because the process is not centered to the target; the process is not considered capable.

7. For the overall performance $PPM = 925.4$, the percentage of the parts that are expected to be defective will be

$$PPM \times \frac{100}{10^6}$$

$$925.4 \times \frac{100}{10^6} = 925.4 \times 10^{-4} = 0.09254$$

Therefore, 0.093% of the parts are expected to be defective.

Process Capability Using Nonnormal Distribution

If the data being analyzed were not normally distributed, an alternative to using a transformation process to run a capability analysis as if the data were normal would be to use the probability distribution that the data actually follow. If, for instance, the data being used to assess capability follow a Weibull or lognormal distribution, it is possible to run a test with Minitab. In these cases, the analysis would not be done using the Z transformation and therefore C_{pk} would not be provided because it is based on the Z formula. The values of Pp and Ppk are not obtained based on the mean and the standard deviation, but rather on the parameters of the particular distributions that the observations follow. In the case of the Weibull distribution, for instance, the shape and the scale of the observations are used to estimate the probability for the event being considered to happen.

Example 2.43   Futa-Toro Electronics manufactures circuit boards. The engineered specification of the failure time of the embedded processors is no less than 45 months. Samples of circuit boards have been taken for testing and they have generated the data in the file *Futa Toro.MPJ*. The data give the lifetime of the processors. The observations have proved to follow a Weibull distribution.

1. Without transforming the data, what is the expected overall capability of the process that generated the processors?
2. What is the expected PPM?

Solution: The process has only one specified limit; because the lifetime of the processors is expected to last more than 45 months, there is no upper specification. The capability analysis will be conducted using the Weibull option.

Open the file *Futa Toro.MPJ*. From the menu bar, click on **Stat**, then select **Quality Tools** and, from the drop down list, select **Capability Analysis,** and click on **Nonnormal**. In the **Capability Analysis (Nonnormal Distribution)** dialog box, select **C1 Lifetime** for the **Single Column** field, select **Weibull** from the **Distribution:** drop down list and type in 45 in the **Lower spec field**. Leave the **Upper Spec** field empty. Then click on the **OK** button.

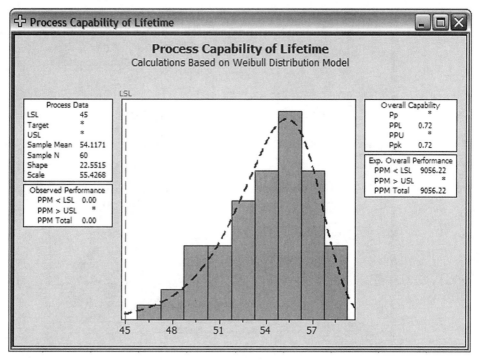

Figure 2.53

Figure 2.53 should pop up.

There is only one specified limit and it is the LSL; therefore, the *Ppk* will be based solely on the PPL, which is equal to 0.72. *Ppk* is a lot lower than the threshold of 1.33. We have to conclude that the process is not capable.

The expected overall PPM is 9056.2.

Example 2.44   The purity level of a metal alloy produced at the Sabadola Gold Mines is critical to the quality of the metal. The engineered specifications have been set to 99.0715% or more. The data contained in the file *sabadola.MPJ* represent samples taken to monitor the production process at Sabadola Gold Mines. The data have proved to be lognormally distributed. How capable is the production process, and what is the overall expected PPM?

Solution: Open the file *sabadola.MPJ*. From the menu bar, click on **Stat,** then select **Quality Tools,** and from the drop down list, select **Capability Analysis,** and click on **Nonnormal.** In the **Capability Analysis (Nonnormal Distribution)** dialog box select **C1** for the **Single Column** field, select **Lognormal** from the **Distribution:** drop down list, and type in 99.0715 in the **Lower spec field.** Leave the **Upper Spec** field empty. Then click on the **OK** button.

Figure 2.54 should appear.

The overall capability is P_{pk} = PPL = 1.02. Therefore, the production process is barely capable and shows opportunity for improvement. The PPM yielded by such a process is 1128.17.

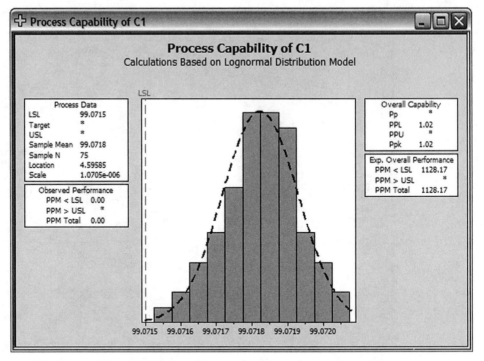

Figure 2.54

Lean Six Sigma Metrics

Once the problem has been clearly defined, its extent has to be measured in order to create a baseline against which the results of any future process changes could be compared. The critical criteria used to evaluate the successful completion of the project should have already been determined in the Define phase and expressed on the charter. However, the charter only gives a rough estimate of the extent of the problem being faced and it determines the objectives based on that estimate.

In the Measure phase, specific metrics are used to exactly assess the gaps in the processes' performance. The first step in this phase is to specifically define the process at a granular level in order to make its critical aspects more visible, and then gather the data needed to estimate the current process performance.

The Measure phase of a project involves assessing the current productivity level of the productive resources in order to determine the baseline. The metrics used in Six Sigma Lean projects apply to both of these methodologies. While the Six Sigma metrics assess the levels of variability of the production processes, the Lean metrics help evaluate the levels of waste generated by the processes.

Six Sigma metrics

A production process is, by definition, a chain of tasks, a sequence of events through which products are transformed from raw material to final goods. The tasks operate as customers and suppliers, with each task being a customer to the task upstream and a supplier to the task downstream. The quality of the product or service generated by each task depends on the variability at the current task and the combined variability of all the tasks upstream. Consequently, the quality of the final product depends on the combined variability that is exhibited by every step in the production process. To better control the quality level of the process yield and reduce operational expenses, it is necessary to control the potential errors at each step of the process and the overall variability of the process itself.

Quality targets are set for each step in a process and metrics are used to assess the performance of the steps and estimate the probability for a unit of product to go through every step free from defects. The metrics that are generally used in Six Sigma to assess quality are the total containment effectiveness (TCE), first time yield (FTY), the defect per unit (DPU), the PPM, the defect per million opportunities (DPMO), and the rolled throughput yield (RTY). These metrics are calculated to determine the probability for a defect to occur at a given step of a production process. The Poisson and normal distributions are primarily used to determine the probability for defects to occur.

Total containment effectiveness (TCE). Since the objective of process improvement is to eliminate defects all across the production systems, the effectiveness of the defects tracking method should reveal the number of defects found at every step and the defects found after the final products have been released. The metric used to measure the effectiveness of the defects tracking method is the TCE. It compares the number of defects and errors prior to the product release to the total number of defects and errors after release.

The difference between errors and defects is that errors are found before the product leaves the processing step where they occurred, while defects are errors caught after the product leaves the step where they occurred.

$$\text{TCE} = \frac{\text{prerelease defects} + \text{errors}}{\text{prerelease defects} + \text{release defects} + \text{errors}} \times 100\%$$

The objective is to minimize the number of defects across the production system but in terms of a defects tracking process, the aim should be to have a high TCE because it would be a sign that more defects and errors are caught before the products are released.

Example 2.45 Table 2.33 contains the errors and a defect found on a product before and after the product was released. Find the TCE.

TABLE 2.33

Processing steps	Defects	Errors
1	2	0
2	3	1
3	7	5
4	3	6
5	2	1
Total	**17**	**13**
Post release	**6**	**0**

Solution:

$$TCE = \frac{\text{prerelease defects} + \text{errors}}{\text{prerelease defects} + \text{release defects} + \text{errors}} \times 100\%$$

$$= \frac{17+13}{17+6+13} \times 100\% = 83.33\%$$

Therefore, 83.33% of the total defects and errors were found before the product was released.

Defect per unit (DPU). A defect is defined as a CTQ that does not meet the specified standard. If the measurement of an object is critical to quality and the specification for the CTQ is that it must fall within the interval [2, 4], any measurement that falls outside that interval will be considered a defect. Most manufactured products have multiple CTQs; therefore, a defective product can have multiple defects.

The DPU is the most basic defects metric. It is the ratio of the number of defects found to the total units of products sent to customers:

$$DPU = \frac{\text{total defects}}{\text{total unit}} = \frac{D}{U}$$

Defects per opportunity (DPO). Every product or service exhibits multiple CTQs and has to be free of defects for those products and services to meet customers' expectations. An opportunity is defined as any part of a product or a service where a defect can be found; in other words, opportunities are those CTQs that the customers expect to receive without any defects.

Total opportunities is the total number of opportunities (O) that are found on a set of units (U):

$$\text{Total opportunities} = O \times U$$

DPO will be equal to the ratio of the defects found on all the units to the total number of opportunities:

$$DPO = \frac{\text{defects}}{O \times U} = \frac{DPU}{O}$$

Defects per million opportunities (DPMO). The DPMO measures the number of defects that the process generates for every million opportunities:

$$DPMO = DPO \times 10^6$$

Example 2.46 A manufacturing plant has five processing lines. Table 2.34 shows the DPMO for each line.

TABLE 2.34

Processing lines	Units	Defects	Opportunities	DPU	Total opportunities	DPO	DPMO
Line 1	125	123	242	0.984	30250	0.004066	4066
Line 2	145	2	442	0.01379	64090	0.000031	31
Line 3	96	132	427	1.375	40992	0.003220	3220
Line 4	142	113	124	0.79577	17608	0.006418	6418
Line 5	93	6	142	0.06452	13206	0.000454	454

First time yield (FTY)

The probability for finding defects on a processing line generally follows a Poisson distribution.

$$P(x) = \frac{\mu^x e^{-\mu}}{x!}$$

If the historic average DPU is known, the density function could be converted into

$$P(x) = \frac{DPU^x e^{-DPU}}{x!}$$

Example 2.47 What is the probability that a product will contain three defects if the historic DPU is 0.9? What is the probability that the process would be free from defects?

Solution:

$$P(3) = \frac{0.9^3 e^{-0.9}}{3!} = \frac{0.729(0.407)}{6} = 0.049$$

The probability that a product will go through a process defect free is called FTY and is obtained using the following formula derived from the Poisson density function.

$$Y = FTY = e^{-DPU}$$

The probability for a product to go through the process defect free is

$$P(0) = \frac{0.9^0 e^{-0.9}}{0!} = e^{-9} = 0.407$$

Based on that formula, the DPU can be estimated if the FTY is known:

$$Y = FTY = e^{-DPU}$$

$$\ln(Y) = \ln e^{-DPU}$$

$$DPU = -\ln(Y)$$

Rolled throughput yield (RTY). While yield measures the probability for a product to go through a step defect free, the RTY measures the probability for the product to go through several steps defect free.

$$RTY = \prod_{i=1}^{n} Y_i = (Y_1)(Y_2)\ldots(Y_n)$$

Suppose that a process has five steps and the yield at each step is given in Table 2.35. Calculate the RTY.

TABLE 2.35

Process steps	1	2	3	4	5
Step yield	0.98	0.95	0.89	0.99	0.91

$$RTY = \prod_{i=1}^{n} Y_i = (0.98)(0.95)(0.89)(0.99)(0.91) = 0.746$$

The probability for a unit of product to go through the five steps defect free is 0.746 and the probability for finding a defect on the product is $1 - 0.746 = 0.254$.

The use of RTY requires investigating the defect rate at every step of the process. Therefore, it enables the producer to determine where the opportunities for improvement are and take action before the products reach the final stages in the process.

The total defect per unit (TDPU) can be estimated from the RTY:

$$TDPU = -\ln(RTY)$$

For the previous example, the TDPU would be

$$TDPU = -\ln(RTY) = -\ln(0.746) = 0.293$$

Lean metrics

The primary purpose of implementing Lean is to identify and eliminate waste from a workplace. The elimination of waste is done through the identification of the activities that do not add value to the customer because the non-value-adding activities increase the time that it takes to complete tasks and therefore reduce the productivities of the three factors of production that determine the efficiency of a process: the workers, the equipment, and the facility.

Resources capacity calculations.   A Lean process provides an efficient utilization of the three main productive resources, which are labor, facility, and equipment. The production processes within an organization are only optimized if the utilization of these three resources is actually optimized. There is a relationship between the efficiencies of these three factors of production. If the labor force is under-utilized, it will affect the productivity of both the facility and the equipment because they will not produce at an optimal level.

Workers capacity calculation.   Employee or worker capacity refers to the amount of time needed for one worker to produce a unit of product. The worker capacity is calculated based on engineered standards developed through observations during regular operating hours. Worker capacity is used to determine the takt time and the cycle time.

The takt time is defined as the maximum amount of time that the producer is allowed to take in order to produce and deliver customer orders on time. It is therefore based on the customers demand, their buying rate; it is the ratio of the available time to the rate of customer demand. A takt time of 5 min means that a maximum of 5 min is needed to produce a unit of product to meet customer expectation.

$$\text{Takt time} = \frac{\text{net available time per period}}{\text{customer demand per period}}$$

Example 2.48   The customer demand for leather chairs is 700 units per day. The company operates with two 8-h shifts per day that include a 30-min lunch for each shift, two 15-min breaks, and 10-min planned downtime per shift.

The total available time is $2[8(60) - 30 - 2(15) - 10] = 820$ min

$$\text{Takt time} = \frac{\text{net available time per period}}{\text{customer demand per period}} = \frac{820}{700} = 1.2 \text{ min}$$

SigmaXL template gives Table 2.36.

The takt time for the processing line is 1.2 min; therefore, the maximum time allowed to produce a unit of leather chair is 1.2 min. Therefore, one chair is completed every 1.2 min.

Cycle time.   The cycle time measures the time that it takes to produce a unit of product. It is different from the takt time because it does not consider the customer current orders; it only considers the capabilities of the available resources. The cycle time for a distribution center, for instance, considers the time that the products stay in a warehouse from the minute that they reach the receiving dock to time that they are staged in inbound inventory to the time they are

TABLE 2.36

SigmaXL Lean Templates: Takt Time Calculator		
Daily Customer Demand:	units per day	700
Scheduled Work:	hours per shift	8
Shifts per Day:		2
Lunch:	minutes per shift	30
Breaks:	minutes per shift	30
Planned Downtime:	minutes per shift	10
Staff/Operator Cycle Time:	minutes per unit	
Available Time:	minutes per day	820.0
Takt Time:	minutes per unit	1.2
Required Number of Staff/Operators:		0.0

transformed and shipped out. The less waste present in a production process in the form of rework, excess inspection, excess inventory, and excess motion, the shorter the cycle time.

A shorter cycle time reduces the time needed to complete orders and enables a more efficient utilization of the productive resources. Even though cycle time is not measured in terms of the resources' ability to meet current customer orders, a short cycle time has a positive effect on time-to-deliver and therefore enables the organization to operate with more flexibility.

$$\text{Cycle time} = \frac{\text{available time to process orders}}{\text{number of units required}}$$

In order to consistently meet customers' expectations, the production processes must be designed in such a way that the cycle times never exceed the takt times.

The workers' daily capacity.   The production process that most effectively enables the optimization of the utilization of productive resources involves what is called in Japanese "*Shojinka*," which translates in English to *flexible work force*. *Shojinka* starts with assigning one person-day to each employee. One person-day is based on engineered standards that determine the exact amount of work that each employee should perform in a given day of regular operating hours. The engineered standards and the one person-day are used to calculate the takt time, the cycle time, and the process efficiency.

The workers' daily capacity is based on the cycle time and is calculated using the following formula:

$$\text{Load time for workers} = \text{cycle time} \times \text{customer orders}$$

If the cycle time is 1.2 min and the customer orders are 3000 units, the load time of workers will be $3000 \times 1.2 = 3600$ min or 60 h.

The amount of workers' time needed to complete the orders will be 60 h.

The rational quota per worker will be

$$\text{Rational quota per worker} = \frac{\text{cycle time} \times \text{customer orders}}{\text{available operating hours}}$$

Facility capacity calculation. While the employee capacity refers to the amount of work that the workers are able to process based on the cycle time and the customer orders, facility capacity refers to the volume of products that a facility can process during regular operating hours.

$$\text{Facility capacity} = \frac{\text{regular operating hours}}{\text{completion time per unit at the bottleneck}}$$

If the completion per unit is 0.005 min and the facility operates 16 regular operating hours every day, then

$$\text{Facility capacity} = \frac{16 \times 60}{0.005} = \frac{960}{0.005} = 192000$$

The facility can process 192,000 units per day.

Let us note that the completion per unit is the time required to complete the processing of one unit of product and that it depends on the processing capabilities of the current bottleneck. The bottleneck is the step in the production process that requires the most time to complete a unit of product.

Equipment utilization. In addition to determining the labor and facility capabilities, a Lean Six Sigma project may also have to assess the equipment utilization efficiency. The efficiency and profitability of the equipment depend on their effective utilization, which is contingent upon their availability. Therefore, all the metrics used to assess the equipment utilization will depend on the extent of their availability. The availability of productive equipment depends on such factors as machine failure, defects due to machine poor performance, loss and waste of products due to machine performance, preventive maintenance that requires production stoppage, and the setup time, that is, the time needed to prepare the equipment for production.

Equipment unavailability can cause an excess of labor and facility capacities, that result in those factors being totally or partially idle because of a lack of equipment. Therefore totally eliminating the factors that impede the equipment availability can help optimize the equipment utilization and boost the facility and labor productivities.

Equipment availability.   The availability of the equipment is measured in terms of the time that the equipment can be utilized during working hours. It is therefore the ratio of the actual operation time to the loading time. The operating time is the loading time during a shift minus unscheduled downtime.

Planned downtime includes shift breaks, planned shift meetings, planned maintenance, and any planned exception time that is scheduled for a shift.

$$\text{Equipment availability} = \frac{\text{operating time}}{\text{loading time}} = \frac{\text{loading time} - \text{downtime}}{\text{loading time}}$$

Example 2.49   A shift is scheduled for 8 h with two 15-min breaks, one 15-min meeting, and 5 min for planned housekeeping. The machines went down during the shift for 45 min. What is the equipment availability?

$$\text{Loading time} = (8 \times 60) - ((2 \times 15) + 15 + 5) = 480 - 50 = 430 \text{ min}$$

$$\text{Equipment availability} = \frac{\text{loading time} - \text{downtime}}{\text{loading time}} = \frac{430 - 45}{430} = 0.895 = 89.5\%$$

The equipment was available 89.5% during the shift.

Equipment performance efficiency.   Each type of equipment is expected to perform at a certain level during operating hours. The speed at which the equipment operates determines not only the equipment's productivity level but also the operator's and the facility's because slow performing machines will lead to a reduction in the productivity of both the workers who use them and the facility in which they operate. Equipment performance efficiency depends on its operating speed rate and the net operating rate.

Equipment performance efficiency = net operating rate × operating speed rate

The operating speed rate.   The operating speed rate measures how fast the machine actually operates, its actual cycle time compared to its specified speed or ideal cycle time. The net operating rate measures the stability of the equipment; it measures the time that the equipment is being used at a constant speed during a specified period of time.

$$\text{Operating speed rate} = \frac{\text{specified cycle time}}{\text{actual cycle time}}$$

Example 2.50 The operating manual of a machine specifies that it should produce five units per minute and its actual cycle time is 0.35 min. What is the machine's operating speed?

Solution: The cycle time is the time that it takes to produce one unit of a product. If the specified production per minute is five units, the specified cycle time would be 1/5 = 0.2 min. It should take the machine 0.2 min to produce one unit.

$$\text{Operating speed rate} = \frac{\text{specified cycle time}}{\text{actual cycle time}} = \frac{0.2}{0.35} = 0.571 = 57.1\%$$

Net operating rate. Net operating rate is the ratio that measures the stability of the equipment. The stability of the equipment can be affected by adjustments made by operators and unexpected stoppages.

It is the ratio of actual processing time to the operating time.

$$\text{Net operating rate} = \frac{\text{actual processing time}}{\text{operating time}}$$

Example 2.51 The actual cycle time for a machine is 1.07 min per unit and the machine processed 387 units during 450 min of operating time. What is the net operating rate?

Solution: The actual processing time is the actual cycle time × processed units = 1.07 × 387 = 414.09 min.

$$\text{Net operating rate} = \frac{\text{actual processing time}}{\text{operating time}} = \frac{414.09}{450} = 0.92 = 92\%$$

$$\text{Equipment performance efficiency} = \text{net operating rate} \times \text{operating speed rate}$$

$$\text{Equipment performance efficiency} = 0.92 \times 57.1 = 52.53$$

Quality rate of products. The quality rate can be estimated in terms of the defect rate. It is the ratio of the defects found out of every unit of product processed.

If 855 units are processed and 59 defects are found, the defect rate would be 59/855 = 0.069 or, in terms of percentage, 6.9%.

The quality rate would be 100% − 6.9% = 93.1%.

In other words, the quality rate would be 93.1%.

Overall equipment efficiency. The overall equipment efficiency (OEE) is the metric that enables us to determine how efficiently the equipment is used. It takes into account the availability of the equipment, its performance efficiency, and the quality rate of the products that it generates.

$$\text{OEE} = \text{availability} \times \text{performance efficiency} \times \text{quality rate}$$

Based on the level of the OEE, the company can determine its overall level of performance, identify bottlenecks, and eliminate waste and reduce operational expenses.

Example 2.52   Based on the information summarized in Table 2.37, find the OEE.

TABLE 2.37

Work hours	8
Breaks	30 min
Meeting	15 min
Unexpected downtime	35 min
Machine's manual spec	1 unit per minute
Actual machine performance	1.05 min per unit
Number of units processed	405
Quality rate	0.97

Solution: Since

$$OEE = availability \times performance\ efficiency \times quality\ rate$$

we need to calculate the availability first.

Availability

Equipment availability =

$$\frac{operating\ time - downtime}{operating\ time} = \frac{(8 \times 60) - (30 + 15) - 35}{(8 \times 60) - (30 + 15)} = \frac{400}{435} = 0.92$$

Equipment performance efficiency   Equipment performance efficiency has two components:

$$Equipment\ performance\ efficiency = net\ operating\ rate \times operating\ speed\ rate$$

$$Net\ operating\ rate = \frac{actual\ processing\ time}{operating\ time} = \frac{1.05 \times 405}{435} = \frac{425.25}{435} = 0.978$$

$$Operating\ speed\ rate = \frac{specified\ cycle\ time}{actual\ cycle\ time} = \frac{1}{1.05} = 0.952$$

$$Equipment\ performance\ efficiency = 0.978 \times 0.952 = 0.931$$

$$OEE = availability \times performance\ efficiency \times quality\ rate$$

$$OEE = 0.92 \times 0.931 \times 0.97 = 0.831$$

Throughput rate—average completion rate   The throughput rate or flow rate measures the rate at which a process generates products. It is the inverse of the process cycle time.

$$Throughput\ rate = \frac{1}{cycle\ time}$$

Since the cycle time is the time necessary for a process to generate one unit of product, throughput rate can be defined as the number of units generated during one unit of time measurement. A unit of time measurement can be hours, minutes, or seconds depending on how the company measures its performance.

Work in Progress (WIP)

Work in progress or WIP can be defined in different ways; it can be defined in terms of time or in terms of inventory.

If a machine produces at a rate of 150 units per hour and it has 1500 units to produce, we say that the machine has 10 h of WIP or we can say that we have 1500 units of WIP inventory.

$$\text{WIP inventory} = \text{throughput rate} \times \text{flow time}$$

$$1500 = 150 \times 10$$

$$\text{Flow time} = \text{inventory/throughput}$$

$$10 = 1500/150$$

The relationship between WIP, throughput, and flow time (or total lead time) is known as Little's law. That principle can be developed further to apply to cycle time because there is a constant relationship between cycle time and throughput.

$$\text{Flow time} = \text{inventory} \times \text{cycle time}$$

If the machine produces at a rate of 150 units per hour, then we can say that the cycle time (the time that it takes to complete one unit of product) is $1/150 = 0.007$ h. If we have 1500 units to process, then

$$\text{Flow time} = \text{inventory} \times \text{cycle time} = 1500 \times 0.007 = 10 \text{ h.}$$

Process cycle efficiency.   The efficiency of a process measures the proportion of work that adds value to the operations. The metric used to determine the process efficiency is the process cycle efficiency (PCE):

$$\text{PCE} = \frac{\text{value-added time}}{\text{flow time}}$$

Analyze

Brainstorming

Most project executions require a cross-functional team effort because different creative ideas at different levels of management are needed in the definition and the shaping of a project. These ideas are better generated through brainstorming sessions. Brainstorming is a tool used at the initial steps or during the Analyze phase of a project. It is a group discussion session that consists of encouraging a voluntary generation of a large volume of creative, new, and not necessarily traditional ideas by all the participants. It is very beneficial because it helps prevent narrowing the scope of the issue being addressed to the limited vision of a small dominant group of managers. Since the participants come from different disciplines, the ideas that they bring forth are very unlikely to be uniform in structure. They can be organized for the purpose of finding root causes of a problem and suggest palliatives. If the brainstorming session is unstructured, the participants can give any idea that comes to their minds, but this might lead the session to stray from its objectives.

A structured brainstorming session provides rules that make the collection of ideas better organized. A form of brainstorming called nominal group process (or nominal group technique) is an effective way of gathering and organizing ideas. At the end of the brainstorming session, a matrix called an affinity diagram helps to arrange and make sense of the many ideas and suggestions that were generated.

Nominal Group Process

Nominal group process (or technique) is a method that provides a reflection group with a framework for a nonthreatening and constraint-free face-to-face discussion where the participants aim at reaching an agreement on a given topic. The objective is to make it easy for every participant to freely express his or her opinion, make suggestions without any pressure, and at the same time prevent a small group of more opinionated participants from taking over the debate.

The first step in the process consists of designating a facilitator. His role will be to conduct the discussions and make sure that the members fully understand the issues being discussed without letting himself take a major part in the generation of substantive ideas. He will have to give every participant a chance to make a suggestion. The suggestions can be made orally or in writing. Every participant is asked to give as many ideas as possible in response to a given question. The facilitator writes down all the ideas on a board for all the participants to see, and then the group discusses them in order to clarify some of the suggestions made before a list of priorities is discussed to determine their merit.

At the end of the discussions, the team assesses the priorities and agrees on the next step. The advantages attached to such a process are that

- A group of people of different horizons reaches a consensus on a given issue.

- A high volume of suggestions is made in a short period of time.

- The ideas generated will usually go far beyond and will be much more creative and seminal than what a meeting of a group of similar minded managers who meet regularly would yield.

- It helps uncover tacit knowledge (knowledge that resides in the employees that has thus far been untapped) by eliminating psychological complexes and freeing repressed ideas.

- It gives a sense of belonging and motivates all the participants to the endeavor.

The disadvantages are

- Because the participants come from different areas and therefore have different backgrounds, the facilitator needs to be competent, knowledgeable, and flexible.

- The more vocal participants may end up having it their way.

- If there are too many participants, the meetings will end up being too long and hard to conduct.

Affinity Diagram

If the ideas generated by the participants to the brainstorming session are few (less than 15), it is easy to clarify, combine them, determine the most important suggestions, and make a decision. However, when the suggestions are too many it becomes difficult to even establish a relationship between them. An affinity diagram or KJ method (named after its author, Kawakita Jiro) is used to diffuse confusion after a brainstorming session by organizing the multiple ideas generated during the session. It is a simple and cost-effective method that consists of categorizing a large amount of ideas, data, or suggestions into logical groupings according to their natural relatedness. When a group of knowledgeable people discusses a subject with which they are all familiar, the ideas they generate should necessarily have affinities. To organize the ideas, perform the following:

1. The first step in building the diagram is to sort the suggestions into groups based on their relatedness and a consensus from the members.

2. Make up a header for the listings of the different categories.

3. An affinity must exist between the items on the same list and if some ideas need to be on several lists, let them be.

4. After all the ideas have been organized; several lists that contain closely related ideas should appear. Listing the ideas according to their affinities makes it much easier to assign deliverables to members of the project team according to their abilities.

Example 3.1   A project was initiated to reduce cycle time in a distribution center. The project team decided to conduct a brainstorming session to find the root causes of the problem and assign deliverables to its different members. To generate as many ideas as possible, the team invited employees from every department in the warehouse to the discussion. The facilitator encouraged each participant to give an idea about what he thinks is causing the orders to take too long to be completed. Each idea was then written on a piece of paper and stuck on a board. The labels in Fig. 3.1 represent the ideas that the team generated.

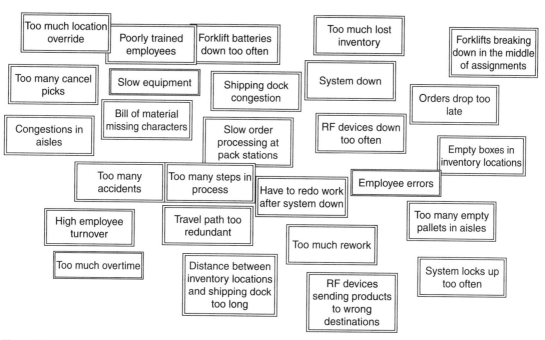

Figure 3.1

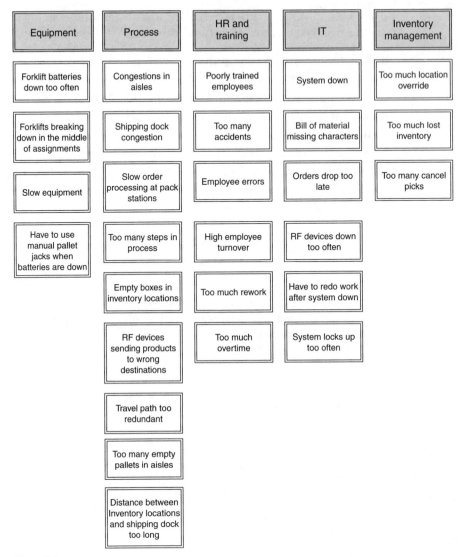

Figure 3.2

After having analyzed all the ideas, the team agreed to organize them according to their affinities in several lists with a heading assigned to each list as shown in Fig. 3.2. According to the listings, the team can easily assign tasks to the different departments that are responsible for their respective problems.

Cause-and-Effect Analysis

The cause-and-effect (C&E) diagram—also known as a fishbone (because of its shape) or Ishikawa diagram (named after Kaoru Ishikawa, its creator)—is used to visualize the relationship between an outcome and its different causes. There is very often more than one cause to an effect in business; the C&E diagram is an analytical tool that provides a visual and systematic way of linking different causes (input) to an effect (output). It shows the relationship between an effect and its first-, second-, and third-order causes.

It can be used in the "Design" phase of a production process as well as in an attempt to identify the root causes of a problem. The building of the diagram is based on the sequence of events. "Sub-causes" are classified according to how they generate "sub-effects," and those "sub-effects" become the causes of the outcome being addressed.

The first step in constructing a fishbone diagram is to define clearly the effect being analyzed. The second step consists of gathering all the data about the key process input variables (KPIV), the potential causes (in the case of a problem), or requirements (in the case of the design of a production process) that can affect the outcome.

The third step consists of categorizing the causes or requirements according to their level of importance or areas of pertinence. The most frequently used categories are

- Manpower, machine, method, measurement, mother nature, and materials for manufacturing
- Equipment, policy, procedure, plant, and people for services

Subcategories are also classified accordingly; for instance, different types of machines and computers can be classified as subcategories of equipment. The last step is the actual drawing of the diagram.

Example 3.2   The average time that it takes Memphis Warehouse to complete a customer order is considered too high and a quality control manager is working to find the root causes for the high cycle time. He collected enough information during a brainstorming session and tabulated the information on a Minitab worksheet found in the file *cycletime.MTW*.

Open the file *cycletime.MTW* and from the menu bar, click on **Stat,** then on **Quality Tools,** and then on **Cause-and-effect**. When the **Cause-and-effect Diagram** box opens up, fill it out as shown in Fig. 3.3 before clicking on the **OK** button.

The diagram in Fig. 3.4 appears.

The fishbone diagram does help visually identify the root causes of an outcome, but it does not quantify the level of correlation between the different causes and the outcome. Further statistical analysis is needed to determine which factors contribute the most to creating the effect. Pareto analysis is a good tool for that purpose but it still requires data gathering. Regression analysis allows the quantification and the determination of the level of association and significance between causes and effects. A combination of Pareto and regression analysis can help not only determine the level of correlation but also stratify the root causes. The causes are stratified hierarchically according to their level of importance and their areas of occurrence.

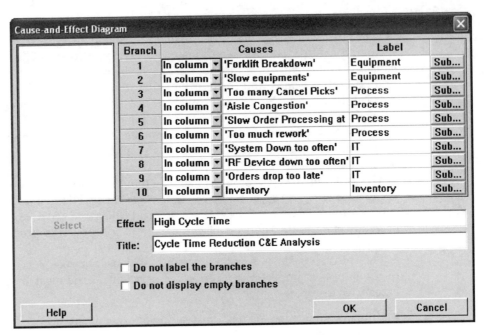

Figure 3.3

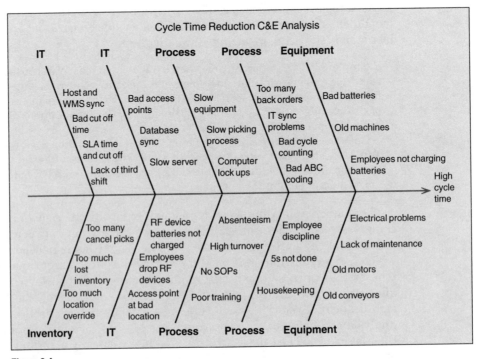

Figure 3.4

Pareto Analysis

Pareto analysis is simple; it is based on the principle that 80% of problems find their roots in 20% of causes. Vilfredo Pareto, a 19th century Italian economist who discovered that 80% of the land in Italy was owned by only 20% of the population, established the principle. Later, empirical evidence showed that the 20/80 ratio was determined to have a universal application. Dr. Joseph Juran discovered in the 1920s in his studies of quality control that a few defects were responsible for the bulk of rejects and rework. He also noted that the same principle applied to employee absenteeism, causes of accidents in a workplace, and many other factors in management. Dr. Juran determined that the Pareto principle did indeed have universal applications.

80% of customer dissatisfaction stems from 20% defects.

80% of the wealth is in the hands of 20% of the people.

20% of customers account for 80% of a business.

When applied to management, the Pareto rule becomes an invaluable cost-effective tool. In the case of problem solving, for instance, the objective should be to find and eliminate the circumstances that make the 20% "vital few" possible so that 80% of the problems are eliminated.

The first step is to define clearly the goals of the analysis. What is it that we are trying to achieve? What is the nature of the problem we are facing?

The next step in the Pareto analysis is data collection. All the data pertaining to factors that can potentially affect the problem being addressed need to be quantified and stratified. In most cases, a sophisticated statistical analysis is not necessary; a simple tally of the numbers suffices to prioritize the different factors. However, in some cases the quantification might require statistical analysis to determine the level of correlation between the causes and the effects. A regression analysis can be used for that purpose. A coefficient of correlation or a coefficient of determination can be derived to estimate the level of association of the different factors to the problem being analyzed.

Then a categorization can be made, and the factors are arranged according to how much they contribute to the problem. The data generated is used to build a cumulative frequency distribution.

The next step is to create a Pareto diagram or Pareto chart in order to visualize the main factors that contribute to the problem, and therefore concentrate on the "vital few" instead of the "trivial many." The Pareto chart is a simple histogram; the horizontal axis shows the different factors, while the vertical line represents the frequencies. Since all the different causes will be listed on the same diagram, it is necessary to standardize the unit of measurement and set the time frame for the occurrences.

The building of the chart requires a data organization. A four-column data summary must be created to organize the information collected. The first column lists the different factors, the second column lists the frequency of occurrence of the effects during a given time frame, the third column records the relative frequencies (in other words, the percentage of the total), and the last column records the cumulative frequencies, bearing in mind that the data are listed from the most important factor to the least.

Example 3.3   A cellular phone service provider was facing a high volume of returned phones from its customers. The quality control manager decided to conduct a Pareto analysis to determine what factors contributed the most to causing customer dissatisfaction. The data in Table 3.1 were gathered from Customer Services during a period of 1 month to analyze the reasons behind the high volume of customers' return of cellular phones ordered online. The table is used to construct a Pareto chart.

TABLE 3.1

Factors	Frequency
Misinformed about the contract	165
Wrong products shipped	37
Took too long to receive	30
Defective product	26
Changed my mind	13
Never received the phone	12
Total	283

The data in Table 3.1 can be reorganized to include the relative and cumulative frequencies (Table 3.2).

TABLE 3.2

Factors	Frequency	Relative frequency	Cumulative frequency
Misinformed about the contract	165	$(165/283) \times 100 = 58\%$	58%
Wrong products shipped	37	$(37/283) \times 100 = 13\%$	$58\% + 13\% = 71\%$
Took too long to receive	30	$(30/283) \times 100 = 11\%$	$71\% + 11\% = 82\%$
Defective product	26	$(26/283) \times 100 = 9.2\%$	$82\% + 9.2\% = 91.2\%$
Changed my mind	13	$(13/283) \times 100 = 4.6\%$	$91.2\% + 4.6\% = 95.8\%$
Never received the phone	12	$(12/283) \times 100 = 4.2\%$	$95.8\% + 4.2\% = 100\%$
Total	283	100%	

Using SigmaXL, open **SigmaXL** and then open the file *cellphone.xls*. Select the area containing the data before clicking on **SigmaXL** from the menu bar. Then click on **Graphical Tools** and then on **Basic Pareto Chart** (Fig. 3.5).

When the **Basic Pareto Chart** appears, it should already have the field **Please select your data** populated. Press the **Next >>** button.

Fill out the next **Basic Pareto Chart** as shown in Fig. 3.6.

You have the choice between pressing the **Finish>>** to obtain a standard chart or pressing **Next >>** to customize the charts.

The diagram itself consists of three axes (see Fig. 3.7). The horizontal axis lists the factors and the left vertical axis lists the frequency of occurrence (it is graded from 0 to at least the highest frequency). The right vertical line is not

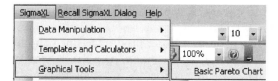

Figure 3.5

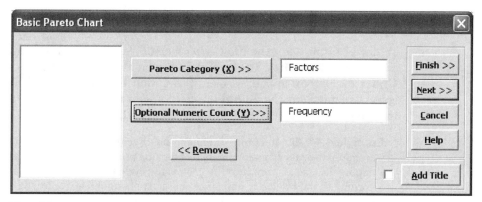

Figure 3.6

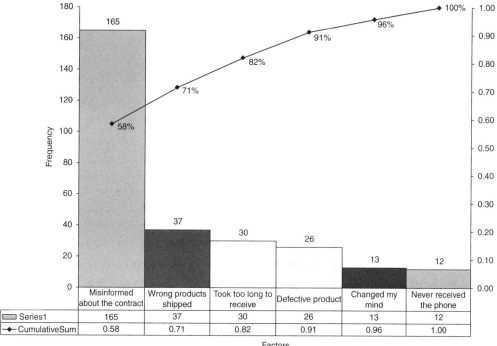

	Misinformed about the contract	Wrong products shipped	Took too long to receive	Defective product	Changed my mind	Never received the phone
Series1	165	37	30	26	13	12
CumulativeSum	0.58	0.71	0.82	0.91	0.96	1.00

Factors

Figure 3.7

always present on Pareto charts; it represents the percentage of occurrences and is graded from 0 to 100%.

The *breaking point* is at around "wrong product." Since the breaking point divides the "vital few" from "the trivial many," the first three factors "misinformed about the contract", "wrong products", and "took too long to receive" are the factors that need more attention. By eliminating the circumstances that make them possible, we will eliminate approximately 71% of our problems.

Using Minitab

To use Minitab, open the file *Cellphone.MTW* and from the menu bar, click on **Stat**, then on **Quality Tools,** and finally on **Pareto Chart**. When the **Pareto Chart** dialog box appears, select the option **Chart defects Table** and then select "Factors" for the **Labels** in field and "Frequency" for the **Frequencies** in field before pressing the **OK** button to get the chart.

Example 3.4 An inventory manager has decided to reorganize the layout of a warehouse in order to make the fastest moving parts more accessible and closer to the shipping dock so as to reduce the processing cycle time. He wants to codify the parts according to their velocity and since there is not enough space by the shipping docks, only a few parts can be transferred there. He tabulates the parts families according to how often they are ordered per day (see Table 3.3).

TABLE 3.3

Part families	Orders per day
Engines	1384
Turbo	1001
Gasket	305
Catalytic converters	298
Washers	287
Pistons	189
Liners	178

Using Minitab, we obtain the graph shown in Fig. 3.8.

The breaking point is clearly at turbo; therefore, the engines and the turbo generators are the two part families that must be moved closer to the dock.

Fault Tree Analysis

Fault tree analysis (FTA) is one of the most methodical tools used for finding the root causes of a problem in reliability studies. FTA is a structured analysis of defects and their causes. It can be used as a reactive tool when a problem has already occurred and the engineers are trying to determine its causes; it can also be used as a proactive tool to prevent a problem from occurring. FTA ties a defect (in the case of a reactive analysis) or an undesirable effect (in the case of a proactive analysis) to its (potential) causes. The causes of defects in a system,

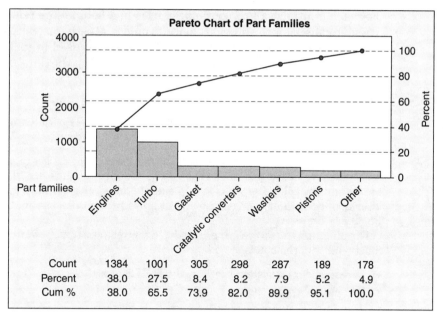

Figure 3.8

process, product, or service have their own causes, which also have causes. Not only does FTA inventory the different causes but it also organizes them according to the factors they affect. Because it does not only consider the first-order causes, but considers all the different factors that could possible cause a defect, FTA is considered as one of the most rigorous and effective tools for root cause analysis. The following steps are used when conduction an FTA.

1. Gather the team conducting the FTA. Just as in the case of a Failure Mode and Effect Analysis (FMEA), an FTA is conducted through a brainstorming session. A group of people knowledgeable about the problem being faced is gathered to determine the several orders of causes of the problem.

2. Specify the main problem or the undesirable effect being analyzed. Be as precise and specific as possible when defining the problem. An example of a specific problem would be "Oil leak draining the sink within an hour of operations" or "RF devices lock up every time an employee scans part number 213242."

3. To make the relationship between defects and their causes visible, draw a relational diagram.

4. Determine the faults that could lead to the main problem. For the examples above, the causes of the problems may be, respectively, "a crack at the bottom of the oil sink" and "Part number 213242 not listed in the database."

5. For each fault, list all the possible causes in a field by the related fault. The possible causes for the previous defects could be "there is a crack at the bottom of the sink because the thickness of the material used to make the sink is too small" and "Part number 213242 is not listed in the database because it is two digits short."

6. Repeat the process until the root of the problem is found.

7. Find a resolution to the problem once the root cause is determined.

Example 3.5   An engineering team is analyzing the root causes of problems found on a newly released notebook model. The stated problems are "The LCD turns blurry and the computer makes a squeaking noise before locking up and shuts off after about 5 min of use." At the end of a brainstorming session, the diagram in Fig. 3.9 is drawn.

The diagram uncovers several existing problems that potentially affect the functionality of the notebook.

- Excess solder on system board by the CPU heat sensors
- Bad base plastic design
- Aluminum metal used as keyboard base is too thin
- Not enough holes on keyboard base
- The cards were mislabeled

The next step for the team should be to figure out how to solve each of these issues.

Seven Types of Waste

Productivity improvement is one of the most significant fundamentals of any endeavor to better business operations and one of the most effective ways to improve productivity is to increase throughput while reducing or maintaining operational resources. Productivity can be defined as the number of items produced for every unit of resources used to produce them; in other words, productivity measures the efficiency of the effort spent on producing the output.

$$P = \frac{\text{number of items produced}}{\text{total cost of production}}$$

Productivity is reduced every time the cost incurred to produce one extra unit is increased. Suppose that the total number of items produced by a process is 2500 and the total cost of production is \$1000. Then the process' productivity is

$$P = \frac{\text{number of items produced}}{\text{total cost of production}} = \frac{2500}{1000} = 2.5$$

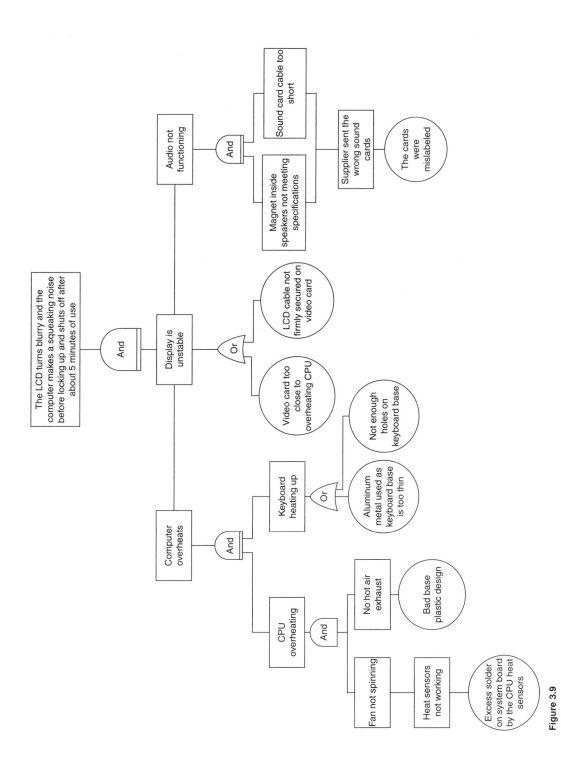

Figure 3.9

For every dollar invested, 2.5 items were produced. If the cost of production is increased by $500 and the output level remains the same, then

$$P = \frac{\text{number of items produced}}{\text{total cost of production}} = \frac{2500}{1000 + 500} = \frac{2500}{1500} = 1.667$$

In other words, for every dollar invested, the process only yields 1.667 units. Because of an increase in the cost of production, productivity has gone down by 2.5 − 1.667 = 0.833 points.

Several factors can contribute to an increase in the cost of production of products or services and a reduction of productivity. Some of those factors add value to the output while others do not. Among the non-value-adding factors, some are necessary to the production processes while others are not. Spending resources to increase the size of a company's parking lot may not directly add value to the products or services that the company produces, but it may be necessary for its operations. Working overtime to increase the volume of inventory of products that are not ready to be shipped to customers would increase operational expenses without adding any value to the products.

Waste in a production process is defined as a factor in the process that unnecessarily increases operational expenses while it does not add value to the products. Identifying the sources of waste and eliminating them is one of the most effective ways of increasing resources' efficiencies and processes' productivity. Despite the significant operational differences within and between industries, Tahiichi Ohno, one of the Toyota Motor Company's quality gurus, has determined the main sources of waste to be similar across industries. He identified and classified the sources of waste in seven categories.

Overproduction

Overproduction is defined as the production of an item that is not ready to be sold to customers (if it is a final product) or to be further processed by the next step in the process (if it is in a transformation process). It can be the result of products being made too early or products being made in excess of the required amount. The cost of overproduction is always prohibitive whether it occurs within a production in progress or under the form of final goods.

When overproduction occurs within a production in progress, it creates clutter and prevents a seamless flow of materials, which lowers productivity. It is generally the result of batch production, inefficient scheduling, and bad coordination. Batch production systems require completed units of products to be held at a step in the process until a predetermined lot is finished before it is moved to the next step. Since a perfectly even and balanced flow of material across a production process seldom occurs because different steps in a process usually require different amounts of resources and time to process the same volume of products, using a batch-and-queue system can reinforce bottlenecks and cause more overproduction between process steps.

Example 3.6 A processing line is made up of five steps and the time that it takes each step to complete its task is given in Table 3.4.

TABLE 3.4

Processing steps	Processing time (min)
1	2
2	2
3	3
4	1
5	2

If a batch-and-queue system is used and steps 1 and 2 continuously process material, inventory will very quickly accumulate at step 3; steps 4 and 5 will remain idle most of the time.

There is a positive correlation between productivity and throughput. An increase in productivity leads to an increase in throughput. Since throughput measures the rate at which a process yields products or services, it depends on the bottleneck, i.e., the step in the process that takes more effort to complete a task. Subsequently, the productivity of a process will be tied to the rate at which the bottleneck executes tasks.

Overproduction of outbound inventory can be the result of a desire to respond quickly to customers' demands. An enterprise produces in excess of customers' current demands and stores the products as outbound inventory waiting for a demand. If the product is highly demanded, the high velocity of the inventory would make the negative impact of high volume of held outbound inventory less visible, but a sudden drop in demand would result in heavy financial losses. The cost attached to overproduction includes

- An increase in unnecessary inventory
- An increase in operational expenses for handling the inventory
- An increase in the defect rate

Overproduction will generally require prohibitive extra floor and labor capacities.

Wait

Waste in the form of wait occurs when productive resources are kept idle while they are waiting for material to process. Waste happens because employees are being paid and resources consumed when they are not producing. Waste in the form of wait also occurs more often because of a lack of coordination and poor scheduling in an unbalanced production process.

In the example in Table 3.4, if a batch-and-queue system is used, the downstream steps from step 3 would be idle most of the time because they processes the same amount of material much faster than step 3; therefore, they would have to wait for more work every time they complete a task. Step 3 would be the bottleneck because it requires more time to process the same volume of material.

Wait increases work in progress (WIP) and cycle time and reduces productivity.

Unnecessary inventory

Inventory can be defined as money invested in the acquisition of products intended for transformation or sales. Not only is it money not readily available for spending and does not generate any interest but it is money that requires extra operational expenses for its maintenance. A volume of inventory will require proportional extra expenses in the form of labor, insurance, and floor capacity. Even though from an accounting perspective inventory is considered an asset, technically, it is a liability.

Unnecessary inventory is generally the result of overproduction. In most organizations, the consequences of unnecessary inventory are not accurately quantified and its impact is seldom made visible. However, unnecessary inventory is largely the root of excessive operational expenses because it is generally responsible for most of the hidden problems that inhibit a seamless flow of material and a clutter-free workplace. It requires more resources for its maintenance in the form of extra labor, extra equipment, more interest, and more insurance.

The way Dell used to manage inventory in the 1980s is a good example of how overproduction can negatively affect a business. By the end of the 1980s, Dell was growing at an unprecedented rate and in order to fulfill quickly ever-growing customers' orders, Dell accumulated a high volume of memory chips among the many computer parts that were bought to meet unanticipated demand. Technology in computer industries evolves very fast and the capacity for memory chips went from 256 KB to 1 MB in a short period, leaving Dell with an outstanding volume of obsolete parts that had to be sold at a fraction of their cost. This resulted in great financial losses for Dell.

Motion

A production process should be designed in such a way that a minimum effort is required from operators to generate a maximum result. Such a design would reduce unnecessary physical movements that would strain resources without adding any value to the product or service. Motion refers to the movements of equipment and labor that do not add any value. If workers have to move around constantly to find the tools that they need to perform their work, this would reduce the time that they spent actually performing work and increase their cycle time. Unnecessary motion increases WIP and deteriorates quality.

Unnecessary transportation

Transportation consists of moving a product from one place to another without altering is composition. Typically, transportation does not add any value to the product but it can be a necessary part of the process because, by definition, material is expected to flow within a transformation process. Unnecessary transportation occurs when production cells are too far apart from one another due to a poorly designed process layout. Excessive transportation unnecessarily increases WIP and cycle time and adds more opportunities for errors. The cost of unnecessary transportation is hard to evaluate accurately because it is not always easy to determine what part of transportation within a process does not add any value. If the non-value-adding part of transportation is determined, the precise and accurate quantification of its implications should go beyond the labor cost associated with moving material from one place to another because unnecessary transportation increases cycle time and defects.

Inappropriate processing

By definition, processing is the activity that every organization performs to generate products and services. However, there are multiple ways to process material to generate the same type of product. Out of a multitude of ways of processing, one must be more efficient in terms of resources consumption. Inappropriate processing is a way of processing that is less than optimal; generally, bad plant layout, poor process design, and poorly maintained equipment generate defects. Inappropriate processing will result in rework, overproduction, unnecessary handling, and defects.

Product defects

A defect is a flaw on a critical to quality (CTQ) of a product that makes it unfit for sale to customers. Among the seven types of waste, the impact of defects on the bottom line is one of the easiest to quantify because the immediate result of defects is either rework or scrap and both of these come with a quantifiable cost. Product defects increase the cost of nonconformance and drastically reduce productivity. There is a correlation between the production of defects and some of the other six types of waste.

- The production of defects increases the wait time for the tasks downstream because the defective products cannot be further processed before the defects are corrected. Further inspection may also be needed to ensure the conformance of the products.

- The defective products may have to be reworked in order to correct the problem. This would not only slow the process but also add to labor costs.

- The rework of the defective products may require additional parts.

- If the defects cannot be corrected, the products are scrapped and both the material and the labor used to produce and scrap them are wasted.

Lean Approach to Waste Reduction

The underlying foundation of Lean manufacturing is the organizational strategy that constantly seeks a continuous improvement through the identification of the non-value-adding activities (known in Japanese as Muda) and their elimination along with the reduction of the time that it takes to perform the value-adding tasks. The identification and elimination of waste is made through the following steps:

1. Specify the exact *value* of each specific product. A definition of the concept of value acceptable to both producers and customers has historically been contentious. The definition of value from the producer standpoint used to be associated with the cost incurred to generate the products or services or the revenue generated from their sale. A product is worth more when the cost incurred to produce it or the return expected from its sales is high. From the customer's perspective, the value of a product depends on its utility or in some cases on its rarity (which confers it a certain status). Since we are discussing mass production, we will consider only the utility functions of the products or services.

   Customers buy a product because of the satisfaction they derive from using it. The value they attach to the product depends on its quality (durability, reliability, comfort, etc.). The higher the quality, the more valuable the product is perceived to be and the more the customers are willing to pay for it. If the producer specifies the value of the product, she would be more likely to do it based on cost of production or revenue. An enterprise that operates based on Lean principles lets the customers specify the value. A good and simple example of an organization that lets its customers specify the value they expect from it is Dell. Until recently, Dell only sold its products directly to customers without a third-party involvement. When a customer wants to buy a Dell computer, he can go online and choose a combination of parts from a multitude of possibilities. If he wants to buy a Latitude D620 for instance, he can select the size of hard drive, the type of CPU, the type of LCD, etc. He specifies the value he wants to acquire from Dell. If he had to go to a retail store to buy a notebook, the choices would be limited.

2. Identify the *value stream* for each product. Each product is manufactured in a unique way. The value stream traces all the steps required to transform the raw materials into the products demanded by the customers. Each step must add value to the product; in other words, the product must be worth more when it leaves a step in the process than when it arrived there. The analysis of the value stream is performed to identify waste and non-value-added steps and reduce the time necessary for the value-added steps. Some non-value-added steps are necessary and inherent to the processes but some are unnecessary, cause clutter, and can be sources of bottlenecks.

   The analysis of the value stream of a Dell notebook starts from the minute a customer places his order to when the product is delivered to him. It studies every single step in the process to detect waste and determine opportunities for reducing the time that it takes the value-adding steps to execute their tasks.

3. Make the value *flow* without interruptions. To eliminate waste and clutter, the producer should put in place a production process that yields a steady and constant flow of products. Therefore, after the value of the products is determined, the value stream mapped, and the clutter removed, the producer should strive to make the production flow relentlessly at a steady pace.

Applying the one-piece-flow principle is one way of doing this and it will eventually lead to the just-in-time (JIT) method: getting the right part, in the right quantity, at the right time, at every step of the process. A one-piece-flow process is only possible if all the steps take the same amount of time to process the materials before they send them to the next step.

4. Let the customer *pull* value from the producer. When the material flows on a steady and constant pace and the just-in-time principles are applied and are working, it becomes easier to predict and plan the work load executions and the deliveries because the time required for the completion of each task is known in advance. Inventory, cycle time, WIP, and complex scheduling are reduced, making it possible to let the customer pull the orders instead of building an excessive outbound stock of products waiting for a potential customer. An excessive stock of product in itself constitutes waste because one cannot know with certitude how long it will stay unsold, and more money will be spent on its maintenance under the form of labor, warranty, and the cost of the space it occupies. Letting the customer pull the products means only producing the products that are ordered by the customers. Let them determine what, when, and in what quantity to produce.

   Toyota Motor Company and Dell have excelled because of letting the customers pull value from them. Until recently, Dell never produced a notebook that had not already been ordered by a customer. This, combined with the fact that the company only orders supplies for products that have already been ordered, keeps its inventory turnover and WIP insignificant compared to the rest of the industry; its working capital (cash + accounts receivable + inventory – accounts payable) is negative. A negative working capital means that Dell generates sales without any investment in working capital; it receives revenue from its customers before it pays its suppliers.

5. Pursue *perfection*. Opportunities for improvements will always be there; therefore, continuous improvement does not end. Once the process flow has started, the company should keep seeking to uncover best practices. There are always possibilities to improve on existing processes by continuously setting higher targets. This in itself will prevent the company from falling back on old ways. A process improvement is made through Kaizen events with the involvement of all concerned employees. A Kaizen event is a small project that targets an area in operations for improvement. It usually lasts less than a month and uses the competencies of all the people who are directly involved in the area.

Cycle Time Reduction

The main purpose of **Lean** is to improve productivity through an elimination of waste. A production process is a sequence of events, a chain of tasks with each link, each step being considered as a customer to the previous task and a supplier to the next. The products being transformed from raw materials to finished goods flow through the process and are gradually transformed at each step. Since the different steps perform different tasks that require different amounts of resources and time to complete, the cycle time at each step is likely to be different from step to step in the process.

Cycle time is defined as the time that it takes to complete a given task in a production process. Within a process, there are generally cycle time variations between the different steps. One way to improve on the overall productivity of a process is to reduce overall process cycle time by first reducing the variations within the process. Balancing the cycle times across a process will lead to an uninterrupted continuous flow of material, which will eventually lead to a reduction in waste.

To improve upon cycle time, Kaizen events are generally organized. When the cycle time variations between the steps within a process are significant, waste will occur in the form of wait and overproduction. Leveling the production process and reducing the cycle time will help reduce waste, improve on delivery time, and increase floor and labor capacities and better capacity planning.

Example 3.7   Suppose that a process is made up of four steps—Receiving, Transfer from Receiving, Packaging and Stocking (Fig. 3.10).

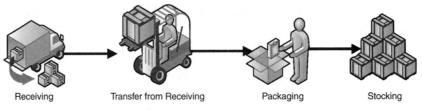

Receiving   Transfer from Receiving   Packaging   Stocking

Figure 3.10

If Receiving takes 10 min to process a unit of material and Transfer from Receiving processes the same unit in 14 min, then an excess of inventory would pile up in front of Transfer from Receiving waiting to be processed. If at the same time Packaging and Stocking process the material faster than Transfer from Receiving, then those steps in the process would end up with idle resources every time they finish processing a piece of material.

The cycle time reduction should take into account the customer demand for delivery time. The time to deliver addresses the takt time. The definition of takt time is close to cycle time but it is different. It is defined as the time needed to meet customer requirements. It is the ratio of the net operating time to the customer requirement.

Takt time

Takt time is the time that it takes to complete customer orders and meet deadlines. When there are cycle time variations within a process, the time to deliver would depend on the step that takes more time to process a piece of material, so that step would be the constraint that would need to be worked on. In the previous example, Transfer from Receiving took more time to process a unit of material; therefore, it was the bottleneck and the takt time depended on it.

Takt time is define as

$$\text{Takt time} = \frac{\text{net available time per period}}{\text{customer demand per period}}$$

Example 3.8   The takt time can help assess the processes' ability to meet customer requirements (see Table 3.5).

TABLE 3.5

	Description	Time
Scheduled work	Hours per shift	8
Number of shifts		3
Lunch	Minutes per shift	45
Breaks	Minutes per shift	30
Expected downtime	Minutes per shift	15
Customer demand	Units per day	120

Total work time = $8 \times 3 = 24$ h
Total work time in minutes = $24 \times 60 = 1440$ min
Total available time = $1440 - [(45 + 30 + 15) \times 3] = 1170$ min

$$\text{Takt time} = \frac{\text{net available time per period}}{\text{customer demand per period}} = \frac{1170}{120} = 9.8 \text{ min}$$

This means that every 9.8 min, a unit of material should be taken from the assembly line. We can use the SigmaXL template to verify the results (Table 3.6).

TABLE 3.6

SigmaXL Lean Templates: Takt Time Calculator		
Daily Customer Demand:	units per day	120
Scheduled Work:	hours per shift	8
Shifts per Day:		3
Lunch:	minutes per shift	45
Breaks:	minutes per shift	30
Planned Downtime:	minutes per shift	15
Staff/Operator Cycle Time	minutes per unit	
Available Time:	minutes per day	1170.0
Takt Time:	minutes per unit	9.8
Required Number of Staff/Operators:		0.0

If the process line is made up of several steps and operates with one-piece flow, the ideal solution would be to have each step take 9.8 min to process each piece.

Example 3.9   A processing line uses a one-piece flow and has six steps. The time that it takes each step to complete a task is shown in Table 3.7.

TABLE 3.7

Processing steps	Processing time (min)
1	10
2	13
3	9
4	10
5	12
6	11

The processing time for each piece depends on step 2 because it is the constraint since it takes more time to complete a task than the rest of the steps. This will result in excess (labor, machine, and/or floor) capacities. Step 1 will complete excess pieces and have materials piled up in front of step 2, and since step 3 and the rest of the downstream steps take much less time to process the material, they will end up being idle every time they run out of work. Therefore, the non-value-added elements on this line come under the form of excess inventory and wait.

If the time that it takes to complete one piece and meet the customers' requirement is 11 min, step 2 will prevent the line from meeting the deadline. Only step 6 really performs to meet the customers' requirements. Therefore, in order to make improvement to the processing line, the improvement team should determine if there are opportunities for improvement within step 2. If there is no way that step 2 can be made to process the material faster, the team can concentrate on finding out how to reallocate resources from steps 1, 3, 4, and 5 to step 2, which may require the redesign of the whole line. Any improvement will depend on how flexible it is for the process to shift resources from one step to another.

Once the takt time has been calculated, the next step is to create a balanced flow of material. Better capacity planning is made by making adjustments on machines, labor, setup time, and floor utilization to balance the workload. If batches are being used, it is better to change the process to a one-piece flow.

Batch versus one-piece flow

In a batch-and-queue process, pieces of material are consolidated in lots after having been processed before being moved to the next steps in the production line, while with one-piece flow every time a piece has been completed it is moved to the next step. The advantage of a one-piece flow over the batch-and-queue process is explained better through an example.

Example 3.10   Suppose in Example 3.9 that batches of 10 pieces have to be completed before being moved to the next step. How much time will it take to complete the first batch of 10 pieces using the batch-and-queue process? How much time will it take to complete 10 pieces using the one-piece flow?

Solution:

TABLE 3.8

Processing steps	Processing time (min)	Time to process 10 pieces
1	10	$10 \times 10 = 100$
2	13	$10 \times 13 = 130$
3	9	$10 \times 9 = 90$
4	10	$10 \times 10 = 100$
5	12	$10 \times 12 = 120$
6	11	$10 \times 11 = 110$
Total		**650**

It will take the processing line 650 min to finish the first batch of 10 pieces. In addition, after that, it will take 130 min to complete every subsequent batch (see Table 3.8). This is caused by the constraint in step 2.

If a one-piece flow were used, it would take 65 min to produce the first pieces and 13 min for every subsequent piece. Therefore, it would take 182 min $(65 + (13 \times 9))$ to complete the first 10 pieces. Compared to using the batch-and-queue process, 468 min would be saved for the first 10 units.

Data Gathering and Process Improvement

In order to determine and evaluate opportunities for improvement, the Kaizen event team needs to gather data at every step and in-between steps in the process to determine the presence of non-value-added elements.

Example 3.11 An audit of the processing line in Example 3.9 provided the information in Table 3.9. Note, all items are in minutes.

TABLE 3.9

Processing steps	Actual value-added work	Rework	Inspection	Wait	Machine adjust	Other	Total
1	6	0	3	0	0	1	10
2	7	1	3	0	1	1	13
3	5	0	1	3	0	0	9
4	4	2	2	0	1	1	10
5	7	1	2	0	1	1	12
6	6	1	1	1	0	2	11
Total	**35**	**5**	**12**	**4**	**3**	**6**	**65**

Based on that information, the Kaizen team can determine the percentage of time that is actually spent on doing value-added work.

Based on Table 3.10, only 53.85% of the overall time spent processing customers' orders is actual value-added work. The rest is spent on inspection, machine adjustments, rework, waiting for work, and other activities. To improve the process cycle time, the team will have to determine how to eliminate the unnecessary non-value-added elements and reduce the time that it takes to perform the value-added work. The improvement process can start with value stream mapping.

TABLE 3.10

Processing steps	Actual value-added work	Total time	Actual value-added work (%)
1	6	10	60
2	7	13	53.85
3	5	9	55.56
4	4	10	40
5	7	12	58.33
6	6	11	54.55
Total	**35**	**65**	**53.85**

Value stream mapping

The purpose of business operations is to produce goods or services that are destined to customers who determine their value through their desire to buy them and through the prices that they are willing to pay for them. For the Lean philosophy, what the customers value is what they are willing to pay for. Since value is what is being sold, the enterprise lets the customers determine the value that they are willing to buy before it creates the value system that will enable it to satisfy the customers' demand. The value system is determined by first identifying all the steps requires to meet customers' expectations, and then putting in place a set of operational mechanisms that enable them to meet customers' expectations.

The set of operations that they determine ranges from how orders are received, to how raw materials are bought from suppliers, to how those raw materials are transformed and the final goods delivered to the customers. The transformation path that the products follow from raw materials to finished goods in the hands of customers is called *value stream*. The analysis of a company's value stream helps trace the flow of material transformation from when the customers place their orders to when they receive their products. For a Lean-oriented organization, the production process should be set in such a way that at every step of the process, value is expected to be added to the material being transformed. Waste occurs every time the material goes through a step without having value added to it and every time it is stored at a given stage of the process, waiting to be further transformed. Non-value-adding elements in a process are:

- Rework
- Scrap
- Excess motion
- Excess transportation
- Excess inventory
- Wait
- Overproduction

Business organizational structures are composed of functional departments that put in place distinct processes and those departments operate in contingent sequences with different operations at different stages and areas of the organization being dependent on each other. Because of this interdependence of the different operations, every time a process at a given department fails to operate to its full potential or take more time to process orders than the other departments, it becomes a constraint, a bottleneck for the entire processing line.

How to map your value stream

Just as in the case of process mapping, the purpose of value stream mapping is to visualize the chain of events that leads to the generation of a throughput in order to determine where the non-value-adding activities are and pinpoint a bottleneck, clutter, or opportunities for improvement. Yet it is necessary to distinguish value stream mapping from process mapping because a value stream map does not only include steps in a process but also includes the flow of detailed and descriptive information that pertain to every step of the process.

Mapping the value stream does not only help visualize the sequence of operations but it also makes it easy to assess the process capabilities and performance with regard to takt time. A company has as many value streams as it has products, so the first step in value stream mapping is separating the products before examining the chain of processes that lead to their production.

A Lean-oriented organization usually uses pull techniques, which consist of subordinating production to the customers' actual orders instead of a forecast-driven production process that consists of building a high volume of outbound inventory that will wait for potential customers. Therefore, the value stream mapping of the current state for a Lean-driven process should start with the customers' demands. The three main players in the production of the demanded products are the customer, the organization's internal operations, and the suppliers of the raw materials and services.

The first step therefore is the assessment of a periodic (weekly, monthly, etc.) volume of orders from the customers that will later be subdivided to determine the daily production requirement. The next step is evaluating the demand for raw materials and services that will be required to satisfy the customers' demands and then determining the company's ability to meet demands along with its productivity, performance, and takt time. How long does it take every operation to process a given number of orders based on the available resources?

The current state of value stream mapping is not intended to describe the ideal state of the company but to visualize what is currently happening. Based on the current map, opportunities for improvement can be determined and actions taken.

The list of the data that need to be collected to give a complete picture of the current situation should include the following:

- Number of operators
- Actual working time (minus breaks and exception time)
- Cycle time—how long it takes each step in the process to complete its tasks

- Set up time—the time it takes to change from producing one type of product to another
- Available time—the time during which the people and the machines can actually perform value adding work
- Production per shift
- How many different types of products can be produced at a given station?
- Scrap rate
- Rate of rework

Example 3.12   Figure 3.11 is a simplified map of the value stream of a soap manufacturer.

Failure Mode and Effect Analysis

When a project team is engaged in a new product or process development or a new system implementation, no matter how well thought out and well conducted the initiative is uncertainty will always be present and it will always involve potential for failure. Being able to foresee the potential impediments to the initiative is a first step toward reducing the probability for their occurrence and the cost attached to future repairs. The techniques commonly used in design reliability analyses are the Ishikawa diagram, the fault tree analysis, and the failure mode and effect analysis (FMEA) or failure mode effects and criticality analysis (FMECA). While a fault tree analysis is an investigative tool that goes from the effects of a failure of a product or process to its root causes and the Ishikawa diagram is a relational diagram that traces effects to their causes, an FMEA is a design reliability analysis technique, a preemptive form of brainstorming that generally follows a process mapping for a product or process design and is usually followed by a Pareto analysis. It is a granular analysis of a process, system, or product design for identifying potential deficiencies. A cross-functional group generally conducts it with all the participants having a stake in or knowledge about the process, system, or product being assessed. While a fault tree analysis is generally conducted to investigate a failure that has already occurred, an FMEA is done prior to a product or process release; it is a preemptive technique. Although the methodology for conducting an FMEA is in general the same, there are minor differences of approach used to carry it out. The differences reside in the collection of the items to be evaluated for potential shortcomings. The collection depends on whether the analysis is done for a product or a process.

An FMEA starts with the gathering of a team of knowledgeable stakeholders who are involved in the design, development, deployment, or marketing of the products or process to be evaluated. The next step is getting all the information pertaining to the subject including diagrams, drawings, and maps that list all the steps (or features) of the process (or product) to be implemented.

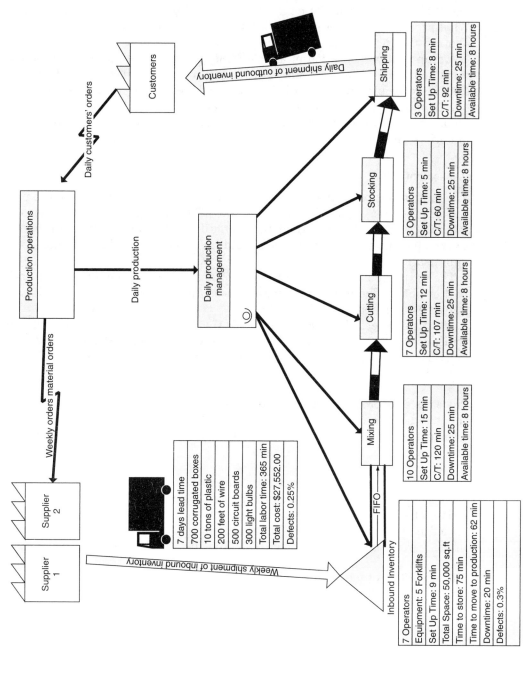

Figure 3.11

169

When the FMEA is done in relation with a new product development, the listing of the items to be assessed will include all the critical parts of the product and their interactions. The next step is determining the extent of the FMEA. What process, system, or product is being studied, what are the critical components in that product or process, and how do they interact?

A graphical presentation of an FMEA is generally a representation of two combined matrices: on one side is the failure mode and effect part, which consists of developing a list of all the causes of the potential failures and their effects on the overall process or product, and on the other side is the action plan, which determines what needs to be done to prevent the failures from materializing.

Failure mode assessment

The first step is to brainstorm and list the critical parts or phases of the process or product at hand. Flow charts and CAD drawings are generally used to map processes and interactions between different components of a product. Every element of the flow chart should be listed on the FMEA matrix for appraisal.

The next step is listing all of the potential failures that might occur to each part or phase. The probable causes of those failures are then listed and their impact established.

A critical aspect of this methodology is the determination of the severity (or criticality) of the failure, how often it is likely to happen, and how easy it is to detect. In general, the levels of severity, occurrence, and detection of each item in the FMEA are ranked from 1 to 10.

Process step or product function.   The process step or product function indicates the item being analyzed. If the FMEA were based on a flow chart, it would be a component in the chart. The flow chart should be divided in such a way that only one stage is analyzed at a time.

Potential failure mode.   The potential failure mode describes the way in which the corresponding item listed in the process step or product function could possibly fail to satisfy its intended purpose. There can be more than one potential failure for a function component. The potential failures may only occur under certain unique circumstance and each one should be listed separately.

Potential effect of the failure.   The potential effect of the failure describes the impact of the failure on the functionality of the product or process. It describes what the customer or process owner will actually notice because of the component's failure. The effect must be a very clear and explicit description of the impact of the noncompliance.

Potential causes of the failure.   This describes the flaws of the design that may lead to a potential failure. Every conceivable ground for failure should be listed in a thoroughly descriptive manner so that the corrective actions can be more promptly and cost effectively taken.

Severity.   The severity measures how critical or serious a potential failure can be on the product or process. If the failure is so serious that it can stop production, it is graded 10 and if it is very easy to correct, it is graded 1.

Detection.   How easy is the failure to detect? Detection is a measure of the ability to detect preemptively the failures. If the potential failure is easy to detect, the grade should be low (1 for very easy to detect and 10 for very hard).

Occurrence.   The occurrence measures how often the failure is likely to happen. Here again, the likelihood of an occurrence is expressed in ranking from 1 to 10 (1 for rare and 10 for often).

Current control.   Current controls are known preventive controls that are currently being used in similar processes or products.

Risk priority number.   The risk priority number (RPN) helps to rank the failures and establish their precedence for problem resolution considerations. It measures the process or product design risk. The RPN is the product of the severity, detection, and occurrence levels. For a failure with a severity of 6, a detection of 3, and an occurrence of 6, the RPN will be 108 ($6 \times 3 \times 6 = 108$). The higher the RPN, the more attention that particular step of the process or that characteristic of the product should get.

The templates used for FMEAs are not always the same, but the items above (severity, detection, occurrence, RPN) should always be present because they are the basis for corrective actions. SigmaXL provides a practical template for conducting an FMEA. We will use it in our example.

Action plan

Since the purpose of an FMEA is to forestall failures, after determining the list of potential failures and their RPNs, the next step is the planning of the actions to take to avert their occurrence. The planned actions to be taken are above all based on the nature of the failures but their presence is contingent upon the RPNs. After finishing the first phase of the FMEA, preventive tasks are assigned to stakeholders according to their aptitude, but the priority of execution should be subjected to the RPN ranking.

Not all FMEAs follow the same pattern of action plan but the following steps are usually considered.

- *Recommended actions.* The recommended preventive actions are generally suggested by the FMEA team during a brainstorming session. It consists of all the suggested proceedings that need to be followed to prevent failures.

The reasons for failures are multifaceted. Every failure can have several causes; that is why recommended preventive actions are better generated by a cross-functional team.

■ *Task owner and projected completion date.* The task owner is the person who has been assigned the task of mending the aspects of the product, process, or design that is subject to failure. Even though the suggested preventive actions are the result of a brainstorming session, the task of executing the actions is performed at an individual or departmental level. A person or a group of people are selected and assigned the task of averting failures. The projected completion date should also be determined to avoid procrastination and enforce accountability.

■ *Severity.* If the actions are taken and conducted according to the suggestions made by the team, by how much are they expected to reduce the potential failure? How would they affect the criticality of the failure? Here again the effects of the actions are ranked from 1 to 10.

■ *Occurrence.* How often will the failures happen if the recommended actions are taken?

■ *Detection.* Detection refers to the ability to identify failures. After improvement, potential failures should be easier to detect than they were before the recommended actions were taken.

■ *RPN.* Here again, the RPN will be the product of the detection, occurrence, and severity. After the improvements have been made, the RPN is expected to be significantly lower than it was before.

Example of an FMEA

The flow chart in Fig. 3.12 represents the future state process map for a new WMS implementation. The map depicts a cross-functional process that involves two interfaced systems' software and a physical process.

The first system software, Compass, is the host; it stores the customer databases, transportation matrices, orders nomenclatures, and warehouse inventory. The WMS is used to manage the daily transactions in the warehouse. The two systems are interfaced in such a way that the host processes the customer orders first before being dropped to the WMS. Once the orders are dropped, picking labels are printed and the warehouse employees are assigned the tasks to physically pick, pack, and ship the products.

After having mapped the future state of the process, an FMEA is conducted to preempt potential failures. The SigmaXL FMEA template summarizes the results of the brainstorming session.

After having evaluated the potential failures, every item on the list becomes a task and each task is assigned to a project team member. The priority for the task completions depends on the RPN numbers. It is clear that "Are SKUs available" has the highest RPN (60); therefore, it must receive special attention.

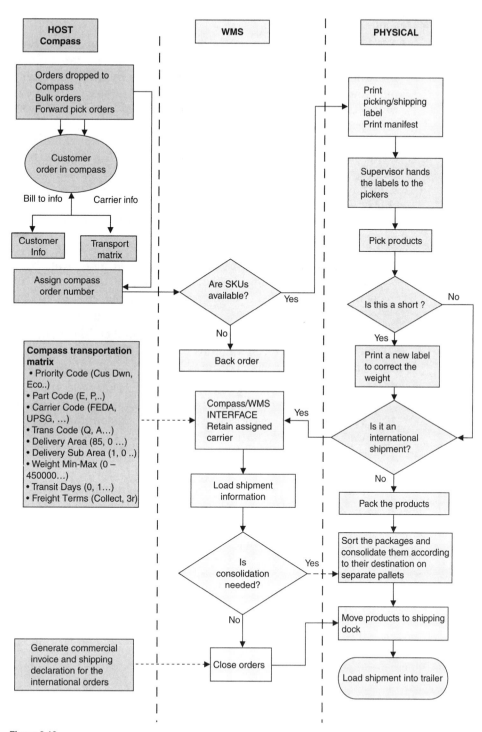

Figure 3.12

Hypothesis testing

A hypothesis is a value judgment made about a circumstance, a statement made about a population. Based on experience, an engineer can, for instance, assume that the amount of carbon monoxide emitted by a certain engine is twice the maximum allowed legally. However, his assertions can only be ascertained by conducting a test to compare the carbon monoxide generated by the engine with the legal requirements.

If the data used to make the comparison are parametric data, that is, data that can be used to derive a mean and a standard deviation, the populations from which the data are taken are normally distributed and they have equal variances. A standard error-based hypothesis testing using the *t*-test can be used to test the validity of the hypothesis made about the population. There are at least five steps to follow when conducting a hypothesis testing

1. *Null hypothesis.* The first step consists of stating the null hypothesis, which is the hypothesis being tested. In the case of the engineer making a statement about the level of carbon monoxide generated by the engine, the null hypothesis is:

   H_0: The level of carbon monoxide generated by the engine is twice as great as the legally required amount.
   The null hypothesis is denoted by H_0.

2. *Alternate hypothesis.* The alternate (or alternative) hypothesis is the opposite of the null hypothesis. It is assumed valid when the null hypothesis is rejected after testing. In the case of the engineer testing the carbon monoxide, the alternate hypothesis would be

   H_1: The level of carbon monoxide generated by the engine is not twice as great as the legally required amount.
   The alternate hypothesis is noted H_1 or in some cases H_a.

3. *Testing the hypothesis.* The objective of the test is to generate a sample test statistic that can be used to reject or fail to reject the null hypothesis. The test statistic is derived from the Z formula if the samples are greater than 30.

$$Z = \frac{\overline{X} - \mu}{\sigma / \sqrt{n}}$$

If the samples are less than 30, the *t*-test is used.

$$t = \frac{\overline{X} - \mu}{s / \sqrt{n}}$$

4. *Level of risk.* The level of risk addresses the kinds of errors that can be made while making an inference based on the test statistics obtained from

the testing. Two types of errors can be made. The experimenter can falsely reject a hypothesis that is true. In that case, we say that he made a Type I error or α error. If he fails to reject a hypothesis that is actually false, he makes a Type II or β error. In the case of the engineer testing the level of the carbon monoxide generated by the engine, if the actual level of carbon monoxide is in fact twice as great as the prescribed level and he rejected the null hypothesis, he would have made a Type I error. If the carbon monoxide generated by the engine is less than the legally prescribed level and the experimenter fails to reject the null hypothesis, he would have made a Type II error, he would have failed to reject a null hypothesis that happened to be false.

5. *Decision rule.* Only two decisions are considered: rejecting the hypothesis or failing to reject it. The decision rule determines the conditions under which the null hypothesis is rejected or failed to be rejected. The decision to reject the null hypothesis is based on the alpha (α) level. Before conducting the test, the experimenter must set the confidence level for the test. He can, for instance, decide to test his hypothesis with a 95% confidence level. That means he would be 95% sure that the decision to reject or fail to reject the null hypothesis is correct. However, 95% confidence level also means that there is a 5% chance that an error will be made.

Example 3.13 A machine used to average a production rate of 245 units per hour before it went for repair. After it came back from repair, over a period of 25 h, it produced an average of 249 units with a standard deviation of 8. Determine if there is a statistically significant difference between the machine's productivity before and after repair at a confidence level of 95%.

Solution: Since the sample is smaller than 30, we will use the t-test

The null hypothesis is the following:

H_0: The productivity before repair = the productivity after repair

The alternate hypothesis should state the opposite:

H_1: The productivity before repair ≠ the productivity after repair

$$n = 25, \quad s = 8, \quad \bar{X} = 249, \quad \mu = 245,$$

We can determine the interval within which we should expect the calculated t to fall in for the null hypothesis not to be rejected. We will determine the critical t-statistic values based on the degree of freedom and the significance level. Since the confidence level is set at 95%, $\alpha = 1 - 0.95 = 0.05$. Since the null hypothesis is stated as equality, we will have a two-tailed curve with each tail covering one-half of α. We would need to find $\alpha/2 = 0.05/2 = 0.025$.

With degree of freedom $df = n - 1 = 25 - 1 = 24$, the t-critical can be obtained from the t table (Table 3.11).

$$t_{\alpha/2, n-1} = t_{0.025, 24} = 2.064$$

TABLE 3.11

df	0.10	0.05	0.025	0.01	0.005
15	1.341	1.753	2.131	2.602	2.947
16	1.337	1.746	2.120	2.583	2.921
17	1.333	1.740	2.110	2.567	2.898
18	1.330	1.734	2.101	2.552	2.878
19	1.328	1.729	2.093	2.539	2.861
20	1.325	1.725	2.086	2.528	2.845
21	1.323	1.721	2.080	2.518	2.831
22	1.321	1.717	2.074	2.508	2.819
23	1.319	1.714	2.069	2.500	2.807
24	1.318	1.711	2.064	2.492	2.797
25	1.316	1.708	2.060	2.485	2.787
26	1.315	1.706	2.056	2.479	2.779
27	1.314	1.703	2.052	2.473	2.771

$$t_{\alpha/2, (n-1)} = t_{0.025, 24} = 2.064$$

Figure 3.13 represents the curve associated with t.

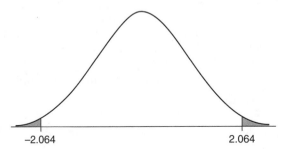

-2.064   2.064

Figure 3.13

If the calculated t-statistic falls within the interval $[-2.064, +2.064]$, we would fail to reject the null hypothesis; otherwise, the null hypothesis would be rejected.

Let us find the calculated t-statistic.

$$t = \frac{\overline{X} - \mu}{s / \sqrt{n}} = \frac{249 - 245}{8 / \sqrt{25}} = 2.5$$

The calculated t-statistic is 2.5. Since it falls outside the interval $[-2.064, +2.064]$, we will have to reject the null hypothesis and conclude that there is a statistically significant difference between the productivity of the machine prior to repair and after repair.

Using the confidence interval method We can use the formula for the confidence interval

$$\overline{X} - t_{\alpha/2,n-1}\frac{s}{\sqrt{n}} \leq \mu \leq \overline{X} + t_{\alpha/2,n-1}\frac{s}{\sqrt{n}}$$

Therefore,

$$249 - 2.064\frac{8}{\sqrt{25}} \leq \mu \leq 249 + 2.064\frac{8}{\sqrt{25}}$$

$$245.698 \leq \mu \leq 252.302$$

The null hypothesis is rejected because the mean μ (245) does not fall within the interval [245.698, 253.302].

Minitab output

```
One-Sample T

Test of mu = 245 vs not = 245


 N    Mean  StDev  SE Mean        95% CI          T      P
25  249.000  8.000   1.600  (245.698, 252.302)  2.50  0.020
```

p-value method

In the previous example, we rejected the null hypothesis because the value of the calculated t-statistic was outside the interval [−2.064, +2.064]. Had it been within that interval, we would have failed to reject the null hypothesis. The reason why [−2.064, +2.064] was chosen because the confidence level was set at 95%, which translates into $\alpha = 0.05$. If α were set at another level, the interval would have been different. The results obtained do not allow a comparison with a single value to make an assessment. Any value of the calculated t-statistic that falls within that interval would lead to a nonrejection of the null hypothesis.

The use of the p-value method enables us not to have to preset the value of α. The null hypothesis is assumed to be true; the p-value sets the smallest value of α for which the null hypothesis has to be rejected. For instance, in the example above, the p-value was 0.020 and $\alpha = 0.05$; therefore, α is greater than the p-value and we have to reject the null hypothesis.

Example 3.14 The monthly electricity bills for a company averages $500. The company decides to cut down on electricity consumption by encouraging employees to open their window shades instead of using lights. Twelve months later, the electricity bills were as shown in Table 3.12.

TABLE 3.12

Jan	Feb	Mar	Apr	May	Jun	Jul	Aug	Sep	Oct	Nov	Dec
479	509	501	470	489	494	445	502	467	489	508	500

Can we say that there has been an improvement in the electricity consumption with a confidence level of 99%?

Solution: The null hypothesis is

H_0: The average electricity bill before changes = the average bill after changes

And the alternate hypothesis is

H_1: The average electricity bill before changes $\neq$ the average bill after changes

where $s = 19.264$
$$\bar{X} = 487.75$$
$$n = 12$$

Because the confidence level is 99%, $\alpha = 1 - 0.99 = 0.01$.

The degree of freedom is 11 and because $\alpha = 0.01$, $\alpha/2 = 0.005$; therefore, from the t-table, the critical t-statistic will be $t_{0.005,11} = 3.106$.

We can determine the confidence interval using the formula

$$\bar{X} - t_{\alpha/2,n-1}\frac{s}{\sqrt{n}} \leq \mu \leq \bar{X} + t_{\alpha/2,n-1}\frac{s}{\sqrt{n}}$$

$$487.75 - 3.106\frac{19.264}{\sqrt{12}} \leq \mu \leq 487.75 + 3.106\frac{19.264}{\sqrt{12}}$$

$$470.477 \leq \mu \leq 505.023$$

The mean $500 falls well within the confidence interval; therefore, we cannot reject the null hypothesis and have to conclude that there has not been a statistically significant change in the level of electricity consumption because of the employees opening their window shades instead of using their lights.

Using SigmaXL   Open SigmaXL and then open the file *Electricitybill.xls*. Select the area containing the data. From the menu bar, click on **SigmaXL** before clicking on **Statistical Tools** and then select **1 Sample t-Test & Confidence Intervals** as indicated in Fig. 3.14.

The **1 Sample t-Test** should appear with the field **Please select your data** already filled out. Press the **Next >>** button. When the dialog box appears (see Fig. 3.15), select the **Unstacked Column Format (2 or More Data Columns)** option, type in "500"

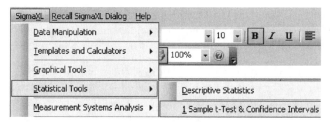

Figure 3.14

Figure 3.15

in the **H0: Mean** = field and for the **Confidence Level** field, type in 99.0. Click on the **Numeric Data Variables (Y)>>** button.

Press the **OK>>** button to get the results shown in Table 3.13.

TABLE 3.13

1 Sample t-test	
H0: Mean (Mu) = 500	
Ha: Mean (Mu) Not Equal To 500	
	Bills
Count	12
Mean	487.75
StDev	19.264
SE Mean	5.561
t	-2.203
P-value (2-sided)	0.0498
UC (2-sided, 99%)	505.02
LC (2-sided, 99%)	470.48

The p-value is equal to 0.0498, which is greater than $\alpha = 0.01$; therefore, we fail to reject the null hypothesis and have to conclude that there has not been any difference after the changes were implemented.

Example 3.15 A food producer claims that the amount of carbohydrate contained in its bread is 247 g of carb per loaf. A consumer group decides to test that claim. It takes a sample of 20 pieces of bread and runs a test. The observations are summarized in Table 3.14. Determine if the amount of carbohydrate is really 247 g of carb at a confidence level of 95%.

TABLE 3.14

283.009	263.273	219.832	236.298
248.934	264.623	275.402	260.859
283.038	235.411	267.245	259.513
272.48	244.518	248.304	277.077
255.197	255.79	250.85	247.606

Solution: The null hypothesis should state that

$$H_0 : \mu = 247 \text{ g}$$

and the alternate hypothesis

$$H_1 : \mu \neq 247 \text{ g}$$

Based on the sample, $n = 20$, $s = 16.636$, $\bar{x} = 257.463$

$$df = 20 - 1 = 19, \quad \alpha/2 = 0.05/2 = 0.025.$$

From the t-table shown in Table 3.15, we determine that $t_{\alpha/2, df} = t_{0.025, 19} = 2.093$.

TABLE 3.15

```
         0.10    0.05  | 0.025
----+-------------------------
15 |    1.341   1.753  | 2.131
16 |    1.337   1.746  | 2.120
17 |    1.333   1.740  | 2.110
18 |    1.330   1.734  | 2.101
19 |    1.328   1.729  | 2.093
```

For the null hypothesis not to be rejected, the calculated t-statistic will have to fall within the interval [–2.093, +2.093] (see Fig. 3.16).

Let us find the calculated t-statistic.

$$t = \frac{\bar{X} - \mu}{s/\sqrt{n}} = \frac{257.463 - 247}{\dfrac{16.636}{\sqrt{20}}} = \frac{10.463}{3.72} = 2.813$$

The calculated t-statistic is equal to 2.813, which falls outside the interval [–2.093, +2.093]; therefore, we have to reject the null hypothesis and conclude that the amount of carbohydrate is not equal to 247 g of carb.

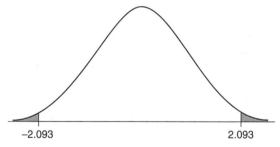

−2.093   2.093

Figure 3.16

Using the confidence interval method

$$257.463 - 2.093\frac{16.636}{\sqrt{20}} \le \mu \le 257.463 + 2.093\frac{16.636}{\sqrt{20}}$$

$$249.677 \le \mu \le 265.249$$

Here again, the mean (i.e. 247 g of carbs) falls outside the confidence interval and we therefore have to reject the null hypothesis.

Using SigmaXL, we obtain Table 3.16.

TABLE 3.16

1 Sample t-test	
H0: Mean (Mu) = 247	
Ha: Mean (Mu) Not Equal To 247	
	Bread
Count	20
Mean	257.46
StDev	16.636
SE Mean	3.720
t	2.813
P-value (2-sided)	0.0111
UC (2-sided, 95%)	265.25
LC (2-sided, 95%)	249.68

The p-value is equal to 0.011, which is much smaller than 0.05; therefore, the null hypothesis has to be rejected.

Nonparametric Hypothesis Testing

In the previous examples, we used means and standard deviations to determine if there were statistically significant differences between samples. What happens if the data cannot yield arithmetic means and standard deviations? What happens if the data are nominal or ordinal?

When we deal with categorical, nominal, or ordinal data, nonparametric statistics are used to conduct a hypothesis testing. A nonparametric test is a test that analyzes data that cannot be converted into parameters such as means and standard deviations. The Chi-square test and the Mann-Whitney U test are examples of nonparametric tests.

Chi-square test

Chi-square goodness-of-fit test.   Fouta Electronics and Touba Inc. are computer manufacturers that use the same third-party call center to handle their customer services. Touba Inc. conducted a survey to evaluate how satisfied its customers were with the services that they receive from the call center. The results of the survey are summarized in Table 3.17.

TABLE 3.17

Categories	Rating (%)
Excellent	10
Very good	45
Good	15
Fair	5
Poor	10
Very poor	15

After having seen the results of the survey, Fouta Electronics decided to find out whether they apply to its customers, so it interviewed 80 randomly selected customers and obtained the results shown in Table 3.18.

TABLE 3.18

Categories	Rating (absolute value)
Excellent	8
Very good	37
Good	11
Fair	7
Poor	9
Very poor	8

To analyze the results, the quality engineer at Fouta Electronics conducts a hypothesis testing. However, in this case, because he is faced with categorical data, he cannot use a t-test since a t-test relies on the standard deviation and the mean, which we cannot obtain from either table. We cannot deduct a mean satisfaction or a standard deviation satisfaction. Therefore, another type of test will be needed to conduct the hypothesis testing. The test that applies to this situation is the Chi-square test, which is a nonparametric test.

Step 1: State the null hypothesis. The null hypothesis will be

H_0 : The results of Touba Inc. survey = the same as the results of Fouta Electronics survey

and the alternate hypothesis will be

H_1 : The results of Touba Inc. survey ≠ the same as the results of Fouta Electronics survey

Step 2: Test statistics to be used. The test statistic used to conduct the hypothesis testing is based on the calculated χ^2, which is obtained from the following formula:

$$\chi^2 = \sum \frac{(f_0 - f_e)^2}{f_e}$$

where f_e represents the expected frequencies and f_0 represents the observed frequencies.

Based on the formula, it is obvious that χ^2 is equal to 0 when there is a perfect agreement between the observed frequencies and the expected frequencies $f_0 = f_e$.

A difference between f_0 and f_e can only result in a positive χ^2 since it is the sum of squares. Chi-square goodness-of-fit test is therefore one-tailed.

The decision to reject or to fail to reject the null hypothesis will be based on a comparison between the calculated χ^2 and the critical χ^2.

The critical Chi-square $\chi^2_{\alpha,df}$ is based on the confidence level and the degree of freedom. The critical $\chi^2_{\alpha,df}$ is obtained from the χ^2 table.

If the calculated χ^2 exceeds the critical $\chi^2_{\alpha,df}$, the null hypothesis is rejected; otherwise, we fail to reject the null hypothesis (see Fig. 3.17).

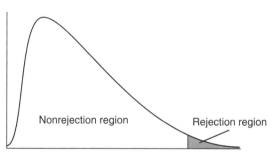

Figure 3.17

Step 3: Calculating the χ^2 test statistic.   To conduct the test, the two tables need to be expressed in the same form. As they stand, one is expressed in the form of percentages while the other is in the form of absolute values.

If a sample of 80 customers were surveyed, the data in Table 3.17 would have resembled the data in Table 3.19 if converted into absolute values.

TABLE 3.19

Categories	Rating (%)	Expected frequencies f_e
Excellent	10	$0.10 \times 80 = 8$
Very good	45	$0.45 \times 80 = 36$
Good	15	$0.15 \times 80 = 12$
Fair	5	$0.05 \times 80 = 4$
Poor	10	$0.10 \times 80 = 8$
Very poor	15	$0.15 \times 80 = 12$
Total	100	80

We can summarize the observed frequencies and the expected frequencies in the same table (Table 3.20).

TABLE 3.20

Categories	Observed frequencies f_0	Expected frequencies f_e	$\dfrac{(f_0 - f_e)^2}{f_e}$
Excellent	8	8	0
Very good	37	36	0.028
Good	11	12	0.083
Fair	7	4	2.25
Poor	9	8	0.125
Very poor	8	12	1.33
Total	80	80	3.816

$$\chi^2 = \sum \frac{(f_0 - f_e)^2}{f_e} = 3.816$$

Now that we have found the calculated χ^2, we can find the critical χ^2 from the table. The critical χ^2 is based on the degree of freedom and the confidence level. Since the number of categories is 6, the degree of freedom is equal to $6 - 1 = 5$. If the confidence level is set at 95%, $\alpha = 0.05$; therefore, the critical χ^2 is equal to 11.070 (see Table 3.21).

$$\chi^2_{0.05,5} = 11.070$$

TABLE 3.21

DF	0.10	0.05	0.025	0.01	0.001
1	2.706	3.841	5.024	6.635	10.828
2	4.605	5.991	7.378	9.210	13.816
3	6.251	7.815	9.348	11.345	16.266
4	7.779	9.488	11.143	13.277	18.467
5	9.236	11.070	12.833	15.086	20.515
6	10.645	12.592	14.449	16.812	22.458
7	12.017	14.067	16.013	18.475	24.322
8	13.362	15.507	17.535	20.090	26.125
9	14.684	16.919	19.023	21.666	27.877
10	15.987	18.307	20.483	23.209	29.588
11	17.275	19.675	21.920	24.725	31.264
12	18.549	21.026	23.337	26.217	32.910
13	19.812	22.362	24.736	27.688	34.528
14	21.064	23.685	26.119	29.141	36.123

Since the critical $\chi^2_{0.05,5} = 11.070$ is much greater than the calculated $\chi^2 = 3.816$, we fail to reject the null hypothesis and have to conclude that the surveys done by Touba Inc. and Fouta Electronics gave statistically similar results (Fig. 3.18).

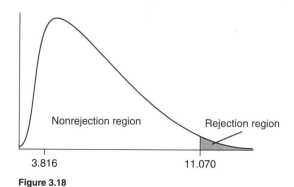

Figure 3.18

Contingency analysis—Chi-square test of independence

In the previous example, we only had one variable, which was the customers' satisfaction about the service they received from the call center. If we have more than one variable with several levels (or categories) to test at the same time, we use the Chi-square test of independence.

Example 3.16   Bakel Digital manufactures computers sold to three groups of customers: businesses, individuals, and governments. The company wants to add a new feature to its products and has conducted a survey to determine if the feature appeals to the three groups in the same way. The results of the survey are shown in Table 3.22.

TABLE 3.22

	Business	Government	Individuals	Total
Not relevant	15	9	33	57
Relevant	13	4	26	43
Somewhat relevant	17	5	23	45
Very relevant	16	9	25	50
Total	**61**	**27**	**107**	**195**

In this case, if there is no difference in the relevance of the new feature to the different groups of customers, there should be no statistically significant difference between categories.

The null hypothesis should state

H_0 : The relevance of the feature is the same for all categories of customers

and the alternate hypothesis should state

H_1 : At least one category of customers is different from the others

The observed frequencies are being compared with the expected frequencies as in the case of the goodness-of-fit test to compute the calculated χ^2. The calculated χ^2 will be compared with the critical χ^2 to determine whether the null hypothesis should be rejected.

The expected frequency for each cell is computed by multiplying the row total by the column total for that cell and then dividing the result by the grand total as shown in Table 3.23.

TABLE 3.23

	Businesses	Governments	Individuals
Not relevant	$(57 \times 61)/195 = 17.831$	$(57 \times 27)/195 = 7.892$	$(57 \times 107)/195 = 31.277$
Relevant	$(43 \times 61)/195 = 13.451$	$(43 \times 27)/195 = 5.594$	$(43 \times 107)/195 = 23.595$
Somewhat relevant	$(45 \times 61)/195 = 14.077$	$(45 \times 27)/195 = 6.231$	$(45 \times 107)/195 = 24.692$
Very relevant	$(50 \times 61)/195 = 15.641$	$(50 \times 27)/195 = 6.923$	$(50 \times 107)/195 = 27.436$

The calculated χ^2 is obtained from the following formula:

$$\chi^2 = \sum \frac{\left(f_e - f_0\right)^2}{f_e}$$

TABLE 3.24

Expected frequencies f_e	Observed frequencies f_0	$\dfrac{(f_e - f_0)^2}{f_e}$
17.831	15	0.449
13.451	13	0.015
14.077	17	0.607
15.641	16	0.008
7.892	9	0.155
5.954	4	0.641
6.231	5	0.243
6.923	9	0.623
31.277	33	0.095
23.595	26	0.245
24.692	23	0.116
27.436	25	0.216
		3.415

$$\chi^2 = \sum \frac{(f_e - f_0)^2}{f_e} = 3.415$$

Now that we have found the calculated χ^2 (Table 3.24), we can determine the critical χ^2 from the χ^2 table. As in the case of the goodness-of-fit, the critical χ^2 depends on the confidence level and the degree of freedom. If the confidence level is set at 95%, $\alpha = 0.05$. The degree of freedom is calculated based on the number of row and columns. We have three columns and four rows. The degree of freedom, $df = (c - 1)(r - 1) = (4 - 1)(3 - 1) = 6$.

The critical $\chi^2_{0.05,6} = 12.592$. Since the calculated $\chi^2 = 3.415$ is much lower than the critical $\chi^2_{0.05,6} = 12.596$, we fail to reject the null hypothesis and have to conclude that the relevance of the feature is independent of the groups of customers.

Using SigmaXL   After having opened **SigmaXL**, open the file *Bakel.xls* and select all the area containing the data. From the menu bar, click on **SigmaXL**, then click on **Statistical Tools,** and then select **Chi-Square Test-Two-Way Table Data**. When the **Chi-Square table data** dialog box appears, click on the **Next>>** button to get the output shown in Table 3.25.

The results shown in the **SigmaXL** output match our calculation and the p-value of 0.7552 indicates that we fail to reject the null hypothesis.

The Mann-Whitney U test

The Mann-Whitney U test is better explained through an example.

Example 3.17   An operations manager wants to compare the number of inventory discrepancies found in two operating shifts. The inventory discrepancies are not normally distributed. The manager takes a sample of discrepancies found over 7 days

TABLE 3.25

Chi-Square Table Statistics			
Observed Counts	Business	Government	Individuals
Not Relevant	15	9	33
Relevant	13	4	26
Somewhat relevant	17	5	23
Very Relevant	16	9	25
Expected Counts	Business	Government	Individuals
Not Relevant	17.831	7.892	31.277
Relevant	13.451	5.954	23.595
Somewhat relevant	14.077	6.231	24.692
Very Relevant	15.641	6.923	27.436
Std. Residuals	Business	Government	Individuals
Not Relevant	-0.670377	0.394291	0.308101
Relevant	-0.123046	-0.800740	0.495142
Somewhat relevant	0.779088	-0.493067	-0.340564
Very Relevant	0.09076759	0.789352	-0.465050
Chi-Square	3.415		
DF	6		
P-value	0.7552		

for the first shift and 5 days for the second shift and tabulates the data as shown in Table 3.26.

TABLE 3.26

First shift	Second shift
15	17
24	23
19	10
9	11
12	18
13	
16	

We can make several observations from this table. First, the sample sizes are small and we only have two samples, so the first thing that comes to mind would be to use the standard error based t-test. However, the t-test assumes that the populations from which the samples are taken should be normally distributed, which is not the case in this example; therefore, the t-test cannot be used. Instead, the Mann-Whitney U test should be used.

The Mann-Whitney U test assumes that the samples are independent and from dissimilar populations.

Step 1: Define the null hypothesis. Just as in the case of the t-test, the Mann-Whitney U test is a hypothesis test. The null and alternate hypotheses are

H_0: The number of discrepancies in the first shift is the same as the one in the second shift

H_1: The number of discrepancies in the first shift is different from the ones in the second shift

The result of the test will lead to the rejection of the null hypothesis or a failure to reject the null hypothesis.

Step 2: Analyze the data. The first step in the analysis of the data consists of naming the groups. In our case, they are already named First Shift and Second Shift. The next step consists of grouping the two columns in one and sorting the observations in ascending order raked from 1 to n. Each observation is paired with the name of the original group to which it belonged.

We obtain Table 3.27.

TABLE 3.27

Observations	Group	Ranks
9	First shift	1
10	Second shift	2
11	Second shift	3
12	First shift	4
13	First shift	5
15	First shift	6
16	First shift	7
17	Second shift	8
18	Second shift	9
19	First shift	10
23	Second shift	11
24	First shift	12

We will call ϖ_1 the sum of the ranks of the observations for group First Shift and ϖ_2 the sum of the ranks of the observations for group Second Shift.

$$\varpi_1 = 1+4+5+6+7+10+12 = 45$$

$$\varpi_2 = 2+3+8+9+11 = 33$$

Step 3: Determining the values of the U statistic. The computation of the U statistics will depend on the samples' sizes.

The samples are small when n_1 and n_2 are both smaller than 10. In that case,

$$U_1 = n_1 n_2 + \frac{n_1(n_1+1)}{2} - v_1$$

$$U_2 = n_1 n_2 + \frac{n_2(n_2+1)}{2} - v_2$$

The test statistic U will be the smallest between U_1 and U_2.

If any or both of the sample sizes are greater than 10, then U will be approximately normally distributed and we could use the Z transformation with

$$\mu = \frac{n_1 n_2}{2}$$

$$\sigma = \sqrt{\frac{n_1 n_2 (n_1 + n_2 + 1)}{12}}$$

and

$$Z = \frac{U - \mu}{\sigma}$$

In our case, both sample sizes are less than 10; therefore,

$$U_1 = 7 \times 5 + \frac{7(7+1)}{2} - 45 = 35 + 28 - 45 = 18$$

$$U_2 = 7 \times 5 + \frac{5(5+1)}{2} - 33 = 35 + 15 - 33 = 17$$

Since the calculated test statistic is the smallest of the two, we will have to consider $U_2 = 17$, so we will use $U = 17$ with $n_2 = 7$ and $n_1 = 5$.

From the Mann-Whitney table, we obtain a p-value equal to 0.5 for a one-tailed graph. Since we are dealing with a two-tailed graph, we have to double the p-value and obtain 1.

Since the p-value is equal to 1, we fail to reject the null hypothesis and have to conclude that the number of discrepancies in the first shift are the same as the ones in the second shift (Table 3.28).

Using SigmaXL   Open **SigmaXL**, and then open the file *Inventorydiscrepancy.xls*. We will stack the data first. Select the area containing the data and from the menu bar, click on **SigmaXL**, then select **Data Manipulation,** and then select **Stack Columns** (Fig. 3.19).

When the **Stack Column** box appears, click on the **Next>>** button. When the second **Stack Column** appears, fill it out as shown in Fig. 3.20.

Press the **OK>>** button to obtain the stacked columns.

Once the **Stacked Columns** appear, click on **SigmaXL** again and select **Statistical Tools**, then **Nonparametric Tests,** and then **2 Sample Mann-Whitney** as shown in Fig. 3.21.

Fill out the **2 Sample Mann-Whitney** box as shown in Fig. 3.22.

Press the **OK>>** button to obtain the output (Table 3.29).

The p-value is the highest it can be; therefore, we cannot reject the null hypothesis. We have to conclude that there is not enough statistical evidence to determine that the two sets of data are not identical.

TABLE 3.28

P-Values for Mann-Whitney U Statistic

$n_2 = 7$	U_o	n_1						
		1	2	3	4	5	6	7
	0	.1250	.0278	.0083	.0030	.0013	.0006	.0003
	1	.2500	.0556	.0167	.0061	.0025	.0012	.0006
	2	.3750	.1111	.0333	.0121	.0051	.0023	.0012
	3	.5000	.1667	.0583	.0212	.0088	.0041	.0020
	4		.2500	.0917	.0364	.0152	.0070	.0035
	5		.3333	.1333	.0545	.0240	.0111	.0055
	6		.4444	.1917	.0818	.0366	.0175	.0087
	7		.5556	.2583	.1152	.0530	.0256	.0131
	8			.3333	.1576	.0745	.0367	.0189
	9			.4167	.2061	.1010	.0507	.0265
	10			.5000	.2636	.1338	.0688	.0364
	11				.3242	.1717	.0903	.0487
	12				.3939	.2159	.1171	.0641
	13				.4636	.2652	.1474	.0825
	14				.5364	.3194	.1830	.1043
	15					.3775	.2226	.1297
	16					.4381	.2669	.1588
	17					.5000	.3141	.1914
	18						.3654	.2279
	19						.4178	.2675
	20						.4726	.3100
	21						.5274	.3552
	22							.4024
	23							.4508
	24							.5000

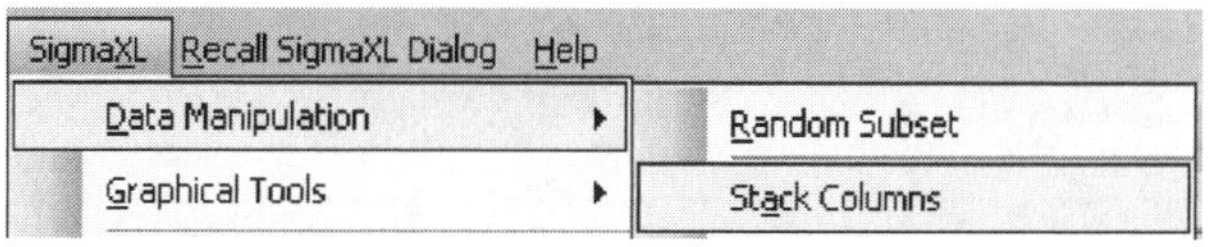

Figure 3.19

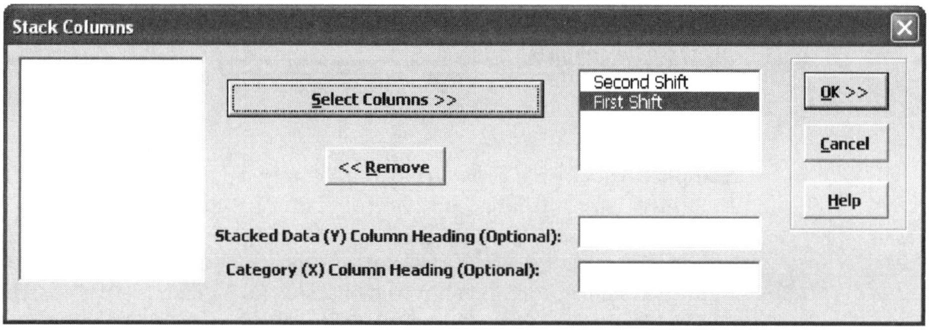

Figure 3.20

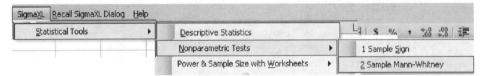

Figure 3.21

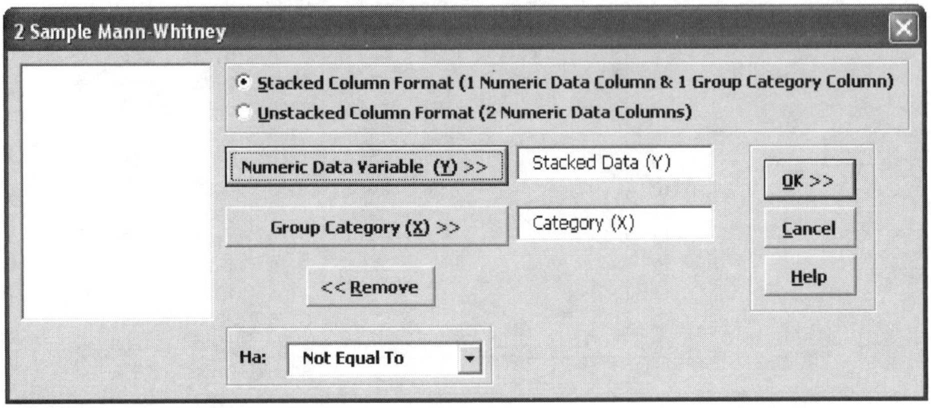

2 Sample Mann-Whitney				✕

● **Stacked Column Format (1 Numeric Data Column & 1 Group Category Column)**
○ **Unstacked Column Format (2 Numeric Data Columns)**

Numeric Data Variable (Y) >>	Stacked Data (Y)	**OK >>**
Group Category (X) >>	Category (X)	**Cancel**
<< Remove		**Help**

Ha: **Not Equal To**   ▼

Figure 3.22

TABLE 3.29

2 Sample Mann-Whitney - Stacked Data (Y)		
H0: Median Difference = 0		
Ha: Median Difference ≠ 0		
	First Shift	**Second Shift**
Count	7	5
Median	15	17
Mann-Whitney Statistic	45.00	
P-value (2-sided, adjusted for ties)	1.0000	

Example 3.18   In the previous example, we used small samples; in this one, we will use large samples. Tambacounda-Savon is a soap manufacturing company. It operates two shifts. The quality manager wants to compare the quality level of the output of the two shifts. He takes a sample at 12 days from the first shift and another sample at 11 days from the second shift and obtains the following errors per 10,000. The data are summarized in Table 3.30. At a confidence level of 95%, can we say that the two shifts produce the same quality level of output?

TABLE 3.30

First shift	Second shift
2	14
4	5
7	1
9	7
6	15
3	4
12	9
13	10
10	17
0	16
11	8
5	

Solution:

Step 1: Define the hypotheses.   The null hypothesis in this case will suggest that there is no difference between the quality levels of the output of the two shifts and the alternate hypothesis will suggest the opposite.

H_0: The quality level of the first shift is the same as the one for the second shift
H_1: The quality level of the first shift is different from the one of the second shift

Step 2: Analyze the data.   Here again, we pool all the data in one column or line and we rank them from the smallest to the highest while still maintaining the original groups to which they belonged (Table 3.31).

TABLE 3.31

Defects	Shift	Rank
0	First shift	1
1	Second shift	2
2	First shift	3
3	First shift	4
4	First shift	5.5
4	Second shift	5.5
5	First shift	7.5
5	Second shift	7.5
6	First shift	9
7	First shift	10.5
7	Second shift	10.5
8	Second shift	12
9	First shift	13.5
9	Second shift	13.5
10	First shift	15.5
10	Second shift	15.5
11	First shift	17
12	First shift	18
13	First shift	19
14	Second shift	20
15	Second shift	21
16	Second shift	22
17	Second shift	23

$$v_{\text{First}} = 1+3+4+5.5+7.5+9+10.5+13.5+15.5+17+18+19 = 123.5$$

$$v_{\text{Second}} = 2+5.5+7.5+13.5+15.5+20+21+22+23 = 152.5$$

We can now find the value of U

$$U_{\text{First}} = 12 \times 11 + \frac{12(12+1)}{2} - 123.5 = 86.5$$

$$U_{\text{Second}} = 12 \times 11 + \frac{11(11+1)}{2} - 152.5 = 45.5$$

The following steps consist of finding the mean and standard deviation.

$$\mu = \frac{n_1 n_2}{2} = \frac{12 \times 11}{2} = 66$$

$$\sigma = \sqrt{\frac{n_1 n_2 (n_1 + n_2 + 1)}{12}} = \sqrt{\frac{132(12+11+1)}{12}} = \sqrt{264} = 16.25$$

Since the U_{Second} is the smallest of the two, it will be used to find the value of Z.

$$Z = \frac{U_{\text{Second}} - \mu}{\sigma} = \frac{45.5 - 66}{16.25} = -1.262$$

What would have happened if we had used U_{First} instead of U_{Second}?

$$Z = \frac{U_{\text{First}} - \mu}{\sigma} = \frac{86.5 - 66}{16.25} = 1.262$$

We would have obtained the same result with the opposite sign.

At a confidence level of 95%, we would reject the null hypothesis if the value of Z is outside the interval [−1.96, +1.96]. In this case, $Z = -1.262$ is well within that interval; therefore, we should not reject the null hypothesis (Fig. 3.23).

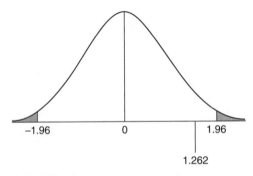

Figure 3.23

Minitab output

> **Mann-Whitney Test and CI: First Shift, Second Shift**
>
> ```
>                N   Median
> First Shift   12    6.500
> Second Shift  11    9.000
>
>
> Point estimate for ETA1-ETA2 is -3.000
> 95.5 Percent CI for ETA1-ETA2 is (-7.001,1.999)
> W = 123.5
> Test of ETA1 = ETA2 vs ETA1 not = ETA2 is significant at 0.2184
> The test is significant at 0.2178 (adjusted for ties)
> ```

The p-value of 0.2178 is greater than 0.05, which suggests that for an alpha level of 0.05, we cannot reject the null hypothesis.

Normality testing

Before conducting a data analysis using the normal distribution, it is essential to make sure that the characteristic being studied is in fact normally distributed. If nonnormal data are analyzed as if they were normal, the interpretations of the results would be misleading.

Both SigmaXL and Minitab offer options to test the normality of data. Minitab uses three main options for normality testing: the Anderson-Darling test, the Ryan-Joyner test, and the Kolmogorov-Smirnov test. All these tests yield the same results and they are all based on a hypothesis testing with the null hypothesis stating that the data are normal and the alternate hypothesis stating that the data are not normal. If the alpha level is set at 0.05, the null hypothesis is rejected if the p-value is less than 0.05. When conducting a hypothesis testing using the Anderson-Darling test, for instance, and the p-value is less than 0.05, the conclusion should be that for $\alpha = 0.05$ the data are not normally distributed.

Example 3.19   The data in Table 3.32 represent the number of defects that were found at a processing line during a 20-day period. The quality assurance manager wants to determine if the daily defect rate is normally distributed.

TABLE 3.32

20	20	22	20	21	21	19	20	20	19	20	20	21	21	20	19	20	19	20	21

To conduct the testing using Minitab, open the file *Defect.MTW* and from the menu bar, click on **Stat**. Then click on **Basic statistics** and then on **Normality test...** When the **Normality Test** box appears, select **Defects** for the **Variables** field and then select the **Anderson–Darling** option before clicking on the **OK** button.

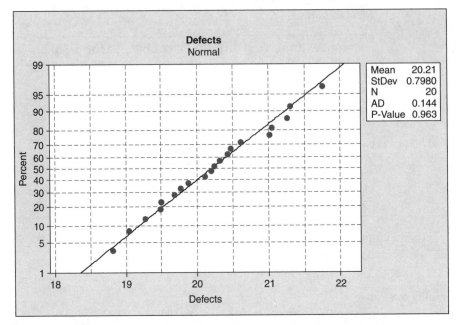

Figure 3.24

The Minitab output in Fig. 3.24 shows a p-value of 0.963, which is greater than 0.05; therefore, we have to conclude that the data are normally distributed.

Normalizing data

When the data being analyzed are not normal and the tools used for the analysis require their normality, then one option would be to normalize them. Normalizing the data means transforming them from nonnormal to normal. This can be done using the Box-Cox transformation, the Johnson Transformation, or the natural logarithm.

Example 3.20   A defective machine is producing pipes with dimensions that exhibit too much variation. The quality engineer took measurements of 40 parts and stored the data in the file *Pipes.MTW*. He wants to determine if the data are normally distributed and if they are not, he will normalize them for further analysis.

Solution:

Testing for normality   Open the file *pipes.MTW* and then, from the menu bar, click on **Stat**, then on **Basic Statistics,** and then on **Normality Testing.** When the **Normality Testing** box appears, select **C1 Length** for the **Variable** field; select the **Anderson-Darling** option for **Tests for normality,** then press the **OK** button.

The graph in Fig. 3.25 appears. It shows that the data points are not randomly clustered around the centerline and the p-value of less than 0.005 indicates that the data are not normal.

Normalizing the data   There are several options for normalizing the data; we can use the Box-Cox transformation or the Johnson Transformation.

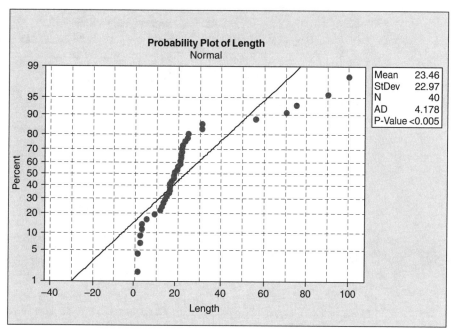

Figure 3.25

From the menu bar, click on **Stat**, then click on **Quality Tools,** and from the dropdown list select **Johnson Transformation**.

When the Johnson Transformation box appears, fill it out as shown in Fig. 3.26.

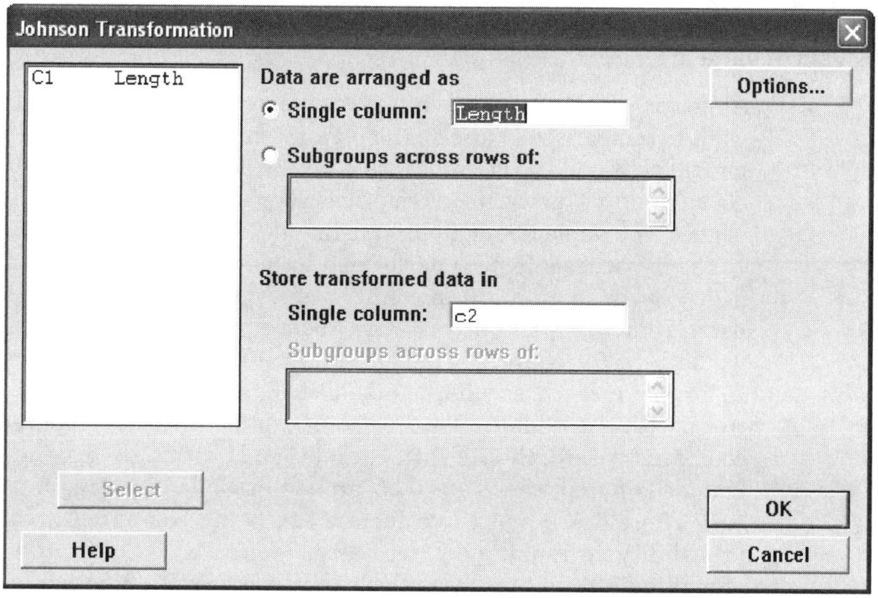

Figure 3.26

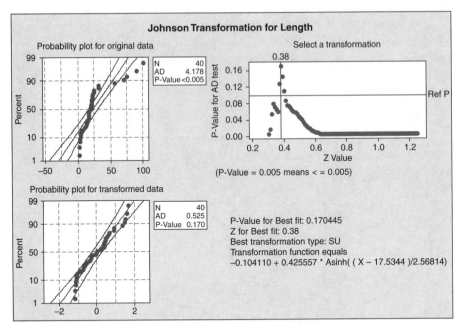

Figure 3.27

Press the **OK** button to obtain the results shown in Fig. 3.27.

The graph shows the probability plots before and after transformation. The probability plot at the bottom left corner shows a p-value of 0.17, which is greater than 0.05. The data after transformation are therefore normally distributed.

Analysis of Variance

A t-test is usually done to compare the means of two treatments. For instance, if we want to compare the performance of a machine before some adjustments are performed on it and the performance after the adjustments are performed, the mean of one sample of products taken prior to adjustments can be compared to the mean of another sample taken after adjustment. In that case, a t-test can be useful.

The hypothesis testing performed based on the t-test is conducted using the degree of freedom and the confidence level, but when two sample means are being compared, there is always room for making errors. If $\alpha = 0.05$, there would be a 5% chance of rejecting a null hypothesis that happens to be true. If, for instance, three sample means A, B, and C are being compared using the t-test with a confidence level of 95%, two factors are compared at a time. A is compared with B, then A with C, and then B with C. Every time two factors are being compared, there are 0.05 probabilities for rejecting a true null hypothesis. Therefore, when three factors are being compared using the t-test, the probability for making a Type I error is inflated. In order to limit the chances of making a Type I error, analysis of variance (ANOVA) can be used instead of the multiple t-tests.

ANOVA is a hypothesis test used when more than two factors' means are being compared.

If k samples are being tested, the null hypothesis will be in the form of

$$H_0 : \mu_1 = \mu_2 = \ldots = \mu_k$$

and the alternate hypothesis will be

H_1 : At least one sample mean is different from the others

The null hypothesis is not rejected if the means of all the samples are equal and it is rejected if at least one mean is different from the rest.

The differences between sample means can be traced to two sources: the variations due to actual differences between the sample means and the variations within the sample (Fig. 3.28).

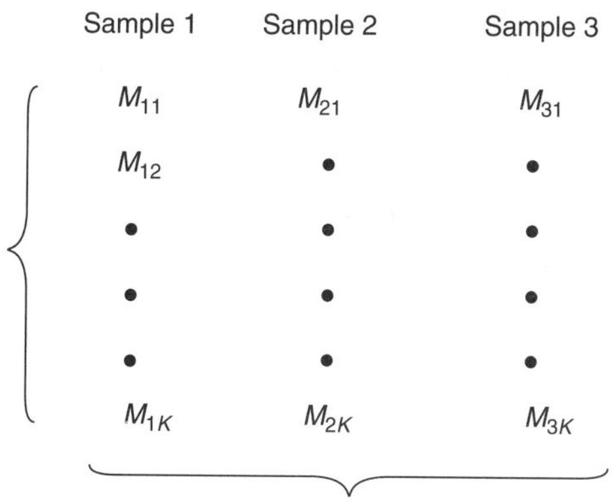

Sample 1	Sample 2	Sample 3
M_{11}	M_{21}	M_{31}
M_{12}	•	•
•	•	•
•	•	•
•	•	•
M_{1K}	M_{2K}	M_{3K}

Figure 3.28

ANOVA examines the variances of the samples to determine if there is a difference between their means. The total sum of squares of the deviations from the mean can be divided into the sum of squares of the deviation between treatments and the sum of squares of the deviation within treatments. The sum of squares of the deviations between samples measures how far each sample mean is from the mean of all the samples. The sum of squares between samples (or treatments) is noted as SST and is obtained from the following formula:

$$\text{SST} = \sum n_j (\bar{x}_j - \bar{\bar{x}})^2$$

The sum of squares of the deviations within treatments measures how far each observation within a treatment deviates from its mean. The sum of squares within a treatment is also called the sum of squares for error (SSE) and it is obtained from the following formula:

$$SSE = \sum_{j=1}^{k} \sum_{i=1}^{n_j} (x_{ij} - \bar{x}_j)^2$$

The total sum of squares is the sum of the sum of squares for treatment (SST) and the sum of squares for error.

$$TSS = SST + SSE$$

Mean square

The next step is to calculate the mean square for the treatments and the mean square for error. The mean squares are obtained by dividing the sums of squares by their respective degrees of freedom.

The mean square for treatment (MST) will therefore be

$$MST = \frac{SST}{df_{\text{Treatment}}}$$

and the mean square for error will be

$$MSE = \frac{SSE}{df_{\text{Error}}}$$

The decision to reject or fail to reject the null hypothesis will be based on the F-statistic or the p-value.

The F-statistic is obtained from the ratio of the mean square for treatments to the mean square for error.

$$F = \frac{MST}{MSE}$$

If the variation between treatments is large, the mean square for treatments will be large and if the variation within treatments is large, the mean square for error will be large. The calculated F-statistic will be large if the mean square for treatments is large and it will be small if the mean square of error is large.

The decision to reject the null hypothesis is made by comparing the calculated F-statistic to the critical F-statistic found in the F table. The null hypothesis is rejected if the calculated F-statistic is greater than the critical F-statistic (Fig. 3.29).

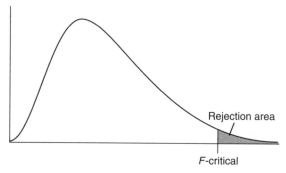

Figure 3.29

Another way to make a decision would be to compare the p-value with the α level. If the p-value is smaller than the α level, the null hypothesis is rejected; otherwise, we fail to reject it.

The formulas used to calculate the F-statistic can be summarized in an ANOVA table (see Table 3.33).

TABLE 3.33   ANOVA

Sources of variation	Sum of square	Degrees of freedom	Mean square	F-Statistic
Between treatments	$SST = \sum n_j (\bar{x}_j - \bar{\bar{x}}_j)^2$	$df_{\text{Treatments}} = k - 1$	$MST = \dfrac{SST}{df_{\text{Treatment}}}$	$F = \dfrac{MST}{MSE}$
Within treatments	$SSE = \sum\limits_{j=1}^{k} \sum\limits_{i=1}^{n_j} (x_{ij} - \bar{x}_j)^2$	$df_{\text{Error}} = n - k$	$MSE = \dfrac{SSE}{df_{\text{Error}}}$	
Total	$SST + SSE$	$df_{\text{Treatments}} + df_{\text{Errors}}$		

Example 3.21   An operations manager is comparing the productivity per hour of four machines. He takes samples of units produced per hour from each machine and tabulates them in Table 3.34. Can we say that the productivities of the three machines are equal with a confidence level of 95%?

TABLE 3.34

Sample 1	Sample 2	Sample 3	Sample 4
23	25	25	23
24	22	25	23
25	22	25	23
24	23	24	23
24	24	24	24
25	22	23	24
26	23	26	25

Solution: The grand mean is $\bar{\bar{x}} = 23.893$

The null hypothesis will be

H_0 : mean of sample 1 = mean of sample 2 = mean of sample 3

and the alternate hypothesis will be

H_1 : At least one sample mean is different from the others

TABLE 3.35   Sum of Square for Treatments

	Sample 1	Sample 2	Sample 3	Sample 4
	23	25	25	23
	24	22	25	23
	25	22	25	23
	24	23	24	23
	24	24	24	24
	25	22	23	24
	26	23	26	25
$\bar{x}_j$	24.429	23.000	24.571	23.571
$\left(\bar{x}_j - \bar{\bar{x}}\right)^2$	0.287	0.797	0.460	0.103

$$\sum(\bar{x}_j - \bar{\bar{x}})^2 = 1.648$$

$$\text{SST} = \sum n_j(\bar{x}_j - \bar{\bar{x}})^2 = 7 \times 1.648 = 11.536$$

TABLE 3.36   The Sum of Square for Error

$$\text{SSE} = \sum_{j=1}^{k}\sum_{i=1}^{n_j}\left(x_{ij} - \bar{x}_j\right)^2$$

	Sample 1	Sample 2	Sample 3	Sample 4
	2.041	4.000	0.184	0.327
	0.184	1.000	0.184	0.327
	0.327	1.000	0.184	0.327
	0.184	0.000	0.327	0.327
	0.184	1.000	0.327	0.184
	0.327	1.000	2.469	0.184
	2.469	0.000	2.041	2.041
$\sum_{i=1}^{n_j}(x_{ij} - \bar{x}_j)^2$	5.714	8.000	5.714	3.714

$$\text{SSE} = \sum_{j=1}^{k}\sum_{i=1}^{n_j}\left(x_{ij} - \bar{x}_j\right)^2 = 5.714 + 8.000 + 5.714 + 3.714 = 23.143$$

$$\text{TSS} = \text{SST} + \text{SSE} = 11.536 + 23.143 = 34.679$$

Table 3.35 shows the sum of the deviations of the treatments' means from the grand mean while table 3.36 shows the sum of square of each observation from its treatment mean.

Degree of freedom

Degree of freedom between treatments: Since there are four treatments, the degree of freedom will be $4 - 1 = 3$.

Degree of freedom within treatments: Since there are 28 observations and 4 treatments, the degree of freedom within treatments will be $28 - 4 = 24$.

Mean squares

$$\text{MST} = \frac{\text{SST}}{df_{\text{Treatment}}} = \frac{11.536}{3} = 3.845$$

$$\text{MSE} = \frac{\text{SSE}}{df_{\text{Error}}} = \frac{23.143}{24} = 0.963$$

Calculated *F*-statistic

$$F = \frac{\text{MST}}{\text{MSE}} = \frac{3.845}{0.963} = 3.993$$

The computed statistics can be summarized in the ANOVA table in Table 3.37.

TABLE 3.37

Sources of variation	Sum of square	Degrees of freedom	Mean square	*F*-statistic
Between treatments	11.536	3	3.845	3.993
Within treatments	23.143	24	0.963	
Total	34.679	27		

We can find the critical *F*-statistic in the *F* table. The critical *F*-statistic is based on the degrees of freedom of the treatments, the degree of freedom for error, and α. Based on the table, $F_{0.05,3,24} = $ (Table 3.38).

TABLE 3.38

		1	2	3	4	5	6	7
	15	4.54	3.68	3.29	3.06	2.90	2.79	2.71
	16	4.49	3.63	3.24	3.01	2.85	2.74	2.66
E r r o r	17	4.45	3.59	3.20	2.96	2.81	2.70	2.61
	18	4.41	3.55	3.16	2.93	2.77	2.66	2.58
	19	4.38	3.52	3.13	2.90	2.74	2.63	2.54
	20	4.35	3.49	3.10	2.87	2.71	2.60	2.51
	22	4.30	3.44	3.05	2.82	2.66	2.55	2.46
	24	4.26	3.40	**3.01**	2.78	2.62	2.51	2.42

Treatments

The critical F-statistic is equal to 3.01, while the calculated F-statistic is equal to 3.993. The calculated F-statistic is larger than the critical F-statistic; therefore, we have to reject the null hypothesis and conclude that at least one sample mean is different from the others. The SigmaXL output shows a p-value of 0.019, which suggests that the null hypothesis has to be rejected at $\alpha = 0.05$ (Table 3.39).

TABLE 3.39

One-Way ANOVA & Means Matrix:				
HO: Mean 1 = Mean 2 = ... = Mean k				
Ha: At least one pair Mean i ≠ Mean j				
	Sample 1	Sample 2	Sample 3	Sample 4
Count	7	7	7	7
Mean	24.429	23	24.571	23.571
Standard Deviation	0.975900	1.155	0.975900	0.786796
UC (2-sided, 95%, pooled)	25.195	23.766	25.337	24.337
LC (2-sided, 95%, pooled)	23.663	22.234	23.805	22.805
ANOVA:				
Pooled Standard Deviation =	0.981981		R-Sq =	33.26%
DF =	24		R-Sq adj. =	24.92%
F =	3.988			
P-value =	0.0194			
Pairwise Mean Difference (row - column)	Sample 1	Sample 2	Sample 3	Sample 4
Sample 1	0	1.429	-0.142857	0.857143
Sample 2		0	-1.571	-0.571429
Sample 3			0	1
Sample 4				0
Pairwise Probabilities	Sample 1	Sample 2	Sample 3	Sample 4
Sample 1		0.0119	0.7878	0.1155
Sample 2			0.0063	0.2871
Sample 3				0.0688
Sample 4				

SigmaXL goes a step beyond just showing that the null hypothesis should be rejected based on the p-value. It also shows in the **Pairwise Mean Difference** section which sample means are different.

Regression Analysis

Businesses are made up of multiple interacting processes within different interconnected functional departments. The decisions made on one process necessarily have an effect on other processes. Being able to quantify accurately the correlations between processes and metrics within an organization can help better assess the first-order consequences of decisions made on operational processes. Some of the questions that an organization must address before considering significant process changes are

How will productivity react to an increase in wages?

How does a reduction in inventory affect time-to-deliver?

Does an increase in marketing expenses affect sales?

How do changes in the prices of a given input affect the cost of the output?

Can a process improvement offset recent increases in the cost of acquisition of supplies?

The regression analysis is the part of statistical analysis that analyzes the relationship between quantitative variables. It helps predict the reaction of a variable when the magnitude of a related variable changes. For instance, a human resources (HR) director can intuitively determine that there is a correlation between wages and employee attrition rate. That intuitive determination does not necessarily provide actionable information that can help assess the optimal level of wages that lead to an acceptable attrition rate. Since the attrition rate depends on wages, the HR director will want to know what attrition rate to expect for each level of wages. Such information can provide better decision making.

The objective of regression analysis is to create a mathematical model under the form of a polynomial function that can help to determine how the predicted or dependent variable y (the variable to be estimated, the attrition rate in the case of the HR director) reacts to the variations in the predicator or independent variables x (wages in the case of the HR director).

The polynomial function is in the form of

$$\text{Attrition rate} = \alpha \text{ wages} + \beta$$

where α and β are constants whose values determine the strength of the relationship between attrition rate and wages.

The first step when conducting a regression analysis should be to determine whether there is any relationship between the independent and dependent variables, and if there is, how important it is. If there is a lack of correlation, there would be no need for creating a model because it would be useless. The covariance, the coefficient of correlation, and the coefficient of determination can determine the existence of that relationship and its level of significance. However, these alone do not help make accurate predictions on how variations of the independent variables affect the dependent variables.

It is obvious that in most cases there is more than one independent variable that can cause the variations in a dependent variable. Our HR director knows that there might be other factors explaining the high attrition rate. Other factors besides wages include the amount of overtime employees have to work, the point system, the number of terminations due to poor performance, the number of terminations due to failure to drug tests, etc.

However, the significance of all these factors in the variation of the dependent variable is disproportional. Therefore, in some cases, it is more beneficial to focus on one factor versus analyzing all the competing factors.

When building a regression model, if more than one independent variable is being considered, we call it a multiple regression analysis. If only one independent variable is being considered, the analysis is a simple linear regression.

In our quest for that model, we will start with the techniques that enable us to find the relatedness between one dependent variable and one independent variable.

Simple linear regression (or first-order linear model)

The simple regression analysis is a bivariate regression in that it involves only two variables: the independent and the dependent variables.

Example 3.22   The HR director believes that the wages are the only significant factor that explains the high attrition rate in her organization. In order to minimize the cost of employee retention, she wants to create a model that will help her know what attrition rate to expect for each level of wages. However, before creating the model, she would like to ascertain her intuitions by actually verifying that there really is a correlation between attrition rate and wages by determining what proportion of changes in attrition rate is explained by changes in wages. She would like to know what attrition rate should be expected if the hourly wages were set at $20.

The first step in any model building is, as always, the data gathering and organization. Therefore, the HR director gathered data from 20 companies that she organized in Table 3.40. She wants the data to be used for the analysis.

TABLE 3.40

Companies	Attrition rate	Wages
A	15	15.7
B	14	14.9
D	15	15.8
E	15	15.3
F	14	14.5
G	17	12
H	19	10
I	20	10.9
J	25	9
K	14	15
L	13	16.5
M	15	15.5
N	12	17
O	9	23
P	7	24
Q	14	16
R	5	25
S	17	15.4
T	3	27
U	15	15

Solution: The first step in the analysis consists of verifying her intuition by determining if there is indeed a relationship between the attrition rate and the wages and if there is, how strong it is.

Coefficient of correlation. The statistic used to determine the measure of relatedness between factors is the coefficient of correlation, generally noted r; it is a number between 1 and +1. When the coefficient of correlation is equal to 1, there is a perfect negative correlation between the factors; in other words, an increase in the magnitude of independent factor will necessarily lead to a decrease in the dependent factor in the exact same proportions and a decrease in the independent factor will lead to an increase in the dependent factor in the exact same proportions. The two factors vary in the same proportions but in opposite directions. If the coefficient of correlation is equal to 0, there is absolutely no relationship between the two factors. When it is equal to +1, there is a perfect positive correlation between the factors; an increase in the independent factor leads to an increase in the independent factor in the same proportions. The two factors vary in the same proportions and in the same direction.

Any value of the coefficient of correlation other than −1, 0, and +1 should be interpreted according to how close it is to these values.

Using Excel Open the file *Attrition.xls* and click on the Insert Function (fx) shortcut button before selecting Statistical for the **Or select a category** field as indicated in Fig. 3.30. Select **CORREL**, and then press **OK.**

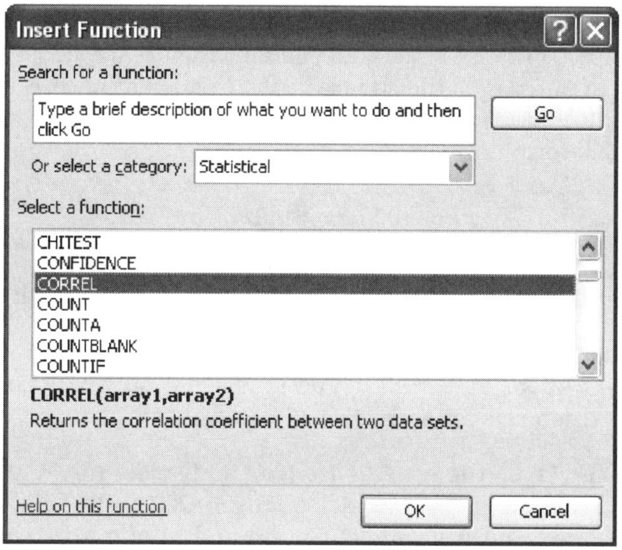

Figure 3.30

When the **Function Arguments** box appears, select the data under **Attrition Rate** for array 1 and the data under **Wages** for array 2. The coefficient of correlation appears as shown in Fig. 3.31.

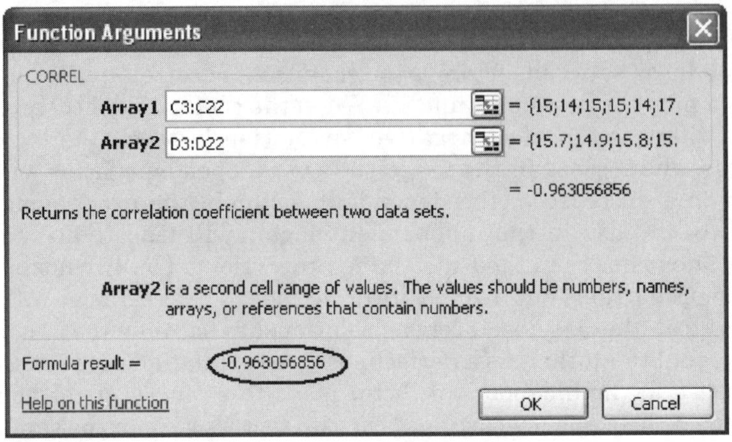

Figure 3.31

The coefficient of correlation is equal to –0.963. It is negative; therefore, we have to conclude that there is a negative correlation between attrition rate and wages. –0.963 is very close to –1, which means that the correlation is very strong. Therefore, an increase in the wages does lead to a reduction in the attrition rate.

Coefficient of determination. While the coefficient of correlation measures the strength of the relation between the two factors, the coefficient of determination shows what proportion in the variations of the dependent factor (attrition rate) is due to the variations in the independent factor (wages). The coefficient of determination noted r^2 is the square of the coefficient of correlation. Now that the coefficient of correlation has been determined, it is easy to determine the coefficient of determination.

$$r^2 = (-0.963)^2 = 0.927 \quad \text{or} \quad 92.7\%$$

92.7% of the changes in attrition rate are due to changes in the wages.

Regression equation. The regression equation will be in the form of $y = f(x)$ or attrition rate = α wages + β. Based on that equation, predictions can be made about the attrition rate for each level of wages.

Using SigmaXL Open the file *Attrition.xls* and select the area containing the data. Then from the menu bar, select **SigmaXL,** select **Statistical Tools**, and then **Regression** from the submenu and finally **Multiple Regression** as shown in Fig. 3.32.

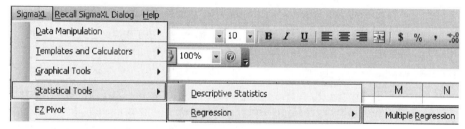

Figure 3.32

The **Multiple Regression** box should come with the **Please select your data** field already filled. Press the **Next >>** button.

Fill out the box as shown in Fig. 3.33 and then press the **OK >>** button.

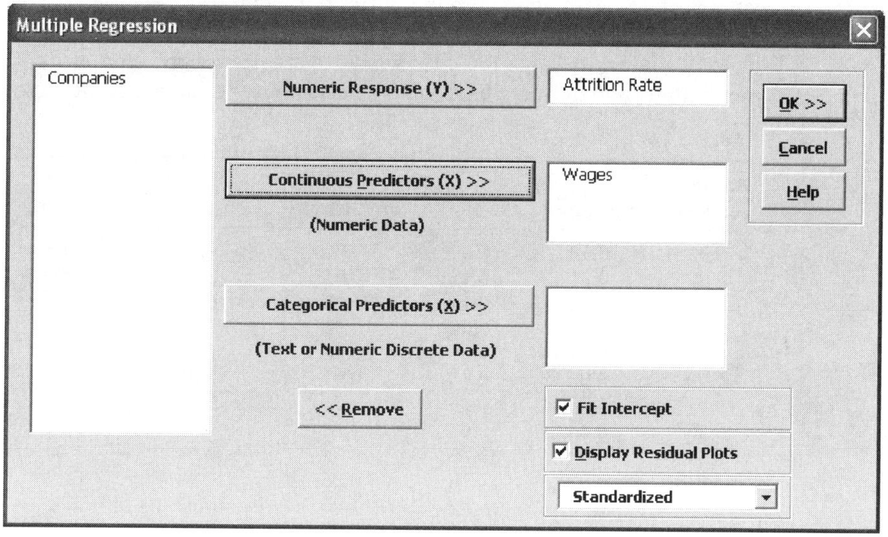

Figure 3.33

SigmaXL output shows the regression equation along with the coefficient of determination (Table 3.41).

TABLE 3.41

Multiple Regression Model: Attrition Rate = (30.338) + (-1.00383) * Wages

Model Summary:		
R-Square	92.75%	Coefficient of determination
R-Square Adjusted	92.34%	
S (Root Mean Square Error)	1.399	

If the wages were set at $20, the attrition rate would be 10.261.

$$\text{Attrition rate} = 30.338 - (1.0038 \times 20) = 10.261$$

Multiple regression analysis

When more than one independent factor is used to explain the dependent factor, a multiple regression analysis is used. The principle followed when conducting a multiple regression analysis is the same as when conducting a simple regression with the difference that more input factors are used. Since multiple independent factors are used, some might be more significant to the model than others might. Therefore, to make the model more fit for use, it might be necessary to eliminate the nonsignificant factors.

Example 3.23   An operations manager has determined the following five factors to be significant in delivering customers' orders on time: (1) inventory accuracy, (2) rate of rework, (3) employees' productivity, (4) delay in the reception of customers' orders, and (5) delays in loading shipments. He collects a sample of 25 days and tabulates the data as shown in Table 3.42. He wants to create a regression model that would help predict on-time delivery based on variations in the independent factors.

TABLE 3.42

On-time delivery	Inventory	Rework	Productivity	Order delay	Loading
0.96	0.98	0.04	0.90	7.00	8.96
0.90	0.94	0.12	0.91	0.00	8.90
0.95	0.94	0.15	0.96	6.95	8.95
0.94	0.93	0.06	0.91	9.00	8.94
0.95	0.94	0.05	0.96	6.95	8.95
0.90	0.95	0.10	0.91	0.00	8.90
0.94	0.93	0.26	0.93	8.00	8.94
0.95	0.98	0.05	0.96	6.95	8.95
0.94	1.00	0.06	0.95	6.94	8.94
0.94	0.93	0.16	0.95	6.94	8.94
0.98	0.97	0.02	1.00	10.00	8.98
0.90	0.93	0.10	0.93	6.90	8.90
0.94	0.93	0.06	0.98	6.94	8.94
0.96	0.95	0.04	0.97	6.96	8.96
0.95	0.94	0.05	0.96	11.00	8.95
0.96	0.95	0.14	0.98	6.96	8.96
0.96	0.95	0.04	0.97	1.00	8.96
0.90	0.94	0.10	0.93	6.90	8.90
0.96	0.95	0.04	0.97	13.00	8.96
0.90	0.95	0.15	0.91	6.90	8.90
0.93	0.92	0.07	0.91	6.93	8.93
0.95	1.00	0.05	0.96	12.00	8.95
0.91	0.92	0.29	0.90	6.91	8.91
0.95	0.94	0.05	0.96	6.95	8.95
0.91	0.95	0.09	0.98	10.00	8.91

Solution: Using SigmaXL, open the file on time *delivery.xls* and repeat the steps used in the previous example to obtain the output shown in Tables 3.43 and 3.44.

TABLE 3.43

Multiple Regression Model: On-Time delivery = (-8.253) + (-0.035989) * Inventory + (0.0141528) * Rework + (-0.045919)
* Productivity + (-1.234E-04) * Order delay + (1.0369636) * Loading

Model Summary:	
R-Square	98.83%
R-Square Adjusted	98.52%
S (Root Mean Square Error)	0.00285164

TABLE 3.44

Parameter Estimates:

Predictor Term	Coefficient	SE Coefficient	T	P	VIF	Tolerance
Constant	-8.253	0.284410	-29.019	0.0000		
Inventory	-0.035988724	0.031950699	-1.126	0.2740	1.306	0.765515
Rework	0.01415284	0.010356768	1.367	0.1877	1.456	0.686857
Productivity	-0.045919302	0.02628011	-1.747	0.0967	1.767	0.566072
Order delay	-1.234E-04	2.065E-04	-0.597798	0.5570	1.226	0.815990
Loading	1.036963599	0.03365807	30.809	0.0000	1.937	0.516263

For an alpha level of 0.05, the p-values in the parameter estimates part of the output show that loading is the only significant factor in the model because it is the only one that has a p-value less than 0.05. Loading is the only input factor that affects on-time delivery.

Improve

Design of Experiments

A design engineer has just finished designing a diesel engine used to run a plodder. After testing the engine, she realized that the level of carbon monoxide that it generates exceeds by far the allowed limits of 30 PPM over 8 h of work.

The engineer decided to determine what part in the engine is causing it to generate such a high level of carbon monoxide. She wants to isolate the factors causing the problem and take corrective actions. After screening all the parts, she decided that only five of them could have contributed to the problem but she does not know with certitude which ones significantly impact the carbon monoxide level. The factors that she isolated are

- The thickness of the oil
- The type of air filter
- The type of fuel filter
- The type of catalytic converter
- The type of oxygen sensors

Based on her analysis, these are the only factors in the engine that could affect the generation of carbon monoxide. However, the extent to which each factor in isolation or in interaction with the other factors contributes to the problem is unknown.

She wants to not only determine the significant factors but also create a regression model that would enable her to predict the level of the carbon monoxide whenever she changes the levels of the factors. She also wants to know with confidence by how much the changes in the level of carbon monoxide are affected by the model that she wants to create.

Since the resources available to run the tests are limited, she chooses to run a design of experiments (DOE) in order to isolate the contributing factors at the lowest possible cost.

The customer satisfaction index (CSI) is used to measure how pleased customers are about the services that they receive from Senegal Bank. The quality engineer at Senegal Bank wants to know what factors have been causing the index to decrease recently to an unprecedented level. He isolated four factors as being potentially significant.

- Customers being kept on hold over the phone for too long
- Long lines at the bank
- Too much paperwork to obtain loans
- High interest rates on short-term loans

The quality engineer knows that some of these factors or their interactions have been detrimental to the CSI but he does not know which ones are. The experiment that can help him determine the significant factors is expensive and time consuming because it involves changes in the interest rate offered by the bank, changes in the volume of paperwork, the opening of special fast lines at the cashiers, and the hiring of new customer service personnel to speed up calls.

To limit the cost involved in the experiment, the engineer chooses to use a DOE because at the end of his experiment he also wants to create a reliable regression model to gauge future fluctuations in the CSI that would result from variations in the significant factors.

The C_{pk} index measures how capable a process is at meeting or exceeding customers' expectations. Lately, the computed C_{pk} has proved to be unacceptably low. The operations manager has isolated the four factors that could contribute to poor performance.

- Raw material from different suppliers
- Machine operator
- Temperature in storage after processing
- Temperature in the trailers that transport the finished product to customers

He wants to test these factor to determine which one of them or what combination of factors is contributing to lowering of the C_{pk}. He wants to isolate the significant factors to take corrective actions. He chooses to use a DOE to limit the cost involved in the testing.

Whether changes are being made to an existing process, product, or system, or a new process, system, or product is being designed, it is always desirable to know the effects of the factors used in the process and the effects of the interactions between those factors on the expected outcome. In other words, a process engineer can optimize his processes if he can quantify the impact of each factor involved in his processes and the effects of the interactions between the factors on the response factor.

The best way to measure the factors' (or main) effects and the interactions' effects of a process is the use of the DOE.

DOEs consist of creating different scenarios of different combinations of (input) factors to test the effects of those combinations on the outcome (the response factor).

Factorial Experiments

Most experiments involve two or more factors. To test the effects of the factors and their interactions, the experimenter can use the same factors but combine them at different levels to see how each combination of factor levels affects the output factor (response variable).

If the experimenter wants to test effects of n factors on the dependent factor with levels $a_1, a_2, \ldots a_n$ the number of configurations he would have would be $a_1 \times a_2 \times a_3 \times \cdots a_n$. Because of the large number of configurations and the large number of samples to take for observations, it is often preferable to consider only the factors at two levels, usually considered high and low or (+) and (−) (or in some cases −1 and +1) with the meaning of high [or (+)] and low [or (−)] depending on the experiment being conducted.

Let us consider the example of an experimenter who is trying to optimize the production of organic fruits. After screening to determine the factors that are significant for his experiment, he narrows the main factors that affect the production of the fruits to "light" and "water." He wants to optimize the time that it takes to produce the fruits. He defines optimum as the minimum time necessary to yield comestible fruits.

To conduct his experiment, he runs several tests combining the two factors (water and light) at different levels. To minimize the cost of the experiment, he decides to use only two levels of the factors: high and low.

In this case, he will have two factors and two levels; therefore, the number of runs will be $2^2 = 4$.

After conducting observations, he obtains the results tabulated in Table 4.1.

TABLE 4.1

	$Light_{high}$	$Light_{low}$
$Water_{high}$	10	20
$Water_{low}$	15	25

Table 4.1 can be rewritten as Table 4.2.

TABLE 4.2

$Water_{high} Light_{high}$	10
$Water_{high} Light_{low}$	20
$Water_{low} Light_{high}$	15
$Water_{low} Light_{low}$	25

The coded form of Table 4.1 is shown in Table 4.3.

TABLE 4.3

Water	Light	Response
+	+	10
+	–	20
–	+	15
–	–	25

The reading we make of the tables is that, when the level of light is high and the level of water is high, it takes the fruits 10 units of time measurement (in this case, days) to become comestible. When the level of water is low and the level of light is low, it takes 25 units of time measurement.

Main effect and interaction effect

Before conducting his test, the experimenter needs to understand and be able to differentiate between the impacts of the main effects and the interaction effects.

Main effect. If a change in the level of an input factor (water or light) leads to a change in the response variable (the time it takes the fruits to become comestible), we should conclude that that factor has an impact on the response variable. For instance, if a change of the level of water from low to high leads to a significant change in the time to produce the fruits, we would consider that water is a significant factor in the production process of the fruits.

The change in the response factor (the time to produce the fruits) that resulted from the change in the input factor (water or light) level (high or low) is called the main effect.

Let us consider the main effects from the fruit production example. We want to determine the average change in the response factor (time it takes the fruit to become comestible) that results from the average change in the input factors (water and light).

$$Water = \frac{25+15}{2} - \frac{10+20}{2} = 20-15 = 5$$

$$Light = \frac{20+25}{2} - \frac{15+10}{2} = 22.5-12.5 = 10$$

A change of the level of water from low to high has resulted in a change of the time it takes the fruits to become comestible by 5 days. A change in the level of light from low to high has resulted in a variation of 10 units. Therefore, 5 and 10 are the main effects of water and light, respectively.

These are the main effects of water and light on the response factor.

Based on the data in Table 4.1, we can plot the graph of light in Fig. 4.1.

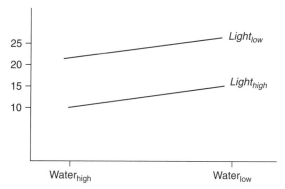

Figure 4.1

Figure 4.1 shows that the lines $Light_{low}$ and $Light_{high}$ are parallel, which suggests that an increase in the response factor to the variations in the level of light is the same at all levels of water and vice versa.

In this case, we conclude that there is an absence of interaction between the two factors (water and light).

Interaction effect.   An interaction is said to occur when a change in the response variable to a variation in one input factor depends on the level of the other input factor. If the factor water can only affect the time that it takes the fruits to become comestible when the variations in the levels of water are accompanied by variations in the levels of light, we conclude that there is an interaction between the two factors. In other words, if the level of water alone were changed, there would not be any change in the response factor if there were interaction between water and light.

Suppose that the observations by the experimenter had generated the data in Table 4.4.

TABLE 4.4

	$Light_{high}$	$Light_{low}$
$Water_{high}$	10	20
$Water_{low}$	15	5

In that case, we would obtain a graph in the form of Fig. 4.2.

When significant interaction is present in an experiment, the effects of the main factors are not consequential and, therefore, are not taken into consideration. Only the interaction effect is considered.

2^k Factorial design

2^k is a simple and special case of factorial design with k factors and only two levels. The levels in this type of factorial design are, in general, designated as

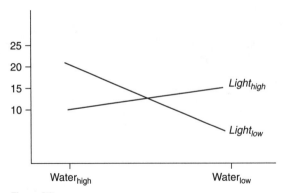

Figure 4.2

"low" and "high" and generally coded with the signs (–) for low and (+) for high. The simplicity of this form of factorial design (only two levels) makes it easier to add more factors.

Let us note that in the case of our fruit producer, the factors are "water" and "light" and the levels are "high" and "low."

The number of trials in that case was $2^2 = 2 \times 2 = 4$. If another factor were to be added to "water" and "light," we would have ended up with $2^3 = 2 \times 2 \times 2 = 8$ trials.

2^2 Two factors and two levels

A 2^2 factorial design will result in four combinations of factors. The different combinations are, in general, represented by the letters $l, a, b,$ and $ab,$ with l being the combination of the two factors at the low level, combination a represents one factor level at the high level and another at the low level, and the letter b represents the opposite. Letters ab represents the combination of both factors at the high levels

Let us suppose that for the sake of ensuring the integrity of his data, the experimenter decides to replicate each observation. Table 4.5a shows the results of his observations.

TABLE 4.5a

Combination	Water	Light	Water × Light	Responses		Total responses
l	–	–	+	13	12	25
a	+	–	–	9	11	20
b	–	+	–	8	7	15
ab	+	+	+	4	6	10

Another way of presenting Table 4.5a is shown in Table 4.5b.

TABLE 4.5b

Water	Light	Responses
−	−	13
+	−	9
−	+	8
+	+	4
−	−	12
+	−	11
−	+	7
+	+	6

These different combinations can be graphically represented by a square for 2^2.

In general, when drawing the square for a 2^2 factorial design, the combination of the two factors at their lower levels represented by the letter l is placed at the lower left corner. The letters a and b are placed at the corners of the sides of the factors they represent when those factors' levels are high. In this case, a represents the combination when the level for water is high and the level for light is low. The square is shown in Fig. 4.3.

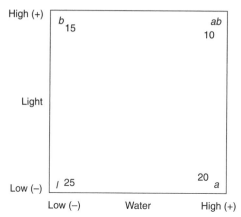

Figure 4.3

To estimate the main effect of the factor water, we will subtract the mean of the left side (where the level of water is low) of the square from the mean of the right side (where the level of water if high) (Fig. 4.4). Figure 4.5 shows the main effect for light.

Since the trials were performed twice, $n = 2$.

Main effect for water

$$= \frac{a+ab}{2^{2-1}n} - \frac{b+l}{2^{2-1}n} = \frac{1}{2n}(a+ab-b-l) = \frac{1}{2(2)}(20+10-15-25) = \frac{-10}{4} = -2.5$$

Main effect for light $= \dfrac{b+ab}{2n} - \dfrac{a+l}{2n} = \dfrac{1}{2n}(b+ab-a-l)$

$$= \frac{1}{2(2)}(15+10-20-25) = \frac{-20}{4} = -5$$

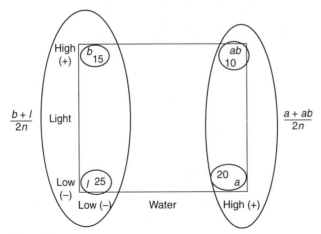

Figure 4.4

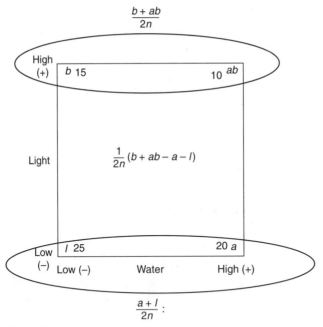

Figure 4.5

The interaction effect is shown in Fig. 4.6.

$$\text{WaterLight} = \frac{ab+l}{2n} - \frac{a+b}{2n} = \frac{1}{2n}(ab+l-a-b) = \frac{1}{2(2)}(10+25-20-15) = \frac{0}{4} = 0$$

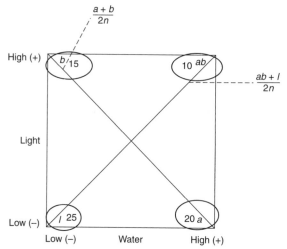

Figure 4.6

The numbers in parenthesis $((a+ab-b-l)$, $(b+ab-a-l)$, and $(ab+l-a-b))$ are measures of contrast. They are very important in the factorial design as they are used to determine the sums of squares for the factors and the interactions when conducting an ANOVA. In fact, the contrasts can be obtained from Table 4.5c by the sum of the product of the run column and the treatment columns.

TABLE 4.5c

Run	Water	Light	Water × Light	Water × Run	Light × Run	Water × Light × Run
l	−	−	+	−l	−l	+l
a	+	−	−	+a	−a	−a
b	−	+	−	−b	+b	−b
ab	+	+	+	+ab	+ab	+ab

$$Water_{Contrast} = a+ab-b-l$$

$$Light_{Contrast} = b+ab-a-l$$

$$Water \times Light_{Contrast} = ab+l-a-b$$

$$SS = \frac{(Contrast)^2}{2^{3-1}n}$$

$$SS_{Water} = \frac{(a+ab-b-l)^2}{2^{3-1}n} = \frac{(-10)^2}{4(2)} = \frac{100}{8} = 12.5$$

$$SS_{Light} = \frac{(b+ab-a-l)^2}{2^{3-1}n} = \frac{(-20)^2}{4(2)} = \frac{400}{8} = 50$$

$$SS_{Water \times Light} = \frac{(ab+l-a-b)^2}{2^{3-1}n} = 0$$

Degrees of freedom

The total number of degrees of freedom will be the number of effects that can be estimated. We have two factors moving from one level to another; in this case, for each factor we are interested in the effect for each movement on the response factor; therefore, the degree of freedom for each factor will be 1. If we had three levels, two degrees of freedoms would have been used. The total number of samples is 8; therefore, we have a total degree of freedom of 7 (8 − 1). The rest of the degrees of freedom will go to the error factor. This gives us an ANOVA table as shown in Table 4.6.

TABLE 4.6

Sources	Degrees of freedom	Sum of squares	Mean squares	F-statistic
Water	1	12.5	12.5	10
Light	1	50	50	40
Water × light	1	0	0	0
Error	4	5	1.25	
Total	**7**	**67.5**		

Using Minitab

We could have used Minitab to obtain the same results. There are several ways that these results could have been obtained using Minitab. The simplest way would be the use of the **General full factorial.**

Open the file *waterlight.mpj* after typing the data in Table 4.5b on a Minitab worksheet as indicated in Fig. 4.7, from the menu bar, click on **Stat**, and then select **DOE**, then **Factorial,** and then **Define Custom Factorial Design** (Fig. 4.8).

The **Define Custom Factorial Design** dialog box pops up. Select Light and Water for **Factors** (Fig. 4.9).

↓	C1-T	C2-T	C3
	Water	Light	Responses
1	-	-	13
2	+	-	9
3	-	+	8
4	+	+	4
5	-	-	12
6	+	-	11
7	-	+	7
8	+	+	6

Figure 4.7

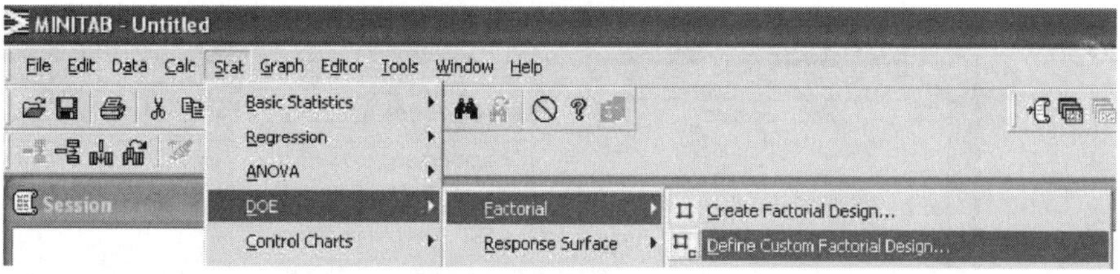

Figure 4.8

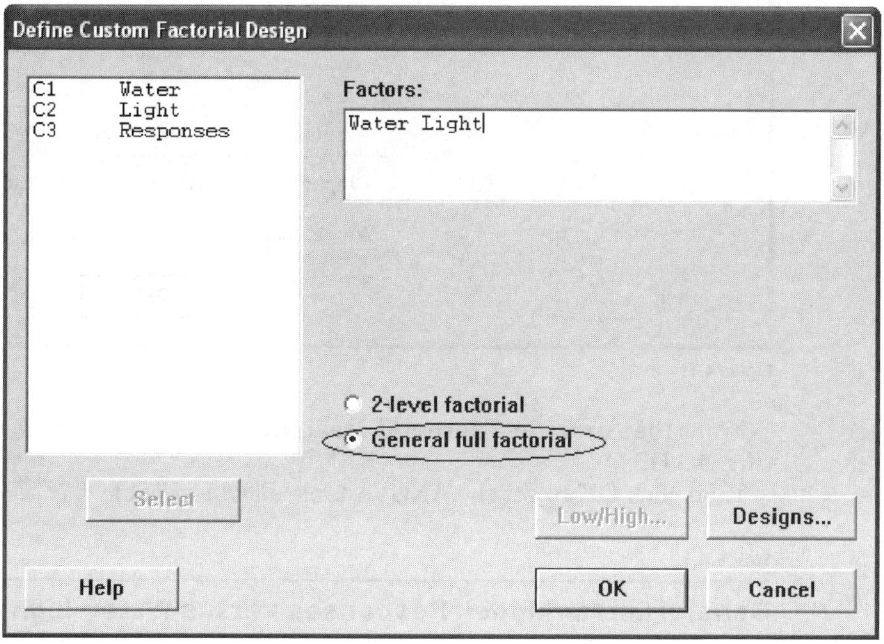

Figure 4.9

Figure 4.10

Select the **General full factorial** option before clicking on **OK.** Then go back to the menu bar and select **Stat**, then select **DOE**, then **Factorial** and the **Analyze Factorial Design** option (Fig. 4.10).

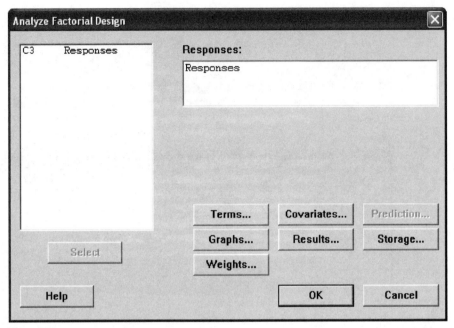

Figure 4.11

From the **Analyze Factorial Design,** select Responses for **Responses** (Fig. 4.11).

Then click **OK** to get the ANOVA table shown in Table 4.7.

TABLE 4.7

General Linear Model: Responses versus Water, Light

```
Factor  Type   Levels  Values
Water   fixed      2   -, +
Light   fixed      2   -, +


Analysis of Variance for Responses, using Adjusted SS for Tests

Source       DF  Seq SS  Adj SS  Adj MS      F      P
Water         1  12.500  12.500  12.500  10.00  0.034
Light         1  50.000  50.000  50.000  40.00  0.003
Water*Light   1   0.000   0.000   0.000   0.00  1.000
Error         4   5.000   5.000   1.250
Total         7  67.500


S = 1.11803   R-Sq = 92.59%   R-Sq(adj) = 87.04%
```

Interpretation

The p-values of both water and light are less than 0.05, so for an alpha level of 0.05, the main effects are significant for the model. In other words, a change in the level for each of them can have an effect on the response factor. The p-value for the interaction effect is equal to 1; therefore, we have to conclude that the interaction effect is insignificant for the model.

Using SigmaXL

From the Excel menu bar, click on **SigmaXL,** select **Design of Experiments** from the drop down list, and then select **2-Level Factorial/Screening Design** from the submenu (Fig. 4.12).

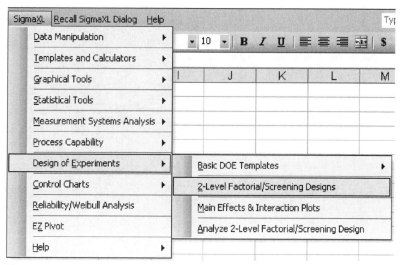

Figure 4.12

The **2-Level Factorial/Screening Design** dialog box appears (Fig. 4.13). You can change the factors names.

Press **OK.**

When the **2-factor DOE** sheet appears, change the values in the **Y1** column as indicated in Table 4.8 (or you can open the file *Waterlight.xls*).

After the screening design, we move on to the analyze phase.

From the menu bar, select **Design of Experiment** and from the submenu, select **Analyze 2-Level Factorial/Screening Design** (see Fig. 4.14).

The **Analyze 2-Level Factorial/Screening Design** dialog box appears as shown in Fig. 4.15.

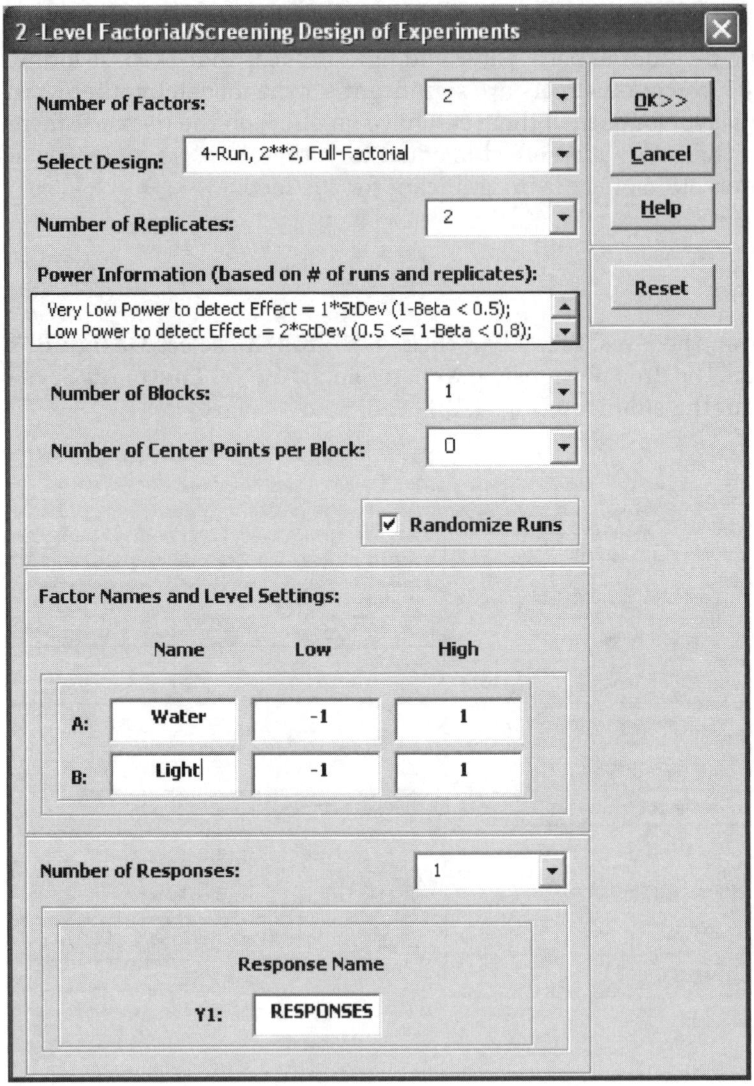

Figure 4.13

Press **OK** to obtain the results in Table 4.9.

Note that in Table 4.9, the *df* and the SS for model are the sums of the degree of freedom and the sum of squares of water and light and their interaction.

Based on the p-values that we have, we can see that the two main effects are significant in the model and that the interaction effect does not have an impact on the response factor.

TABLE 4.8

Design of Experiments Worksheet

Title:	
Date:	
Name of Experimenter:	
Notes:	

Design Type:	2 Factor, 4-Run, 2**2, Full-Factorial
Number of Replicates:	2
Number of Blocks:	1
Number of Center Points per Block:	0
Number of Responses:	1

Run Order	Std. Order	Center Points	Blocks	A: Water	B: Light	RESPONSES
1	8	1	1	1	1	4
2	2	1	1	1	-1	9
3	7	1	1	-1	1	8
4	3	1	1	-1	1	7
5	5	1	1	-1	-1	13
6	6	1	1	1	-1	11
7	4	1	1	1	1	6
8	1	1	1	-1	-1	12

Do not add or delete rows or sort this worksheet.

Aliasing of Effects:
None (Full Factorial Design)

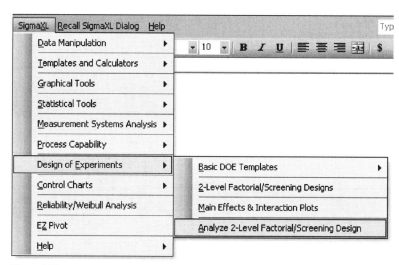

Figure 4.14

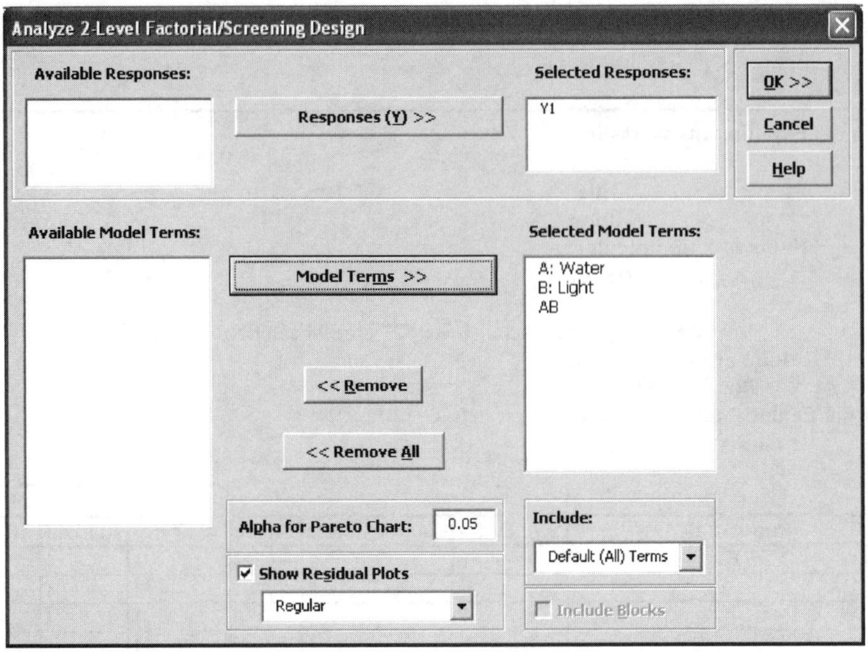

Figure 4.15

TABLE 4.9

Design of Experiments Analysis

DOE Multiple Regression Model: RESPONSES = (8.75) + (-1.25) * A: Water + (-2.5) * B: Light + (0) * AB

Title:	
Date:	
Name of Experimenter:	
Notes:	

Design Type:	2 Factor, 4-Run, 2**2, Full-Factorial
Number of Replicates:	2
Number of Blocks:	1
Number of Center Points per Block:	0
Response:	RESPONSES

Model Summary:

R-Square	92.59%
R-Square Adjusted	87.04%
S (Root Mean Square Error)	1.118

Parameter Estimates:

Term	Coefficient	SE Coefficient	T	P	VIF	Tolerance
Constant	8.75	0.395284708	22.136	0.0000		
A: Water	−1.25	0.395284708	−3.162	0.0341	1	1
B: Light	−2.5	0.395284708	−6.325	0.0032	1	1
AB	0	0.395284708	0	1.0000	1	1

Analysis of Variance for Model:

Source	DF	SS	MS	F	P
Model	3	62.500	20.833	16.667	0.0100
Error	4	5	1.250		
Pure Error	4	5	1.250		
Total (Model + Error)	7	67.500	9.643		

Example 4.1 2^k two levels, two factor with three replicates

A study is being conducted to measure the impact of wind and water on the corrosion of an alloy. Very thin cords of the alloy are used in the experiment. The response factor is determined to be the length of time that it takes the cord to corrode and break. The experimenter conducts the tests by first maintaining the water (denoted by the letter A) level and the wind (denoted by the letter B) level low, then by keeping the water level high and wind level low, then keeping the water level low and wind high, and finally maintaining both water and wind levels high.

To ascertain the validity of the experiment, the experimenter decides to replicate the experiment three times. The results' tests are summarized in Table 4.10a.

Table 4.10a can be converted to Table 4.10b.

TABLE 4.10a

Runs	A	B	Observations I	II	III	Mean	Total
1	−	−	15	15	16	15.33	46
a	+	−	17	16	17	16.67	50
b	−	+	17	17	16	16.67	50
ab	+	+	19	20	20	19.67	59

TABLE 4.10b

A	B	Response
−	−	15
+	−	17
−	+	17
+	+	19
−	−	15
+	−	16
−	+	17
+	+	20
−	−	16
+	−	17
−	+	16
+	+	20

Figure 4.16 is the graphical representation of the data.

Determining the main effects In this case, we have three replicates. Therefore, $n = 3$. We use the same formula to obtain the main effects for water and wind.

$$\text{Main effect for water} = \frac{a+ab}{2^{2-1}n} - \frac{b+l}{2^{2-1}n}$$

$$= \frac{1}{2n}(a+ab-b-l) = \frac{1}{2(3)}(50+59-50-46) = \frac{13}{2(3)} = \frac{13}{6} = 2.167$$

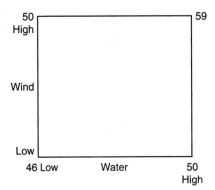

Figure 4.16

$$\text{Main effect for wind} = \frac{b+ab}{2n} - \frac{a+l}{2n}$$

$$= \frac{1}{2n}(b+ab-a-l) = \frac{1}{2(3)}(50+59-50-46) = \frac{13}{2(3)} = \frac{13}{6} = 2.167$$

Both main effects for water and wind are equal to 2.167. This means that a change in the level of water or wind from low to high will result in a change in the response factor by 2.167 units of time measurement.

Interaction effect

$$\text{The interaction effect water} \times \text{wind} = \frac{ab+l}{2n} - \frac{a+b}{2n} = \frac{1}{2n}(ab+l-a-b)$$

$$= \frac{1}{2(3)}(59+46-50-50) = \frac{5}{6} = 0.8333$$

Sums of squares

$$SS = \frac{(Contrast)^2}{2^{3-1}n}$$

$$SS_{Water} = \frac{(a+ab-b-l)^2}{2^{3-1}n} = \frac{(13)^2}{4(3)} = \frac{169}{12} = 14.083$$

$$SS_{Wind} = \frac{(b+ab-a-l)^2}{2^{3-1}n} = \frac{(13)^2}{4(3)} = \frac{169}{12} = 14.083$$

$$SS_{Water \times Wind} = \frac{(ab+l-a-b)^2}{2^{3-1}n} = \frac{(5)^2}{12} = \frac{25}{12} = 2.083$$

ANOVA tables The ANOVA tables for a 2^k factorial design can be presented in at least three ways, depending on the objectives of the experimenter and the type of software used. The ANOVA table can be presented in a way that the different main effects are separated as in the case of the general linear model. It can also be presented in a way that combines the main effects and shows the interaction effect separately as in the case of the Minitab 2-level factorial, or we can have the model that combines the main effects and the interaction effect as in the case of the SigmaXL output in Table 4.11.

TABLE 4.11

Source	Degrees of freedom	Sums of square	Mean square	F-statistic
Water	1	14.083	14.083	42.25
Wind	1	14.083	14.083	42.25
Interaction effect	1	2.083	2.083	6.26
Residual error	8	2.667	0.333	
Total	11			

Using Minitab Open the file *Alloy Corrosion.mtw*, and then click on **Stat**, then click on **DOE**, then on **Factorial,** and then select **Define Custom Factorial Design.** The dialog box shown in Fig. 4.17 appears. Fill it out as shown and make sure to select **2-level factorial.**

Then press **Low/High** button. The dialog box shown in Fig. 4.18 appears. Fill it out as indicated.

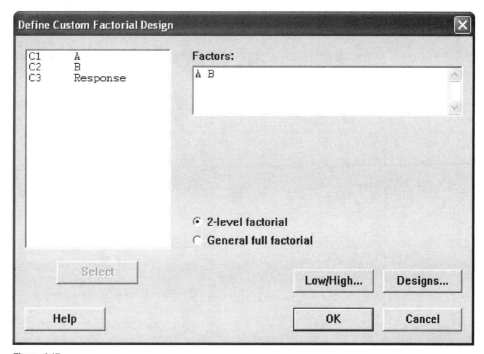

Figure 4.17

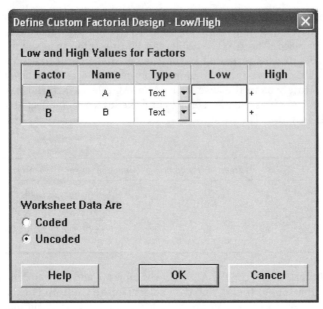

Figure 4.18

Press **OK** to get back to the **Define Custom Factorial Design** dialog box. Four new columns will appear, C4 to C7. Click on **Stat,** then click on **DOE,** then on **Factorial** and select **Analyze Factorial Design.** The Analyze **Factorial Design** dialog box pops up (Fig. 4.19). Select "Responses" for the **Responses** field.

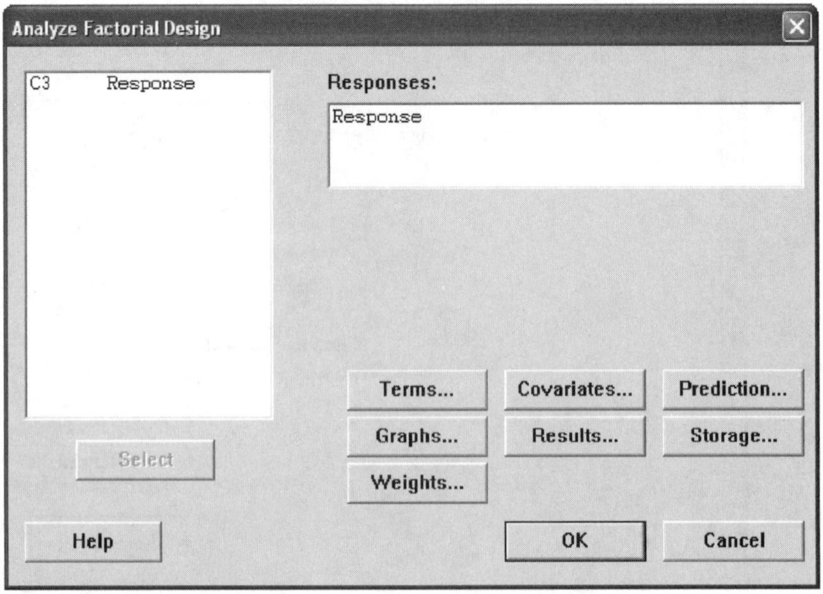

Figure 4.19

Press **OK** again to get the output shown in Table 4.12.

Table 4.12 shows that the p-value for the main effects is null and the one for the interactions is smaller than 0.05. Therefore, for an alpha level of 0.05, the interaction effect is significant; we therefore have to take into account only the interaction effect.

TABLE 4.12

Results for: ALLOY CORROSION.MTW

Factorial Fit: Response versus A, B

Estimated Effects and Coefficients for Response (coded units)

Term	Effect	Coef	SE Coef	T	P
Constant		17.0833	0.1667	102.50	0.000
A	2.1667	1.0833	0.1667	6.50	0.000
B	2.1667	1.0833	0.1667	6.50	0.000
A*B	0.8333	0.4167	0.1667	2.50	0.037

S = 0.577350   R-Sq = 91.90%   R-Sq(adj) = 88.86%

Analysis of Variance for Response (coded units)

Source	DF	Seq SS	Adj SS	Adj MS	F	P
Main Effects	2	28.167	28.167	14.0833	42.25	0.000
2-Way Interactions	1	2.083	2.083	2.0833	6.25	0.037
Residual Error	8	2.667	2.667	0.3333		
Pure Error	8	2.667	2.667	0.3333		
Total	11	32.917				

Using SigmaXL Open the file *WaterWind.xls* (SigmaXL should be open already). The data should already be there. Click on **SigmaXL,** then select **Design of Experiments** and select **Analyze 2-Level factorial/screening designs** to obtain the ANOVA in Table 4.13.

To visualize the main effects and the interaction effect on the response factor, let us use SigmaXL to plot them.

From the menu bar, click on **SigmaXL,** then select **Design of Experiments,** and then select **Main effect and Interaction plots** to obtain the plots in Fig. 4.20.

Even though the two lines in Fig. 4.21 do not intersect, they are not parallel and show that there is a slight interaction between the two factors.

TABLE 4.13

Number of Blocks:	1
Number of Center Points per Block:	0
Response:	RESPONSES

Model Summary:

R-Square	91.90%
R-Square Adjusted	88.86%
S (Root Mean Square Error)	0.577350

Parameter Estimates:

Term	Coefficient	SE Coefficient	T	P	VIF	Tolerance
Constant	17.08333333	0.166666667	102.50	0.0000		
A: Water	1.083333333	0.166666667	6.500	0.0002	1	1
B: Wind	1.083333333	0.166666667	6.500	0.0002	1	1
AB	0.416666667	0.166666667	2.500	0.0369	1	1

Analysis of Variance for Model:

Source	DF	SS	MS	F	P
Model	3	30.250	10.0833333	30.250	0.0001
Error	8	2.667	0.333333		
Pure Error	8	2.667	0.333333		
Total (Model + Error)	11	32.917	2.992		

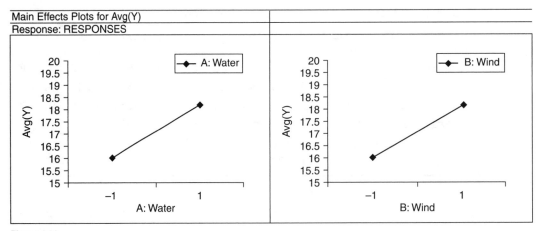

Main Effects Plots for Avg(Y)
Response: RESPONSES

Figure 4.20

Regression Model

DOE is a collection technique that enables the experimenter to determine what factors are significant for his experiment. Once the factors are determined, the experimenter will create a model, a linear equation that will allow him to predict what will happen to his response factor when one or more factors are varied. The best way to create that model is with regression analysis.

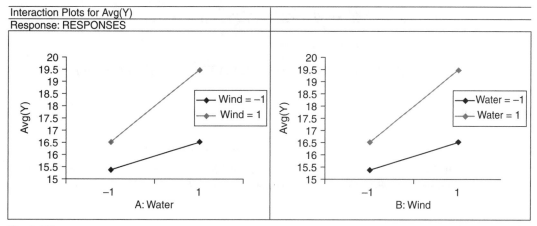

Figure 4.21

The multiple regression model that will be derived from the analysis should ideally contain only the factors that have been determined to be significant by the DOE.

A typical regression equation is

$$\vec{Y} = \beta_0 + \beta_1 x_1 + \beta_2 x_2 + \cdots + \beta_3 x_1 x_2 \ldots x_i \ldots x_n$$

where β_0 is the mean of all the observations, β_i is the ith coefficient, x_i is the ith factor, and $x_1 x_2 \ldots x_i \ldots x_n$ is the highest interaction.

In our alloy corrosion example,

$$\vec{Y} = \beta_0 + \beta_1 Water + \beta_2 Wind + \beta_3 WaterWind$$

β_0 is the mean of all the observations.

$$\beta_0 = \frac{15.33 + 16.67 + 16.67 + 19.67}{4} = \frac{68.34}{4} = 17.085$$

$$\beta_1 = \frac{main\ effect\ for\ water}{2} = \frac{2.167}{2} = 1.0835$$

$$\beta_2 = \frac{main\ effect\ for\ wind}{2} = \frac{2.167}{2} = 1.0835$$

$$\beta_1 = \frac{Interaction\ Effect}{2} = \frac{0.8333}{2} = 0.417$$

$$\vec{Y} = 17.085 + 1.083Water + 1.083Wind + 0.417WaterWind$$

SigmaXL output

Design of Experiments Analysis

DOE Multiple Regression Model: $RESPONSES = (17.08333333) + (1.083333333)*$ A: Water $+ (1.083333333)*$ B: Wind $+ (0.41666666)*$AB: WaterWind

Residual analysis

Now that we have the regression equation, we can conduct a residual analysis to obtain the predicted values for each combination level. The residual is the difference between the predicted value and the actual value (Table 4.14).

$$17.083333 - 1.0833333 + 1.0833333 + 0.4167 = 16.667$$

TABLE 4.14

A	B	Response	Predicted Y		Response − Predicted Y	Residual
−	−	15	17.083333 − 1.0833333 − 1.083333 + 0.4167 =	15.333	15 − 15.333	−0.333
+	−	17	17.083333 + 1.0833333 − 1.083333 + 0.4167 =	16.667	17 − 16.667	0.333
−	+	17	17.083333 − 1.0833333 + 1.083333 + 0.4167 =	16.667	17 − 16.667	0.333
+	+	19	17.083333 + 1.0833333 + 1.083333 + 0.4167 =	19.667	19 − 19.667	−0.667
−	−	15	17.083333 − 1.0833333 − 1.083333 + 0.4167 =	15.333	15 − 15.333	−0.333
+	−	16	17.083333 + 1.0833333 − 1.083333 + 0.4167 =	16.667	16 − 16.667	−0.667
−	+	17	17.083333 − 1.0833333 + 1.083333 + 0.4167 =	16.667	17 − 16.667	0.333
+	+	20	17.083333 + 1.0833333 + 1.083333 + 0.4167 =	19.667	20 − 19.667	0.333
−	−	16	17.083333 − 1.0833333 − 1.083333 + 0.4167 =	15.333	16 − 15.333	0.667
+	−	17	17.083333 + 1.0833333 − 1.083333 + 0.4167 =	16.667	17 − 16.667	0.333
−	+	16	17.083333 − 1.0833333 + 1.083333 + 0.4167 =	16.667	16 − 16.667	−0.667
+	+	20	17.083333 + 1.0833333 + 1.083333 + 0.4167 =	19.667	20 − 19.667	0.333

TABLE 4.15

$17.083333 - 1.0833333 - 1.0833333 + .04167 = 15.333$

A: Water	B: Wind	RESPONSES	Predicted (Fitted) Values	Residuals	Standardized Residuals	Studentized (Deleted t)	Cook's Distance (Influence)	Leverage	DFITS
1	1	19	19.667	-0.666667	-1.414	-1.528	0.250000	0.333333	-1.08012345
-1	-1	15	15.333	-0.333333	-0.707107	-0.683130	0.0625	0.333333	-0.483046
-1	1	17	16.667	0.333333	0.707107	0.683130	0.0625	0.333333	0.483046
-1	1	17	16.667	0.333333	0.707107	0.683130	0.0625	0.333333	0.483046
-1	-1	16	15.333	0.666667	1.414	1.528	0.250000	0.333333	1.08012345
1	-1	17	16.667	0.333333	0.707107	0.683130	0.0625	0.333333	0.483046
1	-1	16	16.667	-0.666667	-1.414	-1.528	0.250000	0.333333	-1.08012345
1	-1	17	16.667	0.333333	0.707107	0.683130	0.0625	0.333333	0.483046
1	1	20	19.667	0.333333	0.707107	0.683130	0.0625	0.333333	0.483046
-1	-1	15	15.333	-0.333333	-0.707107	-0.683130	0.0625	0.333333	-0.483046
1	1	20	19.667	0.333333	0.707107	0.683130	0.0625	0.333333	0.483046
-1	1	16	16.667	0.666667	-1.414	-1.528	0.250000	0.333333	-1.08012345

$15 - 15.333 = -0.3333$

Using the SigmaXL residual table (Table 4.15).

Example 4.2 An experimenter is trying to determine the significant factors that impact the gas consumption for a new engine that she has designed. She estimates that the gas consumption (which is the response factor in this case) depends on the engine's RPM and the sizes of the vehicle's tires. She decides to replicate the experiment twice and obtains the following data shown in Table 4.16.

1. Determine the significant factors that affect gas consumption.
2. What can we say about the interaction effect?
3. Create a regression model for only the significant factors.
4. Create the residual table.
5. Use SigmaXL to verify your results.

TABLE 4.16

Runs	RPM	Tire sizes	Observations I	Observations II
1	–	–	75	77
a	+	–	90	82
b	–	+	76	94
ab	+	+	89	99

2^k Two levels with more than 2 factors

In Example 4.1, only two factors were considered at two levels, high and low. If the design were to be extended to three factors, we would have 8 runs (factors' combinations). Adding an extra factor to the two previous ones will lead to a change in the graphical representation of the combinations of factors.

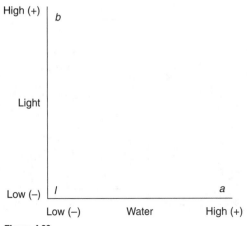

Figure 4.22

The graphical representation of the two-factor-based combinations is really created from two-dimension coordinates (Fig. 4.22).

Let us suppose that the experimenter decided to do some further screening and concludes that "fertilizer" should also be tested for significance in the experiment. He decides to add that factor to "water" and "light." Therefore, we end up with a three-dimension graph as shown in Fig. 4.23.

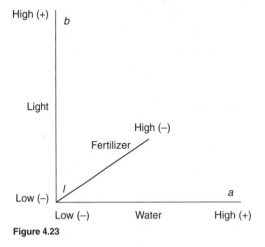

Figure 4.23

Figure 4.23 does not account for the interactions' effects of the three factors. In the case of the two factors with two levels, we were interested only in one interaction effect, water × light. Now that we have three factors, we will need to consider four interactions' effects in addition to the main effects. The interactions' effects in which we are interested are: water × light, water × fertilizer, fertilizer × light, and water × light × fertilizer. When the interactions' effects are added to the graph, we obtain a cube (Fig. 4.24).

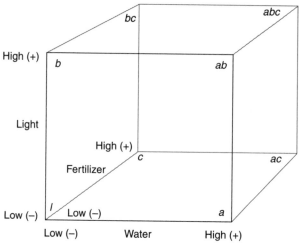

Figure 4.24

Main effects for 2^3—two levels with three factors

The main effect for each factor should also account for the other two factors and all the interactions. For each factor, we will subtract the values at the corners of the "wall" that contains l from the opposite "wall" where that factor is positive, and divide the result by $n2^{3-1}$ which is equal to $4n$.

The main effect for fertilizer is therefore (Fig. 4.25):

$$Fertilizer\ effect = \frac{(c+ac+bc+abc)-(l+a+b+ab)}{n2^{k-1}}$$

$$= \frac{c+ac+bc+abc-l-a-b-ab}{2^{3-1}n}$$

$$= \frac{c+ac+bc+abc-l-a-b-ab}{4n}$$

The main effect for water will be (Fig. 4.26):

$$Water\ effect = \frac{(a+ac+ab+abc)-(l+b+c+bc)}{2^{3-1}n} = \frac{a+ac+ab+abc-l-c-bc-b}{4n}$$

and the main effect for light will be (Fig. 4.27):

$$Light\ effect = \frac{(b+ab+bc+abc)-(l+a+ac+c)}{2^{k-1}n}$$

$$= \frac{b+ab+bc+abc-l-a-ac-c}{4n}$$

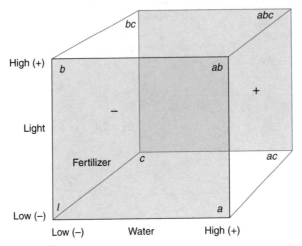

Figure 4.25

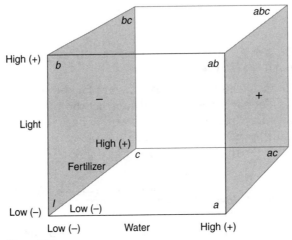

Figure 4.26

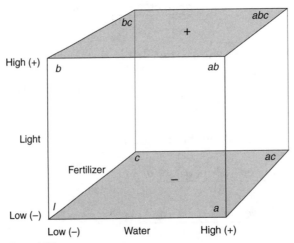

Figure 4.27

Example 4.3 Let us suppose that the experimenter decides to replicate the test four times for each level. Therefore, the number of responses for each level will be four. The results of the experiments are summarized in Table 4.17a.

TABLE 4.17a

Notation	Water	Light	Fertilizer	Observed responses				Total responses	Mean responses
l	−	−	−	19	18	18	20	75	25
a	+	−	−	20	20	27	20	87	29
b	−	+	−	18	27	20	19	84	28
c	−	−	+	20	27	16	30	93	31
ab	+	+	−	20	20	21	20	81	27
ac	+	−	+	21	15	19	20	75	25
bc	−	+	+	16	18	11	18	63	21
abc	+	+	+	12	12	8	13	45	15

Table 4.17a can be expanded to account for the interactions between factors (Table 4.17b).

TABLE 4.17b

Notation	Water	Light	Fertilizer	Water × Light	Water × Fertilizer	Fertilizer × Water	Water × Fertilizer × Light	Observed responses				Total responses	Mean responses
l	−	−	−	+	+	+	−	19	18	18	20	75	25
a	+	−	−	−	−	+	+	20	20	27	20	87	29
b	−	+	−	−	+	−	+	18	27	20	19	84	28
c	−	−	+	+	−	−	+	20	27	16	30	93	31
ab	+	+	−	+	−	−	−	20	20	21	20	81	27
ac	+	−	+	−	+	−	−	21	15	19	20	75	25
bc	−	+	+	−	−	+	−	16	18	11	18	63	21
abc	+	+	+	+	+	+	+	12	12	8	13	45	15

Notice that ab represents the combination where A and B are at a high level and C is at a low level; ac is the combination where A and C are high and B is low; bc represents the combination where B and C are high and A low; and abc represents the combination of A, B, and C when they are all at the high level.

The signs for the combination treatments for each level are obtained by multiplying the signs of the main factors at that level. The sign of the combination water × fertilizer at level ab is negative because, at that level, the sign for the factor water is positive and the sign for the factor fertilizer is negative.

Table 4.17b can be rewritten as

TABLE 4.18

Water	Light	Fertilizer	Responses
A	B	C	
−	−	−	19
+	−	−	20
−	+	−	18
−	−	+	20
+	+	−	20
+	−	+	21
−	+	+	16
+	+	+	12
−	−	−	18
+	−	−	20
−	+	−	27
−	−	+	27
+	+	−	20
+	−	+	15
−	+	+	18
+	+	+	12
−	−	−	18
+	−	−	27
−	+	−	20
−	−	+	16
+	+	−	21
+	−	+	19
−	+	+	11
+	+	+	8
−	−	−	20
+	−	−	20
−	+	−	19
−	−	+	30
+	+	−	20
+	−	+	20
−	+	+	18
+	+	+	13

Main effects (Figs. 4.28, 4.29, and 4.30)

$$Water\ effect = \frac{(a + ac + ab + abc) - (l + c + bc + b)}{n2^{3-1}}$$

$$= \frac{(87 + 75 + 81 + 45) - (75 + 93 + 63 + 84)}{4(4)}$$

$$= \frac{-27}{16} = -1.688$$

$$SS_{Water} = \frac{(Contrast)^2}{n2^k} = \frac{(-27)^2}{4 \times 2^3} = \frac{729}{32} = 22.78$$

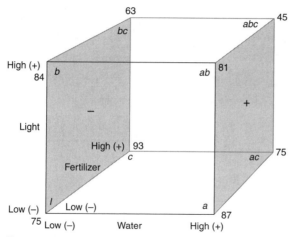

Figure 4.28

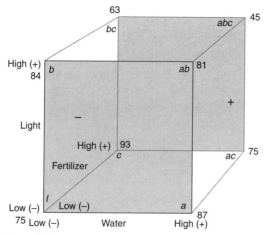

Figure 4.29

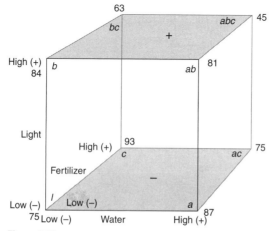

Figure 4.30

243

$$Fertilizer\ effect = \frac{(c+ac+bc+abc)-(l+a+b+ab)}{2^{3-1}n}$$

$$= \frac{c+ac+bc+abc-l-a-b-ab}{4n}$$

$$= \frac{(93+75+63+45)-(75+87+84+81)}{4\times4} = \frac{-51}{16} = -3.188$$

$$SS_{Fertilizer} = \frac{(Contrast)^2}{n2^k} = \frac{(-51)^2}{4\times2^3} = \frac{2601}{32} = 81.28$$

$$Light\ effect = \frac{(b+ab+bc+abc)-(l+c+a+ac)}{2^{3-1}n}$$

$$= \frac{(84+81+45+63)-(75+87+75+93)}{4\times4} = \frac{-57}{16} = -3.563$$

$$SS_{Light} = \frac{(Contrast)^2}{n2^k} = \frac{(-57)^2}{4\times2^3} = \frac{3249}{32} = 101.53$$

Interaction effects

Water-fertilizer effect   Average effect of water at the high and low levels of fertilizer:
Mean water effect at high fertilizer level

$$\frac{(abc-bc)+(ac-c)}{2n}$$

Mean effect of water at the low level of fertilizer

$$\frac{(ab-b)+(a-l)}{2n}$$

The interaction effect will be the difference between the two effects divided by 2 (Fig. 4.31).

$$Water \times Fertilizer\ effect = \frac{(l+ac+abc+b)-(a+ab+bc+c)}{n2^{k-1}}$$

$$= \frac{abc-bc+ac-c-ab+b-a+l}{n2^{3-1}}$$

$$= \frac{45-63+75-93-81+84-87+75}{2^{3-1}(4)}$$

$$= \frac{-45}{16} = -2.81$$

$$SS_{Water \times Fertilizer} = \frac{(Contrast)^2}{n2^k} = \frac{(-45)^2}{4(2^3)} = \frac{2025}{32} = 63.281$$

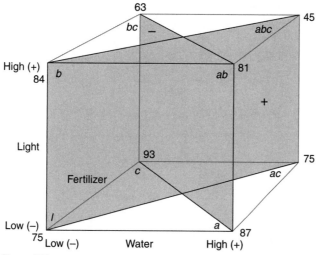

Figure 4.31

Water-light effect (Fig. 4.32)

$$Water \times Light\ effect = \frac{(l + c + abc + ab) - (b + bc + a + ac)}{n2^{k-1}}$$

$$= \frac{(75 + 93 + 45 + 81) - (63 + 84 + 87 + 75)}{16} = \frac{-15}{16} = -0.94$$

$$SS_{Water \times Light} = \frac{(Contrast)^2}{n2^k} = \frac{(-15)^2}{4 \times 2^3} = \frac{225}{32} = 7.031$$

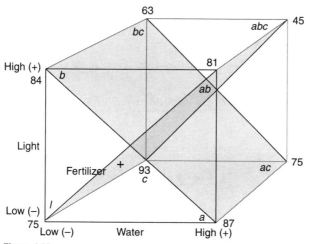

Figure 4.32

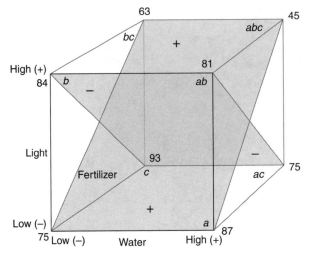

Figure 4.33

Light-fertilizer effect (Fig. 4.33)

$$Light \times Fertilizer\ effect = \frac{(bc+abc+a+l)-(ac+c+b+ab)}{n2^{k-1}}$$

$$= \frac{(63+45+87+75)-(84+81+75+93)}{4 \times 2^{3-1}} = \frac{-63}{16} = -3.94$$

$$SS_{Light \times Fertilizer} = \frac{(Contrast)^2}{n2^k} = \frac{(-63)^2}{4 \times 2^3} = \frac{3969}{32} = 124.03$$

Water-light-fertilizer interaction effect

$$Water \times Light \times Fertilizer = \frac{(abc-bc)-(ac-c)-(ab-b)+(a-l)}{n2^{k-1}}$$

$$= \frac{abc-bc-ac+c-ab+b+a-l}{n2^{3-1}}$$

$$= \frac{a+b+c+abc-ab-ac-bc-l}{4n}$$

$$= \frac{87+84+93+45-81-75-63-75}{16}$$

$$= \frac{15}{16} = 0.94$$

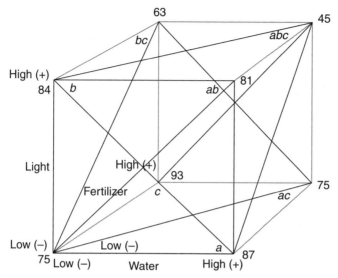

Figure 4.34

Sums of squares (Fig. 4.34)

$$SS_{Water} = \frac{(Contast)^2}{2^k n} = \frac{(-27)^2}{8 \times 4} = \frac{729}{32} = 22.78$$

$$SS_{Light} = \frac{(Contrast)^2}{n2^k} = \frac{(-57)^2}{4 \times 2^3} = \frac{3249}{32} = 101.53$$

$$SS_{Fertilizer} = \frac{(Contrast)^2}{n2^k} = \frac{(-51)^2}{4 \times 2^3} = \frac{2601}{32} = 81.28$$

$$SS_{Water \times Light} = \frac{(Contrast)^2}{n2^k} = \frac{(-15)^2}{4 \times 2^3} = \frac{225}{32} = 7.031$$

$$SS_{Water \times Fertilizer} = \frac{(Contrast)^2}{n2^k} = \frac{(-45)^2}{4 \times 2^3} = \frac{2025}{32}$$

$$= 63.28$$

$$SS_{Light \times Fertilizer} = \frac{(Contrast)^2}{n2^k} = \frac{(-63)^2}{4 \times 2^3} = \frac{3969}{32}$$

$$= 124.03$$

$$SS_{Water \times Light \times Fertilizer} = \frac{(Contrast)^2}{n2^k} = \frac{(15)^2}{4 \times 2^3} = \frac{225}{32} = 7.03$$

Degrees of freedom   We are running a full factorial design with two levels and three factors. This gives us 8 runs. Since we have replicated the test four times, we end up with 32 samples (8×4). Analyzing the experiments requires 1 degree of freedom for the grand average. We then end up with 31 ($32 - 1$); 3 degrees of freedom for the main effects, 3 degrees of freedom for the second order interaction, and 1 degree of freedom for the third order interaction. In all, 8 degrees of freedom have been used up, leaving 24 for the error term.

The mean square of the factors and their interactions

$$MS_{Water \times Light} = \frac{SS_{Water \times Light}}{df_{Water \times Light}} = \frac{7.031}{1} = 7.031$$

$$MS_{Water \times Fertilizer} = \frac{SS_{Water \times Fertilizer}}{df_{Water \times Fertilizer}} = \frac{63.28125}{1} = 63.28$$

$$MS_{Light \times Fertilizer} = \frac{SS_{Light \times Fertilizer}}{df_{Light \times Fertilizer}} = \frac{124.03}{1} = 124.03$$

$$MS_{Water \times Light \times Fertilizer} = \frac{SS_{Water \times Light \times Fertilizer}}{df_{Water \times Light \times Fertilizer}} = \frac{7.03}{1} = 7.03$$

$$MS_{Water} = \frac{SS}{df} = \frac{22.78}{1} = 22.78$$

$$MS_{Light} = \frac{SS}{df} = \frac{101.53}{1} = 101.53$$

$$MS_{Fertilizer} = \frac{SS}{df} = \frac{81.28}{1} = 81.28$$

$$MS_{Error} = \frac{281.25}{24} = 11.719$$

F-statistics

$$F - Stat_{Water} = \frac{MS_{Water}}{MS_{Error}} = \frac{22.78}{11.719} = 1.941$$

$$F - Stat_{Light} = \frac{MS_{Light}}{MS_{Error}} = \frac{101.53}{11.719} = 8.66$$

$$F - Stat_{Light \times Fertilizer} = \frac{MS_{Light \times Fertilizer}}{MS_{Error}} = \frac{124.0}{11.719} = 10.58$$

Table 4.19 represents the ANOVA table while table 4.20 shows the relative contributions of the treatment effects.

TABLE 4.19

Sources	Sum of squares	Degrees of freedom	Mean square	F-statistic
Water	22.78	1	22.78	1.941
Light	101.53	1	101.53	8.66
Fertilizer	81.28	1	81.28	6.94
Water × light	7.03	1	7.031	0.60
Water × fertilizer	63.28	1	63.28	5.40
Light × fertilizer	124.03	1	124.03	10.58
Water × light × fertilizer	7.03	1	7.03	0.60
Error	281.25	24	11.719	
Total	688.21	31		

TABLE 4.20

Model treatments	Treatment effects	Sum of squares	Contribution (%)
Water	−1.688	22.78	0.06
Light	−3.563	101.53	0.25
Fertilizer	−3.188	81.28	0.20
Water × light	−0.94	7.03	0.02
Water × fertilizer	−2.81	63.28	0.16
Light × fertilizer	−3.94	124.03	0.30
Water × light × fertilizer	0.94	7.03	0.02

Coefficient of determination R^2   The coefficient of determination in a DOE measures proportion in the total variation that is due to the model. It is therefore the ratio of the sum of squares of the model to the total sum of squares.

$$R^2 = \frac{SS_{model}}{SS_{Total}} = \frac{22.78 + 101.53 + 81.28 + 7.03 + 63.28 + 124.03 + 7.03}{688.21}$$

$$= \frac{406.96}{688.21} = 0.59133$$

Therefore, the proportion of the total variability due to the model is equal to 59.133%.

Adjusted R^2   The adjusted R^2 accounts for the magnitude of the model. The magnitude of the model is affected by the number of factors in it. If the number of factors increases, R^2 will also increase. If some of the factors in the model are insignificant, R^2 will tend to be arbitrarily inflated. Adjusted R^2 helps remove the inflation. It does so by taking into account the degrees of freedom for both the error term and the total sum of squares.

$$R^2(adj) = 1 - \frac{SS_e/df_e}{SS_{Total}/df_{Total}} = 1 - \frac{281.25/24}{688.21/31} = 1 - \frac{11.7185}{22.200} = 1 - 0.5279 = 0.4721$$

Using Minitab   We can use Minitab to verify the results of the test. Open the file *3-Factor fruit.mpj* (Fig. 4.35).

↓	C1	C2	C3	C4
	Water	Light	Fertilizer	Responses
1	-1	1	1	16
2	1	1	-1	20
3	1	1	-1	20
4	1	-1	-1	20
5	1	-1	-1	20
6	-1	1	1	18
7	-1	1	-1	18

Figure 4.35

Click on **Stat**, and then select **DOE,** then **Factoria**l, and finally **Define Custom Factorial Design** (Fig. 4.36).

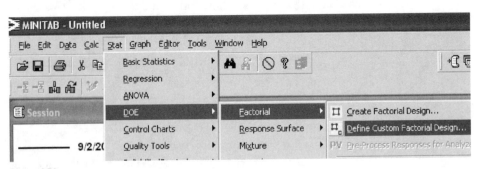

Figure 4.36

Select "Water," "Light," and "Fertilizer" for the **Factors field** and select **General full factorial** before clicking on **OK** (Fig. 4.37).

Then click on **Stat**, then **DOE**, then **Factorial,** and then **Analyze Factorial Design** (Fig. 4.38).

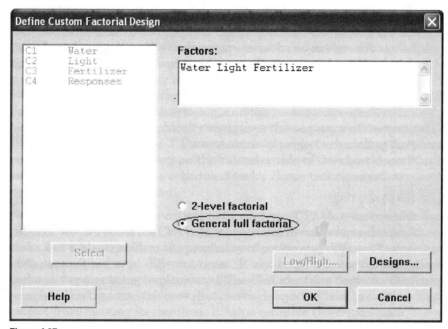

Figure 4.37

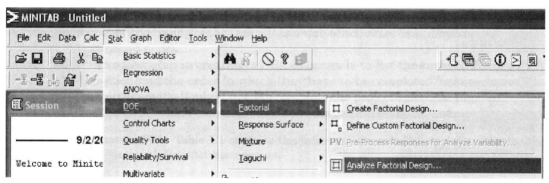

Figure 4.38

The **Analyze Factorial Design** dialog box pops up. Select "Responses" for the **Responses** field (Fig. 4.39).

Press **OK** to obtain the Minitab results shown in Table 4.21.

Interpretation   The ANOVA table matches what we found. The interpretation that we make of the results is that for an alpha level of 0.05, factor B (which represents light)

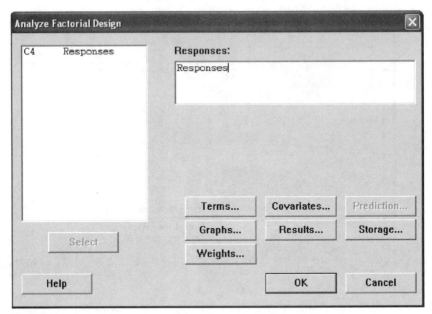

Figure 4.39

TABLE 4.21

Analysis of Variance for Responses, using Adjusted SS for Tests

Source	DF	Seq SS	Adj SS	Adj MS	F	P
Water	1	22.78	22.78	22.78	1.94	0.176
Light	1	101.53	101.53	101.53	8.66	0.007
Fertilizer	1	81.28	81.28	81.28	6.94	0.015
Water*Light	1	7.03	7.03	7.03	0.60	0.446
Water*Fertilizer	1	63.28	63.28	63.28	5.40	0.029
Light*Fertilizer	1	124.03	124.03	124.03	10.58	0.003
Water*Light*Fertilizer	1	7.03	7.03	7.03	0.60	0.446
Error	24	281.25	281.25	11.72		
Total	31	688.22				

$S = 3.42327$ R-Sq = 59.13%   R-Sq(adj) = 47.21%

and C (fertilizer) are highly significant with p-values of 0.007 and 0.015, respectively. The interactions "light × fertilizer" and "water × fertilizer" are also significant with p-values of 0.003 and 0.029, respectively. The interaction effect of "water × light × fertilizer" is negligible.

Blocking

In the previous design, the experimenter tested Water, Light, and Fertilizer for significance in the length of time it takes fruit to become comestible. If the runs were not conducted in homogeneous conditions, i.e., if all the conditions under which the different runs were performed were not identical, some uncontrolled variables could have affected the results of the experiments. If, for instance, the quality of the soil used in all the runs were not identical, one of the input factors might have seemed to be significant when in fact it was not. For instance, light has been determined to be significant in the design but its significance might be the result of a difference in the quality of the soil.

Since the experimenter had elected to replicate the tests four times, in order to avoid the negative impact of having the nonhomogeneous types of soil affect the experiment, he can create blocks of soil and run each replicate in a different block so that every combination is represented in every type of soil.

Table 4.18 can be converted into Table 4.22 to accommodate the blocking.

TABLE 4.22

	Water	Light	Fertilizer	Response	
l	−	−	−	19	Block I
a	+	−	−	20	
b	−	+	−	18	
c	−	−	+	20	
ab	+	+	−	20	
ac	+	−	+	21	
bc	−	+	+	16	
abc	+	+	+	12	
l	−	−	−	18	Block II
a	+	−	−	20	
b	−	+	−	27	
c	−	−	+	27	
ab	+	+	−	20	
ac	+	−	+	15	
bc	−	+	+	18	
abc	+	+	+	12	
l	−	−	−	18	Block III
a	+	−	−	27	
b	−	+	−	20	
c	−	−	+	16	
ab	+	+	−	21	
ac	+	−	+	19	
bc	−	+	+	11	
abc	+	+	+	8	
l	−	−	−	20	Block IV
a	+	−	−	20	
b	−	+	−	19	
c	−	−	+	30	
ab	+	+	−	20	
ac	+	−	+	20	
bc	−	+	+	18	
abc	+	+	+	13	

To conduct the analysis using Minitab, taking into account the block effect, open the file *WaterBlocking.mtw*. From the menu bar, click on **Stat**, then on **DOE**, and then on **Factorial**. In the **Factorial** drop down list, select **Define Custom Factorial Design**. In the **Define Custom Factorial Design** dialog box, select *Water*, *Light*, and *Fertilizer* for the **Factors**, then select the option **General full factorial**. Press the **Designs**... button. The **Define Custom General Factorial Design–Design** dialog box appears. Under **Blocks**, select **Specify by column** and select *Block* for that field, then click on **OK**. Then click on **OK** again.

Go back to the menu bar and click on **Stat**, then on **DOE**, and then click on **Factorial**. From the drop down list, select **Analyze factorial design**. Select *Response* for the **Responses** field, and then click on the **OK** button to get the results shown in Table 4.23.

TABLE 4.23

Analysis of Variance for Response, using Adjusted SS for Tests

Source	DF	Seq SS	Adj SS	Adj MS	F	P
Block	3	32.84	32.84	10.95	0.93	0.446
Water	1	22.78	22.78	22.78	1.93	0.180
Light	1	101.53	101.53	101.53	8.58	0.008
Fertilizer	1	81.28	81.28	81.28	6.87	0.016
Water*Light	1	7.03	7.03	7.03	0.59	0.449
Water*Fertilizer	1	63.28	63.28	63.28	5.35	0.031
Light*Fertilizer	1	124.03	124.03	124.03	10.49	0.004
Water*Light*Fertilizer	1	7.03	7.03	7.03	0.59	0.449
Error	21	248.41	248.41	11.83		
Total	31	688.22				

The ANOVA table shows a p-value of 0.446 for the effect of the blocks. This is very insignificant for the design. The difference in the quality of the soil is not affecting the model. Notice that including the blocks in the design has had very insignificant impacts on the p-values of the main factors and their interactions.

Confounding

In the previous example, the experimenter was able to have every treatment combination in each block. That is because he had enough soil to accommodate all the different combinations in different blocks. Under some circumstances, the experimenter may not be able to have every combination in each block. Our experimenter could well have found himself short of soil and would not be able to have blocks wide enough to accommodate every replicate in each block. Had the experimenter been faced with the impossibility of replicating every treatment combination in each block, he could have used a technique called confounding.

Confounding designs occur when a full factorial design is run in blocks and the block sizes are smaller than the number of different treatment combinations.

Confounding is a special case of blocking. It is a technique that consists of dividing the experiment's treatment combinations into blocks containing subsets of the total number of the combinations. Confounding causes some of the treatment effects to be mixed up with the block effects. When the treatment effects and the blocking effects are confounded, the treatment effects of the main factors (or the interaction effects) are estimated by the same linear combination of the experimental observations as the blocking effects.

Example 4.4 Let us consider the case of our experimenter. Suppose that he is still using a 2-level, 3-factor experiment with only a single replicate.

Table 4.24 shows the combinations of the different coded factors. Consider that the experimenter wants to confound the third level interaction Water × Light × Fertilizer with blocks. From Table 4.24, we sort the treatment Water × Light × Fertilizer and assign the combinations with a plus sign to Block I and those with a minus sign to Block II.

TABLE 4.24

	I	Water	Light	Fertilizer	Water × Light	Water × Fertilizer	Light × Fertilizer	Water × Light × Fertilizer	Response
a	+	+	−	−	−	−	+	+	20
b	+	−	+	−	−	+	−	+	18
c	+	−	−	+	+	−	−	+	20
abc	+	+	+	+	+	+	+	+	12
l	+	−	−	−	+	+	+	−	19
ab	+	+	+	−	+	−	−	−	20
ac	+	+	−	+	−	+	−	−	21
bc	+	−	+	+	−	−	+	−	16

In Fig. 4.40, Block I corresponds to the corners with the dots.

Confounding happens every time a fractional factorial design is conducted instead of a full factorial design as shown in Table 4.25.

TABLE 4.25

Block I	Block II
a	l
b	ab
c	ac
abc	bc

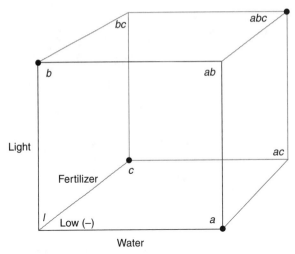

Figure 4.40

2^{k-1} **Fractional factorial design**

In the previous example, we analyzed a design with two levels, three factors, and four replicates. That particular design required 32 samples. If we had added an extra replicate, we would have needed 40 samples, and if instead of adding an extra replicate, we had added an extra factor, we would have ended up with a two-level, four-factor design with four replicates and therefore needing 64 samples, that is, $n2^k$ where k is the factor and n is the number of replicates.

That design would have generated four main factor treatments, six 2-factor interaction treatments, four 3-factor interactions, and one 4-factor interaction. In all, the total degrees of freedom for the treatments would have been 15, of which only four would pertain to the main factors while the other 11 would pertain to the interaction treatments and the error.

Every time a new factor is added to the design, the number of the samples needed would double and the proportion of the degrees of freedom that are associated with the main factors to the total degrees of freedom would shrink.

If the time that it takes to collect the samples is long and the resources needed to conduct an experiment are expensive, taking this many samples would be exorbitant. There are several ways to reduce the cost of collecting samples. First, replicating the experiment three times would not necessarily make a difference in the results of the experiment. One of the assumptions for conducting an experimental design is that the populations from which the samples are taken are normally distributed. Therefore, the probability for rejecting a null hypothesis that happened to be true solely based on the fact that the experiment was not replicated three times would be infinitesimal if the normality condition is met, unless in the extreme case of all the samples coming from outliers. As a result, replicating a design three times would be excessive.

The experimenter can also reduce the number of samples collected and still obtain results that are statistically close to what would have been obtained with a full factorial design by using the method called *fractional factorial design*. In a fractional factorial design, the experimenter considers only a subset of the total number of combinations. The purpose of a factorial experiment is to determine the main factors and the interactions that have an effect on the response variable at the lowest possible cost. If some treatments can rationally be deemed to have insignificant effects on the response factor, then they can be omitted from the design.

Since the selection of the factors to include in the design is made prior to conducting the experiment, the question that arises would be "how do we determine the treatments to be considered?"

The hierarchical significance of the treatments' effects on the response factor has been determined to decrease gradually from the main factors through the interaction treatments. If we have four factors, A, B, C, and D, we would end up with four main effects, six 2-factor interaction effects, four 3-factor interaction effects, and one 4-factor interaction effect. Since the higher the interaction order, the more negligible it is for the design, those interactions could be unnecessarily burdensome and can therefore be dropped from the model (Table 4.26).

TABLE 4.26

Main factors				2-Factor interactions						3-Factor interactions				4-Factor interactions
4				6						4				1
A	B	C	D	AB	AC	AD	BC	BD	CD	ABC	ACD	ABD	BCD	ABCD

In Table 4.24, we divided the design into two blocks. Block I contained the treatment combinations *a, b, c,* and *abc,* while Block II contained *l, ab, ac,* and *bc*. Block I corresponds to the top part of Table 4.24 where the treatment Water × Light × Fertilizer has plus signs. In that design, if only Block I is considered for the experiment, we would be conducting a half-factorial design. A half-factorial design is noted 2^{k-1} because the number of runs for 2^{k-1} equals half of 2^k. In the example for Table 4.24, we had two levels and three factors, 2^3 which generates 8 runs. If only a half-factorial design is being considered, we would end up with $2^{3-1} = 2^2 = 4$ runs.

Let us examine some properties of Table 4.24.

The determinant factor in the building of Block I and Block II was the plus and the minus signs associated with the treatment Water × Light × Fertilizer. For that reason, Water × Light × Fertilizer is called the generator. Column I with all plus signs is called the identifier. Multiplying any column by itself would generate the identifier.

$$\text{Water} \times \text{Water} = \text{Water}^2 = I$$

The product of any column with the identifier equals the column itself.

$$\text{Water} \times I = \text{Water}$$

The product of any two main factor columns would generate an interaction treatment column.

$$\text{Water} \times \text{Light} = \text{Water} \times \text{Light}$$

The sums of the pluses and minuses of each main factor column or interaction column is equal to zero; therefore the factors are said to be orthogonal. For Block I, Water × Light × Fertilizer = I

2^{3-1} Fractional factorial design

Since only the Block I portion of Table 4.24 will be used, let us isolate it and examine it in Table 4.27.

TABLE 4.27

Run	I	Water	Light	Fertilizer	Water × Light	Water × Fertilizer	Light × Fertilizer	Water × Light × Fertilizer	Response
a	+	+	−	−	−	−	+	+	20
b	+	−	+	−	−	+	−	+	18
c	+	−	−	+	+	−	−	+	20
abc	+	+	+	+	+	+	+	+	12

One of the first things that we notice is that each main factor column has a pattern of the plus and minus signs that is identical to an interaction treatment column and the pattern of the identifier (I) is identical to the one of the highest interaction orders of Water × Light × Fertilizer.

When two treatment factors are identical, their effects on the response factor cannot be separated. The effect of the main factor water, for instance, will be exactly the same as the effect of the interaction of Light × Fertilizer because they are identical.

For that reason, we say that Water and Light × Fertilizer are aliases. Since the effects of treatments that are aliases are confounded, considering both of them in the model would be redundant. Since the objective of a factorial experiment is to get the most with a minimum of effort, only one of the aliased treatments should be considered.

We can verify that the treatments are confounded by using their contrasts. The contrasts once again are the sums of the products of the treatment columns with the run column.

Notice that each main factor is aliased with an interaction treatment that involves only the other two main factors.

$$Contrast_{Water} = a - b - c + abc$$

$$Contrast_{Light \times Fertilizer} = a - b - c + abc$$

$$Contrast_{Light} = -a + b - c + abc$$

$$Contrast_{Water \times Fertilizer} = -a + b - c + abc$$

$$Contrast_{Fertilizer} = -a - b + c + abc$$

$$Contrast_{Water \times Light} = -a - b + c + abc$$

To determine what interaction treatment is aliased with a main factor, all we need to do is multiply that main factor with the generator treatment. To find what interaction effect is aliased with water, for instance, we can multiply water with the generator Water × Light × Fertilizer.

$$Water \times Water \times Light \times Fertilizer = Water^2 \times Light \times Fertilizer$$

$$= I \times Light \times Fertilizer = Light \times Fertilizer$$

When determining the effect of water on the response factor, both the effect of water and the effect of its alias Fertilizer × Light is what is actually being determined.

$$Water + Fertilizer \times Light$$

$$Light + Water \times Fertilizer$$

$$Fertilizer + Water \times Light$$

Thus far, all the results that we obtained are based on the fact that we selected Block I for our design. What would have happened if we had chosen Block II, which corresponds to the combinations with minus signs at the generator column?

Block II is called the alternate or complementary half fraction as opposed to Block I, which is the principal fraction. We can see that the identifier column is the opposite of the generator column.

$$I = -Water \times Light \times Fertilizer$$

TABLE 4.28

Run	I	Water	Light	Fertilizer	Water × Light	Water × Fertilizer	Light × Fertilizer	Water × Light × Fertilizer	Response
1	+	−	−	−	+	+	+	−	19
ab	+	+	+	−	+	−	−	−	20
ac	+	+	−	+	−	+	−	−	21
bc	+	−	+	+	−	−	+	−	16

From Table 4.28, we can also see that

$$Water = -Light \times Fertilizer$$

$$Light = -Water \times Fertilizer$$

$$Fertilizer = -Water \times Light$$

2^{4-1} Factorial design

The conduct of a half-fractional factorial design with two levels and four factors is similar to a 2^{3-1} factorial design with the difference being that the number of interaction orders increases and the nature of the aliases changes. Let us suppose that our experimenter deems the factor heat to be potentially significant for his design and adds it to the experiment. Based on what we have learned thus far, we know that the number of factorial interactions will increase and the pattern of confounding treatments will change. For the sake of convenience, we will use alphabetical letters to rename the factors.

$$A = Water$$

$$B = Fertilizer$$

$$C = Light$$

$$D = Heat$$

Table 4.29 summarizes the different treatments and combinations. The creation of the table follows the same patterns as the one for Table 4.24 with the difference being that an extra factor has been added. In Table 4.29, the generator is ABCD and it is confounded with I. We still have two blocks with the upper part of the table being the principal half fraction and the lower part being the complementary half fraction.

Design resolution

Notice that in Table 4.28, with the 2^{3-1} half-factorial design, the main factors were confounded with the 2-factor interaction treatments. In the case of Table 4.29, they are aliased with the 3-factor interactions and the 2-factor interactions are confounded among themselves.

TABLE 4.29

Notation	I	A	B	C	D	AC	AB	BC	AD	BD	CD	ABC	ABD	ACD	BCD	ABCD	Response
1	+	−	−	−	−	+	+	+	+	+	+	−	−	−	−	+	76
ab	+	+	−	+	−	+	−	−	−	+	−	−	+	−	+	+	81
ac	+	+	+	−	−	−	+	−	−	−	+	−	−	+	+	+	80
ad	+	+	−	−	+	−	−	+	+	−	−	+	−	−	+	+	85
bc	+	−	+	+	−	−	−	+	+	−	−	−	+	+	−	+	86
bd	+	−	−	+	+	−	+	−	−	−	+	+	+	−	−	+	70
cd	+	−	+	−	+	+	−	−	−	+	−	+	−	+	−	+	79
abcd	+	+	+	+	+	+	+	+	+	+	+	+	+	+	+	+	85
a	+	+	−	−	−	−	−	+	−	+	+	+	+	+	−	−	85
b	+	−	−	+	−	−	+	−	+	+	−	+	−	+	−	−	80
c	+	−	+	−	−	+	−	−	+	−	+	+	+	−	+	−	79
d	+	−	−	−	+	+	+	+	−	−	−	−	+	+	+	−	80
abc	+	+	+	+	−	+	+	+	−	−	−	+	−	−	−	−	82
abd	+	+	−	+	+	+	−	−	+	−	+	−	−	+	−	−	83
acd	+	+	+	−	+	−	+	−	+	+	−	−	+	−	−	−	85
bcd	+	−	+	+	+	−	−	+	−	+	+	−	−	−	+	−	87

The design resolution shows the confounding structure, it shows how the effects are confounded with one another. Knowing the confounding patterns prior to starting the experiment can help in the screening of the factors to be included in the experiment. It therefore helps reduce the cost of the experiment. Roman numbers are used to denote the types of resolutions. The most commonly used resolutions are III, IV, and V.

- *Design Resolution III.* The main effects are confounded with 2-factor interactions as in the case of 2_{III}^{3-1} (Table 4.28).

- *Design Resolution IV.* The main effects are aliased neither with themselves nor with the 2-factor interactions. As in the case of 2_{IV}^{4-1} (Table 4.29), the main effects are aliased with the 3-factor interactions, while the 2-factor interaction treatments are confounded among themselves.

- *Design Resolution V.* With the 2_V^{5-1} Design Resolution V, I = ABCDE. The 2-factor interaction effects are not confounded with each other, but instead with the 3-factor interaction effects.

Example 4.5 Wurossogui Ceramics is a manufacturer of ceramic tiles. The company has been having many returns of the Dakar model because the tiles of that model crack too easily. A design engineer is charged with the task of finding out the reasons for the cracks and to take corrective actions.

The design engineer believes that the cracks are related to the strength of the tiles, which is measured in terms of PSI. The PSI depends on three factors: the heat level used to harden the molded tiles, the type of clay, and the talc used as raw materials. The company has two suppliers of raw materials, Sine and Saloum.

To conduct his experiment, the engineer is given limited resources, so he decides to conduct a fractional factorial design using three factors (heat, clay, and talc) at two levels for each factor. He considers Sine as a low level (−1) and Saloum as a high level (+1).

Because he is running a Resolution III half-fractional factorial design, he will only need the samples that he summarized in Table 4.30.

TABLE 4.30

Heat	Clay	Talc	Response
-1	-1	1	120
1	-1	-1	190
-1	1	-1	260
1	1	1	200
-1	1	-1	270
1	-1	-1	200
1	1	1	270
-1	-1	1	110

Find the main effect   We already know from our discussion that because we are running a Resolution III half-fractional factorial design we will not need to find the interaction effects since all the 2-order interaction effects are confounded with the main effects and the 3-order interaction effect is confounded with the identifier.

Alias structure

$$I + (Heat \times Clay \times Talc)$$

$$Heat + (Clay \times Talc)$$

$$Clay + (Heat \times Talc)$$

$$Talc + (Heat \times Clay)$$

TABLE 4.31

Heat	Clay	Talc	Response
1	-1	-1	390
-1	1	-1	530
-1	-1	1	230
1	1	1	470

$$Contrast_{heat} = a - b - c + abc = 390 - 530 - 230 + 470 = 100$$

$$Contrast_{clay} = -a + b - c + abc = -390 + 530 - 230 + 470 = 380$$

$$Contrast_{talc} = -a - b + c + abc = -390 - 530 + 230 + 470 = -220$$

Main effects

$$Main\ Effect_{heat} = \frac{Contrast_{heat}}{n2^{3-1}} = \frac{100}{4} = 25$$

$$Main\ Effect_{clay} = \frac{Contrast_{clay}}{n2^{3-1}} = \frac{380}{4} = 95$$

$$Main\ Effect_{talc} = \frac{Contrast_{talc}}{n2^{3-1}} = \frac{-220}{4} = -55$$

Regression coefficients   The regression coefficients are obtained by dividing the main effects by 2.

$$Constant\ t = \frac{195 + 265 + 115 + 235}{4} = 202.5$$

$$Coef_{heat} = \frac{25}{2} = 12.5$$

$$Coef_{clay} = \frac{95}{2} = 47.5$$

$$Coef_{talc} = \frac{-55}{2} = -27.5$$

Regression equation

$$PSI = (Coef_{heat} \times heat) + (Coef_{clay} \times clay) + (Coef_{talc} \times talc) + Constant$$
$$PSI = 12.5\ heat + 47.5\ clay - 27.5\ talc + 202.5$$

Using SigmaXL   SigmaXL is very practical when it comes to a 2-level DOE. From the SigmaXL menu, select **Design of Experiments** and then select **2-Level Factorial/Screening Design** from the drop down list. The **2-Level Factorial/Screening Design** dialog box appears. Fill it out as indicated in Fig. 4.41.

Press the **OK>>** button. The **Design of Experiments Worksheet** appears. Fill out the Response column with the data gathered in Table 4.30. (The data can be found in the file *Ceramic Tiles.xls*). Do not just paste the data in the column, make sure that the numbers match the combination levels; otherwise, the results of the test will be wrong (Table 4.31).

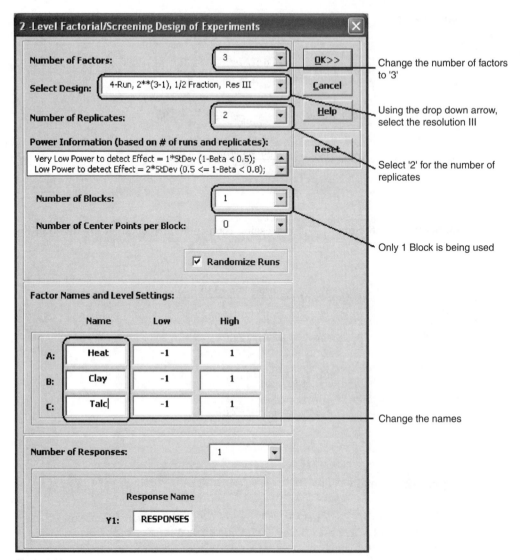

Figure 4.41

Go back to the SigmaXL menu and select **Analyze 2-Level Factorial/Screening Design.** The **Analyze 2-Level Factorial/Screening Design** box will appear already filled in; just press the **OK>>** button to obtain the results shown in Table 4.32.

The output will show among other graphs the Pareto chart (Fig. 4.42), which shows the order of importance of the coefficients for the response.

The residual table will also appear along with the probability plot (Table 4.33 and Fig. 4.43).

TABLE 4.32

Design of Experiments Analysis

DOE Multiple Regression Model: RESPONSES = (202.5) + (12.5) * A: Heat=BC + (47.5) * B: Clay=AC + (-27.5) * C: Talc=AB

Regression equation

Title:	
Date:	
Name of Experimenter:	
Notes:	

Design Type:	3 Factor, 4-Run, 2**(3-1), 1/2 Fraction, Res III
Number of Replicates:	2
Number of Blocks:	1
Number of Center Points per Block:	0
Response:	RESPONSES

Coefficient of determination

Model Summary:

R-Square	90.70%
R-Square Adjusted	83.72%
S (Root Mean Square Error)	25.495

Regression coefficients   P-values

Parameter Estimates:

Term	Coefficient	SE Coefficient	T	P	VIF	Tolerance
Constant	202.5	9.013878189	22.465	0.0000		
A: Heat=BC	12.5	9.013878189	1.387	0.2378	1	1
B: Clay=AC	47.5	9.013878189	5.270	0.0062	1	1
C: Talc=AB	-27.5	9.013878189	-3.051	0.0380	1	1

Main factors and their aliases

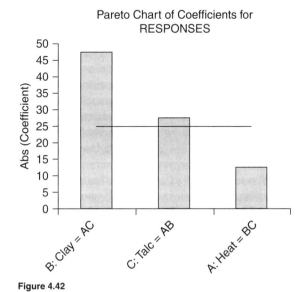

Pareto Chart of Coefficients for RESPONSES

Figure 4.42

TABLE 4.33

A: Heat	B: Clay	C: Talc	Response	Predicted (fitted) values	Residuals	Standardized residuals
1	−1	−1	190	195	−5.000	−0.277350
1	−1	−1	200	195	5.000	0.277350
−1	−1	1	120	115	5.000	0.277350
−1	1	−1	260	265	−5.000	−0.277350
1	1	1	200	235	−35.000	−1.941
−1	1	−1	270	265	5.000	0.277350
1	1	1	270	235	35.000	1.941
−1	−1	1	110	115	−5.000	−0.277350

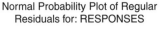

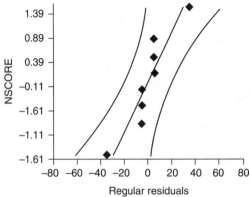

Figure 4.43

Using Minitab   Open the Minitab file *Ceramic tiles.mtw.* From the menu bar, click on **Stat**, then select **DOE**, then **Factorial**, and then **Create factorial design.** From the **Create factorial design** box, select '3' for the **Number of Factors** (Fig. 4.44).

Press on the **Display available designs** button. Select Resolution III for 3 factors 4 runs (Fig. 4.45).

Press the **Design** button to get the dialog box shown in Fig. 4.46. Select **½ fraction 4 runs and Resolution III** and for the **number of replicates for corner points,** select "**2**".

Press **OK** and then **OK** again. Minitab will display the alias structure, the design generator, and a new worksheet containing factor treatments denoted A, B, and C. The response column will not be added. We will have to add it. See Fig. 4.47.

Go back to the menu bar and select **DOE** again and then **Factorial.** This time click on **Analyze Factorial Design.**

Select the responses column for the field **"Responses".**

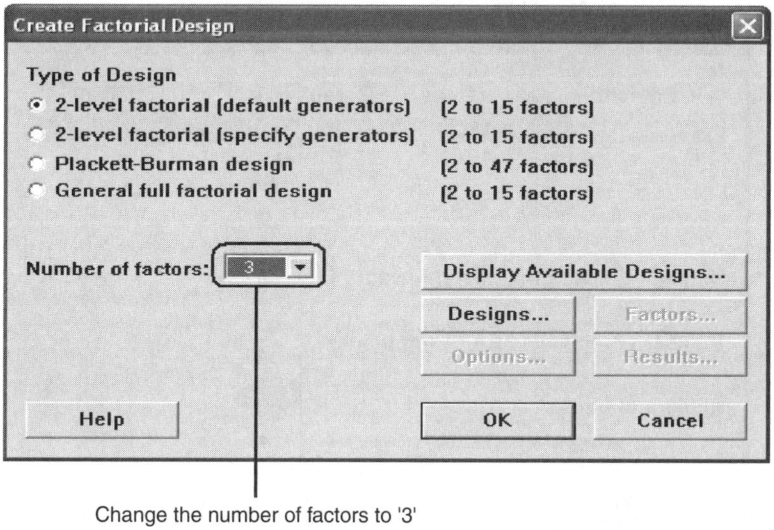

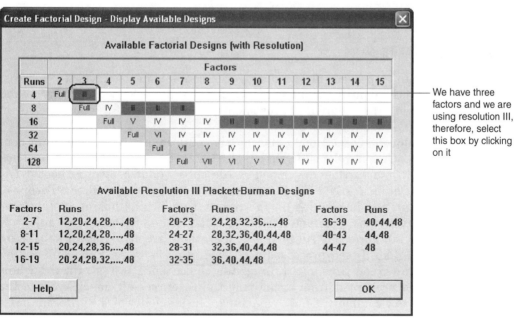

Change the number of factors to '3'

Figure 4.44

Figure 4.45

Press **Terms** to get the **Analyze Factorial Design–Terms** box (Fig 4.48). Select "1" for **Include terms in the model up through order** drop down list.

Press **OK** and then **OK** again to obtain the results in Table 4.34.

The design shows that at an alpha level of 0.05, the factors Talc and Clay are significant in the model but heat, with a p-value of 0.238, is not.

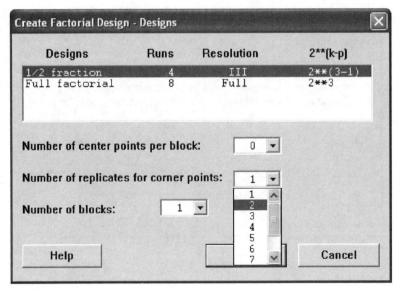

Figure 4.46

The Theory of Constraints

Continuous improvement is a managerial concept that emphasizes the need to constantly identify the areas of a company that need special attention and proceed with the required improvement for the benefit of the business as an entity.

This concept is well embodied by the philosophy of the theory of constraints (TOC), developed in the mid-1980s by Eliyahu Goldratt in his book *The Goal* (and further developed in his subsequent bestseller books, *The Critical Chain, It's Not Luck, The Haystack Syndrome,* and the *Theory of Constraints*). It is founded on the notion that in any business structure, at any given time, one factor tends to impede the organization's ability to reach its full potential. All business operations are structured like a chain of events, like linked processes with each process being a dependent link and at any given time, one link on the chain tends to restrain the whole chain and prevent it from optimizing the resources efficiencies. Since the objective of a company is not to maximize the efficiency of the different parts that compose it, but to maximize the overall efficiency of the business as an entity, it becomes necessary to identify the constraint and proceed with the needed improvement. The improvement of the system requires the focus on the weakest link. Disregarding the constraint, the weakest link, and improving any other aspect of a business will eventually lead to a worsening of the problems. The improvement process itself requires some steps, the first of

Fractional Factorial Design

Factors:	3	Base Design:	3, 4	Resolution:	III
Runs:	8	Replicates:	2	Fraction:	1/2
Blocks:	1	Center pts (total):	0		

* NOTE * Some main effects are confounded with two-way interactions.

Design Generators: C = AB

Alias Structure

I + ABC

A + BC
B + AC
C + AB

Change the main factors names   Add the response column

Worksheet 1 ***

C1	C2	C3	C4	C5	C6	C7	C8	C
StdOrder	RunOrder	CenterPt	Blocks	Heat	Clay	Talc	Response	
1	1	1	1	1	-1	-1	190	
2	2	1	1	1	-1	-1	200	
3	3	1	1	-1	1	-1	260	
4	4	1	1	-1	1	-1	270	
5	5	1	1	-1	-1	1	120	
6	6	1	1	-1	-1	1	110	
7	7	1	1	1	1	1	200	
8	8	1	1	1	1	1	270	

Figure 4.47

which is the identification of the constraint. Once the constraint is identified, it becomes necessary to determine what changes need to take place and how they are to be brought about.

To make the necessary changes, the company first needs to answer the following three questions:

1. What do we need to change?

2. What do we need to change to?

3. How do we make the change happen?

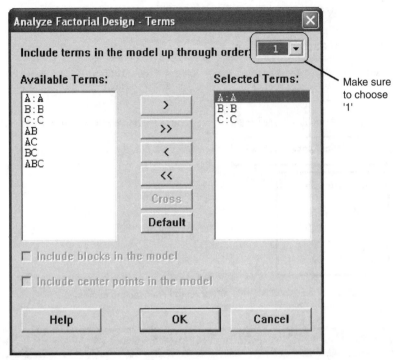

Figure 4.48

TABLE 4.34

Factorial Fit: Response versus Heat, Clay, Talc

Estimated Effects and Coefficients for Response (coded units)

```
Term       Effect    Coef   SE Coef      T      P
Constant            202.50    9.014  22.47  0.000
Heat       25.00    12.50    9.014   1.39  0.238
Clay       95.00    47.50    9.014   5.27  0.006
Talc      -55.00   -27.50    9.014  -3.05  0.038
```

S = 25.4951   R-Sq = 90.70%   R-Sq(adj) = 83.72%

Analysis of Variance for Response (coded units)

```
Source          DF   Seq SS   Adj SS   Adj MS      F      P
Main Effects     3  25350.0  25350.0  8450.0  13.00  0.016
Residual Error   4   2600.0   2600.0   650.0
  Pure Error     4   2600.0   2600.0   650.0
Total            7  27950.0
```

Alias Structure
I + Heat*Clay*Talc
Heat + Clay*Talc
Clay + Heat*Talc
Talc + Heat*Clay

The changes that need to be made must address the area of the business that constitutes the bottleneck. Overlooking the interactions between the different departments in a company and only improving on areas that are perceived to constitute a problem might address the symptoms and in some cases worsen the problems.

The process throughput is tied to the bottleneck

The ideal production process that would eliminate the waste that comes in the form of excess inventory or idle machines would be the balanced production process based on the concept of one-piece flow. The ideal process does not build inventory, and uses both labor and machines to their full potential. For this to happen, every step in the process should take the same amount of time to complete a task. Yet in the real world, this seldom happens. Because of the complexity involved in manufacturing processes and the usually uneven process capabilities of the operating resources that result in different cycle times for different processing steps, a one-piece flow process will make it hard to use all the factors of production to their full potential at the same time. The task of management therefore becomes to allocate resources in such a way that the production process is balanced and even.

Example 4.6 For the sake of our argument let us consider the operations in a fictitious soap manufacturing company. The operations are simplified to the following steps.

- Reception and stocking of the raw material
- Mixing the raw material to produce the soap
- Cutting the bar soap and packing the pieces
- Stocking the products at outbound inventory
- Shipping the product to the customers

The time spent by the different steps to process 5 tons of material is: 1 h for reception and stocking the inventory, 2 h for mixing, 3 h for cutting and packing, 2 h for outbound stocking, and 1 h for shipping (Fig. 4.49). Goldratt suggests five steps for the improvement process.

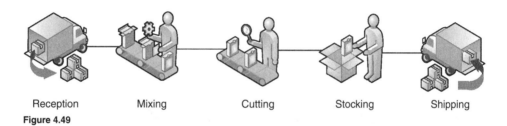

| Reception | Mixing | Cutting | Stocking | Shipping |

Figure 4.49

1. *Identify the constraint.* This scenario clearly shows the "cutting and packing" step to be the critical constraint for the processing line as a whole because no matter how well all the other steps perform it will be impossible for the business to process 5 tons of soap in less than 3 hours. The bottleneck caused by the cutting and packing step translates into two forms of waste: overproduction generated by the upstream operations because the cutting and packing step cannot process all the output that comes from the mixing step (if it operates at its full potential) and underused resources downstream.

   An improvement in any aspect of the processing line other than the cutting and packing step will lead to an increase in either excessive idle inventory after the mixing step or more idle machines and higher cycle time.

   Once the constraint has been identified, the next step is proceeding with the improvement.

2. *Exploit the constraint.* A constraint is not a step in the process that is unnecessary and needs to be eliminated. The bottleneck is defined as a resource whose capacity is equal to or is less than the demand placed on it. It is the slowest performing area in a process and it determines the level of output generated by that process. Analyze the constraint to ensure that it is only performing the tasks for which it was meant. Decide on what can be done to enhance its performance. Eliminate all the clutter and the non-value-adding activities that might be slowing the constraint. In our example, the exploitation of the constraint should include the inspection of the cutting machine to ensure that it is performing to expectation, an audit and time studies to determine if the employees working at the cutting station are following the process, and an analysis to determine if there is any quality problem that makes it necessary to perform rework or scrap part of the products.

3. *Subordinate all activities to the needs of the constraint.* In this step, the improvement team focuses all its effort on improving the performance of the constraint. If there is too much rework done at the constraint level, determine what needs to be done to eliminate it. If there is any adjustment that needs to be performed on the machines or too much motion that reduces the employees' performance that needs to be addressed, the improvement team should find a way to make the necessary improvement.

4. *Elevate the constraint.* If the process is flexible to allow labor capacity to be moved from the other steps to the constraint, workers from stocking and shipping, for instance, could be allocated to the cutting step to alleviate it since stocking and shipping would remain idle most of the time.

   Removing operational capacity from other processing steps and assigning it to the cutting step will increase the cutting throughput, and since the performance of the overall production process is tied to the constraint, the process yield will be increased.

5. *Restart the process without letting inertia become the system's constraint.* Now that capacity has been added to the constraint and its performance increased, the improvement process must be restarted again from step 1 until the current constraint is no longer the constraint. The improvement process must be repeated without letting inertia set in. The way we can tell that the current constraint is no longer the constraint is when further changes are made on the current constraint and they do not positively impact the production process throughput. Therefore,

another step in the process has become the weakest link, the new constraint, and it needs improvement.

TOC Metrics

The performance measurements used by the TOC are different from the traditional cost accounting systems. Goldratt borrows some commonly used business terms but he gives them a different meaning. Three of the most important of those are throughput, inventory, and operational expenses, which he defines as follows.

1. *Throughput (T)*. The rate at which an organization generates money through sales. It is based on the total sales minus the variable cost, which includes the cost of materials used to produce the goods sold.
2. *Inventory (I)*. Money invested in purchasing things intended for sales. This includes the building used for operations. It is money currently tied up in the production system.
3. *Operational expenses (OE)*. Money spent to turn inventory into throughput. It includes salaries and wages paid to workers, rent, utilities, etc.

Some of the derivatives of these metrics are the throughput per unit and the throughput per unit of the constraining factor.

1. *Throughput per unit* = throughput/(units of product)
2. *Throughput per unit of the constraining factor* = (throughput per unit)/(units of the constraining factor required to produce each unit of product)
3. *Net profit* = T − OE
4. *Return on asset* = (T − OE)/I
5. *Productivity* = T/OE
6. *Asset turn* = T/I

To maximize its throughput, the company must concentrate on improving the sales of the products that provide the highest throughput per unit of the constraining factor. This is because the bottleneck determines the overall process throughput.

Thinking Process

The TOC is primarily a thinking process founded on the notion that, in general, all of the problems encountered in an organization can be traced to a single cause. In a manufacturing process, the throughput is obviously constrained by the bottleneck. However, constraining factors are not limited to a bottleneck in a manufacturing process. Constraining factors do exist in any business, even those that do not exhibit linear flows of production processes. Even in those businesses, Goldratt believes that only a single factor inhibits them from achieving their ideal potential. Finding the root cause of the problems and solving it will translate to the resolution of all its undesirable effects at a low cost. The thinking process of the TOC starts with the definition of the problems being confronted. A problem can be defined as a gap, a contradiction between the current realities and a desired reality; it can also be defined in terms of a conflict between the rational requirements and their respective prerequisites that can help feel the gap. After having determined the objective to be attained, Goldratt suggests the use of an evaporated cloud to analyze the contradicting requirements to reach it.

The Goldratt Cloud

When dealing with a problem, we usually have an intuition about ways to solve it but because by definition a problem is a made up of conflicting alternatives, the options to resolve the problem are contradictory. To solve the contradiction, Goldratt suggests that we verbalize all of the components of the problem: the current situation, the contradicting resolution requirements, and their prerequisites. The verbalization is better done through a group discussion and the use of what Goldratt calls the *evaporating cloud*. The cloud helps better organize and visualize the intuition. The objective in the resolution of the problem is not to choose one option over another but to find a synergy, a win-win solution. In order to build the cloud, the following steps must be followed:

- Clearly define the objective to be reached.

- State the contradicting requirements for the objective to be reached.

- State the prerequisites for the requirements to be satisfied.

- Analyze the prerequisites to determine possibilities for a solution that satisfies both of them. In other words, determine what arrow can be broken so that the contradictions are eliminated.

The cloud is made up of blocks and arrows and it should be read as: "In order to reach the objective, one of the requirements must be satisfied, and in order to satisfy the requirement, its prerequisite must be satisfied." (See Fig. 4.50.)

Example 4.7 A distribution plant had initiated a labor productivity improvement plan that included the use of a software suite to track the transactions that the employees perform in their daily activities. The consulting firm that produces and sells the software suite asked for $950,000 to have consultants come to the plant, configure the software, and install it. The alternative to that option would be for the distribution plant to buy the software and have its employees do the work themselves, but because they do not have the expertise, it would take them too much time to study and implement it, the cost of which would have been excessive.

After a verbalization session, the cloud shown in Fig. 4.51 was created.

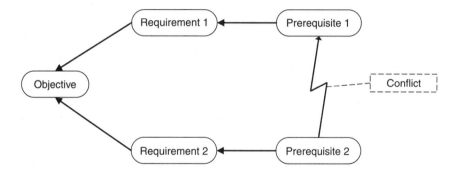

Figure 4.50

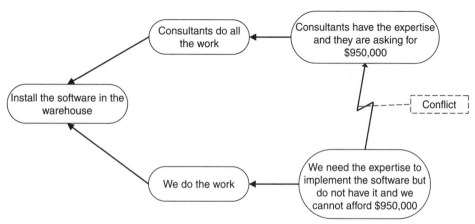

Figure 4.51

During the brainstorming session, it was determined that for the consulting firm to do all the implementation work, it would have to send two of its engineers to the plant and have them work full time on the project. The solution to the problem was found by having the consulting firm send an engineer to the plant 1 day a week for 3 months. The engineer would train the employees on how to implement the software, give them assignments, and verify the tasks accomplished each time he comes to the plant. The total implementation cost was brought down from $950,000 to $230,000 and the project was completed on time.

The Goldratt Reality Trees

The evaporating cloud is a systemic visualization of the alternative ways of solving a problem. But looking at the cloud alone would not solve the problem. Goldratt suggests that the cloud is analyzed through a verbalization of the problem. The verbalization enables us to understand the current realities and to determine the best ways to reach an ideal situation.

Current reality tree

In his book, *It's Not Luck,* Goldratt shows that in most cases, what we think to be problems in an organization are not really problems, they are nothing but the undesirable effects (UDEs) of *one* actual problem. If a manager is confronted with a multitude of poor performing areas in his operations that end up making his scorecards red, chances are that all the problems that he thinks he is facing are nothing but the UDEs of a single actual problem. Unless the actual problem is found and corrected, any attempt to solve the UDEs would only result in costly temporary fixes that would never totally eliminate the problem and instead would put management in a constant fire-fighting mode. Goldratt suggests that the actual problem can be found by building what he called a current reality tree (CRT).

The CRT is a relational diagram that uses a syllogistic reasoning to analyze all the existing problems in an organization and seeks to find their root causes. The idea is to find relatedness between UDEs by using syllogisms, major premises, minor premises, injections, and conditional reasoning to come to a common cause of all the UDEs.

A CRT is built at a brainstorming session. It starts with the listing of all the UDEs before using a cause-and-effect reasoning to conduct the analysis.

Example 4.8   Ziguinchor Distribution Center is a forward logistics warehousing storage facility that specializes in receiving inventory from suppliers, storing the products, and shipping them to its customers when orders are received. The performance of the warehouse's operations is tracked using scorecards and the metrics in the scorecards have not been meeting expectations. A brainstorming session was organized to find the root cause of the situation. The members of the team involved with the session started by listing the poor performing areas. Their objective is to find the root cause.

1. High employee turnover
2. Failure to deliver customers' orders on time
3. Loss of inventory
4. Poor quality
5. Low inventory turnover
6. High lead time
7. Too many work-related accidents
8. Low productivity
9. High operational expenses
10. Low morale
11. Loss of customers
12. Too many customer complaints
13. Customer services too costly
14. Employee absenteeism
15. Inefficient production processes

After having inventoried all the UDEs, the team started by finding affinities and causalities between them. Then it created the first branches of the tree. The following branch should be read as: "*If* the company sends poor quality products *and* it fails to

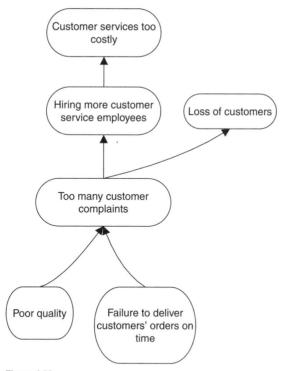

Figure 4.52

deliver customer orders on time, *then* there will be too many customer complaints. *If* there are too many customer complaints, *then* the company will lose customers and will have to hire more customer service employees. *If* the company hires more customer service employees, *then* customer services will end up being more costly." (See Fig. 4.52.)

The second branch should be read as: "*If* new employees are hired and trained too often, *then* they will not have enough expertise to do the job. *If* they do not have enough expertise, *then* their productivity will be low and they will have too many accidents. *If* their productivity is too low, *then* lead time will be too high. *If* the lead time is too high, *then* the company will tend to increase its inventory to avoid customer back orders. *If* inventory stock is increased, *then* inventory turnover will be lowered and opportunities for inventory loss will be increased and more employees will be needed for inventory audit and cycle counting. *If* inventory turnover is decreased and inventory loss is increased, *then* the production processes will become inefficient." (See Fig. 4.53.)

Notice that some new injections that were not on the initial list have been added to make the linkages more explicit. *If* the morale in the warehouse is too low, *then* employee absenteeism will be too high. *If* the absenteeism is too high, *then* the level of employee termination will be too high. *If* there are too many employees being terminated, *then* the turnover will be too high. *If* the employee turnover is too high, *then* new employees will be hired and trained too often. *If* employees are hired too often, *then* the cost of hiring and training new employees and the cost of customer services will lead to higher operational expenses.

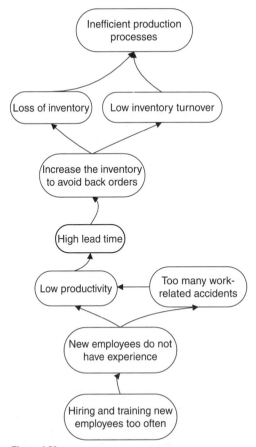

Figure 4.53

Therefore, we will have to conclude that the only problem that the company faces is the low morale and that all the other issues on the scorecards are nothing but UDEs of the low morale. To solve the problem at the lowest possible cost, the causes of the low morale need to be found and addressed (Fig. 4.54).

Future reality tree

Based on the finding obtained from the CRT, a future reality tree (FRT) can be built to remedy the problem. The building of the FRT follows the same patterns in structure as the CRT with the difference being that we are inventorying the necessary injections to realize the ideal vision. In the previous example, we determined the low morale to be the root cause of the UDEs. The building of the FRT will consist of determining all the necessary conditions that elevate the morale of the employees and help retain them in their jobs so that employee turnover is

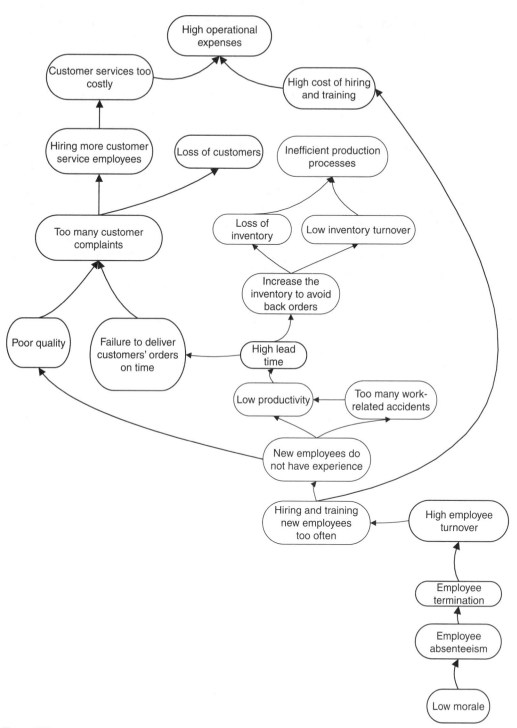

Figure 4.54

reversed and quality and productivity are improved. Of course, the causes of low morale will have to be determined before remedies are considered.

5s

5s is derived from five Japanese words: seiri, seiton, seiso, seiketsu, and shitsuke. In English, these words closely mean, respectively, sort, set in order, shine, standardize, and sustain. 5s represents a five-step process aimed at reducing waste, streamlining operations, and ultimately increase productivity. It is a process of organizing a workplace in order to optimize the efficiency of the resources' utilization by keeping it tidy. It is about having the right products and tools (and only the right products and the right tools) at the right places at the right time. It is a process that eliminates clutter, reduces workers' motion, and increases floor capacity by reducing the size of the space needed for operations. By assigning every tool or product a known and identified location close to where it is normally used, confusion is reduced, the time spent looking for things is reduced and, consequently, productivity is improved. The execution of a 5s process is done by enforcing a discipline of cleanliness and orderliness that leads to an improvement in productivity and quality and a better control over inventory. But 5s is not just a form of housekeeping; it is a methodology of organizing and developing a productive work environment.

The benefits derived from the implementation of a 5s program can be very tangible. A company with which the author worked saved approximately $1.5 million in one fiscal year and drastically reduced customers' complaints after implementing 5s. The company was suffering from a great deal of disorganization in its shipping docks, which resulted in very often systemically shipping customers' orders but not physically loading the shipments in the trailers. As a result of that, customers' packages would be found days or weeks later somewhere in the warehouse after the customers repeatedly called to complain about not receiving their orders. The cost incurred to correct the mistakes was exorbitant because not only would the plant face more rework to fill the orders again but the shipping cost was also doubled even though the first packages were not delivered. Implementing 5s helped a great deal to correct the problem by not only reducing customers' complaints but also by eliminating orphan packages and therefore reducing inventory discrepancies. It also increased floor capacity and improved cycle time by improving productivity.

The five steps of the 5s are as follows:

Step 1: Seiri or sorting. In this phase, the team implementing the 5s methodology inventories all the tools, material, and equipment that are currently in the plant. Then the team separates the tools and equipment that are presently needed for regular operations from those that are not needed. The products that are not needed are removed and taken away from the space used for daily operations to a temporary location where they will be recycled, reassigned, or disposed of. Sorting enables the operations to claim a great deal of space.

Step 2: Seiton or set in order. Arrange and label the needed tools in such a way that they are handy and easy to use. Every item needs to be at a specific place and only that item needs to be there. The workers should not have to travel far to search for the tools that they need to do their work. The storage locations should be clearly identified and the workers should know exactly where to find the tools that they need. This process helps improve efficiency by reducing motion.

Step 3: Seiso or shine. Once the unneeded items have been removed, the clutter and the clog disposed of, and the needed tools and equipment reorganized, labeled, and stored at specified locations, the next step is to steadily clean the workplace. A clean environment renders unexpected items visible. At the plant mentioned previously, customers' packages were left at places where they were not supposed to be and as a result, they were not being loaded in the trailers. No one was alerting the supervisors because nothing indicated that the packages were not supposed to be where they were found. After the 5s process was implemented, the entire warehouse floor was painted and all the locations identified using labels and placards. By keeping the plant clean and tidy, every time a package is left at a wrong location it is very quickly noticed and removed.

Step 4: Seiketsu or standardize. The standardization step consists of creating a sense of consistency in the ways that the tasks are performed. Standardization makes it easy to move workers around without losing out on productivity as well as making it easy to control tools, equipment, and material.

Step 5: Shitsuke or sustain. The last step consists of making the four previous steps a habit for all the employees to follow the procedures. This is the most difficult step in a 5s process because if the four previous steps are not translated into a culture for the organization, it is easy to fall back to their old ways.

5

Control

Statistical Process Control (SPC)

Once the improvements have been made and the best practices defined, management should take control measures to ensure that the production processes do not deviate from the defined standards.

In the past, the most widely used technique to ensure quality in products that were bought or sold by most businesses was acceptance sampling. Acceptance sampling is a technique that consists of randomly selecting samples from batches for testing to determine their conformance to predetermined standards. The samples are taken after production has been completed. Once the products have been manufactured, a quality controller inspects some samples to sift out the nonconforming products that would be either discarded or sent for rework. Quality was inspected in the products and the inspection only took place when the products were about to be delivered to customers. It was a quality management system that only enforced localized actions on the output instead of taking actions on the overall processes that generated the output.

That process was not only time consuming but also it was costly because it negatively affected productivity without necessarily leading to a defect-free production system. The purpose of improving the quality of the products delivered to customers is not just to satisfy customer; it is also to reduce cost by improving productivity. Improving the quality of the products shipped to the end users by solely relying on inspections at the end of production lines can be extremely exorbitant because it requires extra quality controllers, more equipment and space for the inspections. It would eventually decrease the overall productivity of the business. If special circumstances were to create alterations to the production process and result in the production of a considerable amount of defects, this would translate into waste and the output would have to be either discarded or reworked. The alternative to scrapping or reworking those defective products is enduring customer complaints and the cost attached to them.

Productivity is a business metric that measures the resources' efficiency. The productivity of a process is calculated as the ratio of the output to the amount of resources used to produce it, or as the number of items generated for each dollar invested. Expressed as a formula, where P = productivity, O = output, and C = the cost incurred to produce O, productivity would be

$$P = O/C$$

There is a positive correlation between productivity of a process and good quality. In a defect-free process, the cost of production is the cost of producing all the goods that are actually used by the customer. In a process that generates defects, the cost of production includes the cost of the defects (inspection costs, returns from customers, rework in the process flow, etc.).

One alternative to inspections at the end of production lines is the use of SPC. SPC is a preemptive quality control technique that does not aim at inspecting quality in the products themselves but rather at instilling quality in the process that leads to the production of those products by making sure that the production process is stable and in control. SPC is more about managing the quality of a production process than it is about sifting defective products from a process output.

The objective of SPC is to avoid waste in the form of scrap and rework by not producing defective products in the first place. It is, therefore, a tool that improves both quality and productivity.

Control charts are the most widely used tools in SPC. They were invented by Dr. Walter Shewhart in the 1920s and later further developed by Dr. Edward Deming.

Some of the important benefits of using control charts are

- To keep the production process in statistical control and stable
- To detect special causes of variations before it is too late
- To determine if there is a need for machine maintenance or operator retraining
- To improve productivity while improving quality
- To prevent defects

Variation Is the Root Cause of Defects

Customers are the ultimate judges of the quality of products and services; their judgments about the quality levels of their purchases depend on certain characteristics of the products that they buy. Every product or service exhibits certain characteristics, some of which are critical to its quality. The critical to quality (CTQ) characteristics are those whose absence or nonconformity to customers' expectations result in the product being deemed of poor quality.

Before producing the goods and services intended for customers, the design engineers predetermine their CTQ characteristics and they include them in the designs of the products with the expectation that each item produced will meet the predetermined standards.

When the production process for the goods and services is in progress, the manufacturer tries to generate identical products that meet the engineered standards. But producing perfectly identical products is not possible because variations are an inherent part of every production process. Variations in production are defined as the difference in magnitude between the CTQ characteristics of the items produced.

When we buy a case of Pepsi, chances are that all the cans in the case were manufactured by the same machine and within a short time frame. The expectations are that the cans are identical. They might indeed seem identical but a closer observation is likely to reveal differences in the CTQs between them; no two cans are likely to be identical.

Some of the CTQs are explicitly written on the cans. For instance, the sugar content is 41 g, the total carbohydrates is 41 g, the total fat is 0, the total protein is 0, the total amount of sodium is 30 mg, the caffeine content is 38 mg, the total calories is 150 g, and the volume of drink inside each can is set at 355 mL. These are very explicit and clearly written CTQs but they are not the only ones. Some CTQs such as taste are implicit and obviously expected to be present in each can. If any of these CTQs deviates too far from these preset targets, the customers would consider the products as being of poor quality. For instance, the customers expect the sugar level to be 41 g. If they buy a can and the sugar level happens to be 30 g or 50 g, they would consider this CTQ to be too far from target of 41 g and, consequently, the can of Pepsi would have failed to meet their expectations. However, if the sugar level was 40.8 g or 41.2 g, the difference from the target of 41 g would be so insignificant that it might go unnoticed by the customers. Therefore, the manufacturer can consider cans with sugar levels between 40.8 g and 41.2 g as being good enough to be shipped to customers, but cans with sugar levels of 43 g would be considered as nonconforming to standards and therefore defective.

In manufacturing, the variability can be the results of several factors including differences in raw materials, differences in operators' skills, and poor adjustments of machines.

Variations can be short lived as in the case of piece-to-piece variations that result from a lack of attention to details on the part of an operator. But they can also be the long-term effect of an unattended machine that is gradually deteriorating over time.

The determination of the nature of the causes of variation depends on the extent of the variations.

Assignable (or special) causes of variation

When the causes of variation generate differences in the CTQs of the output that are significantly important, they are called assignable causes (by Shewhart) or special causes (by Deming) and are considered sources of defects. When they occur, the production process is stopped and the causes of variations investigated. Those causes can range from a badly calibrated machine, to an operator's lack of attention to details, to poor quality raw materials from suppliers. Assignable causes of variation can lead the cans of Pepsi to only contain 30 g of sugar.

Common (or chance) causes of variation

These causes of variation are often referred to as *background noise* or chance causes (by Shewhart) or common cause (by Deming). They are multiple, relatively small, uncontrollable factors that are inherent to the production process. Their effects on the CTQs of the products are relatively insignificant but detectable. They occur randomly and cannot be eliminated from the process. Because they are relatively insignificant, they are considered acceptable. The causes of variation that resulted in the sugar contained in the can of Pepsi being between 40.8 g and 41.2 g would be an example of common causes.

When only common causes of variation are present in the production process, the process is said to be stable and in control, it is predictable, and all of its output will follow the same probability distribution over time. Special causes of variation will lead to the instability of the process, the impossibility to make predictions on the production process and they will, in the end, eventually alter its distribution.

Being able to distinguish between common causes of variation and assignable causes of variation is very important. According to Forrest Breyfogle (*Implementing Six Sigma*), one important mistake that many organizations make is to confuse the two and focus too much resources trying to correct variations that are the result of common causes, not knowing that those are inherent to the production process and cannot be eliminated. Suppose that in the case of the manufacturing process of the Pepsi cans, every time a can fails to be 355 mL full, the process is stopped for investigation. In that case, the operational expenses would end up being utterly exorbitant because common causes of variation cannot be eliminated.

In order keep from producing defective products, the most commonly used tools to monitor a production process in progress are

- Control charts
- Histograms
- Pareto charts
- Checklists
- Scatter diagrams
- Defect-concentration diagrams
- Ishikawa diagrams
- A3 reports

The most widely used of these tools is the control chart. The other tools are investigative tools that can be used to determine the sources of variations but are not monitoring tools and cannot determine on their own whether the production process is stable and in control, nor can they help assess the processes' capabilities, in other words, their ability to generate goods or services that meet or exceed the customers' expectations.

Since variations are an intrinsic part of every production process and they can be either common or assignable, producers have to define how much variation they are willing to tolerate to keep their processes stable and in control. These specifications are addressed in the way the control charts are built.

How to build a control chart

Control charts are built to determine if a production process is stable and in statistical control and to eventually reduce the common causes of variability to an infinitesimal level. They are built for a production process in progress; samples of the output are randomly taken at predetermined time intervals for inspection and the samples' means are determined and plotted on a chart.

A control chart is generally made up of three horizontal lines and one vertical line as shown in Fig. 5.1. The vertical line measures the levels of the sample means while the horizontal center line represents the process mean. The upper horizontal line is the UCL and the lower horizontal line is the LCL.

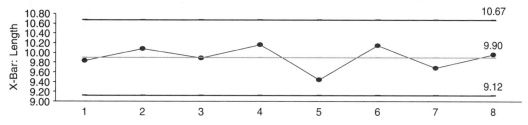

Figure 5.1

If a sample of n items is taken and the value plotted on the chart, and that sample's CTQ is expressed in continuous measurements, $\overline{X} = \frac{\sum X}{N}$ would be that value.

The control limits are calculated in a manner that they are 3 standard deviations away from the center line.

$$UCL = \mu + 3\sigma$$

$$CL = \mu$$

$$LCL = \mu - 3\sigma$$

Therefore, the range of the tolerance for the common causes of variation would be equal to 6σ because

$$UCL - LCL = (\mu + 3\sigma) - (\mu - 3\sigma) = 6\sigma$$

If all the plots of the sample means fall within the control limits and they follow a random pattern as in the case of Fig. 5.1, the production process is considered as being in statistical control. If a sample is taken and its mean plots outside of the control limits, this would indicate that a special cause of variation has occurred and that the production process needs to be stopped so that the causes of the variation can be investigated.

Example 5.1   A production process is supposed to generate pistons that weigh 10 lb each. Samples of 5 pistons are taken at time intervals of 45 min for inspection and the mean weights of the samples are plotted on the control chart depicted in Fig. 5.2. Every sample mean has fallen within the control limits until sample 33 was plotted. The control chart has shown a stable and in control pattern until sample 33. Since the mean of the sample is way outside of the control limits, something special must have happened to cause that sample to be out of control. The production process needs to be stopped so that the special causes for that variation can be investigated and corrections made. The causes of that variation can be anything from a defective or unattended machine to an operator's mistake. Once the corrections are made, the process should be back in control.

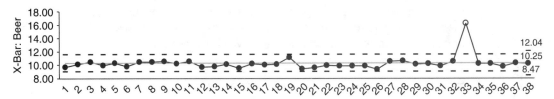

Figure 5.2

Rational subgrouping

One of the main reasons for using control charts is to be able to detect the special causes of variation and to reduce gradually the common causes of variation to an infinitesimal level. Since the purpose of the control charts is to detect the signal factors (the factors that signal to the production supervisor the presence of a special cause of variation), the control charts must be built in such a way that only the noise factors (common causes of variation) determine the levels of the control limits.

The building of control charts is based on samples taken from the processing lines. The sizes of the samples taken and the time intervals between samples can impact the configurations of the charts and thus their interpretation.

According to Dr. Walter Shewhart, samples should be taken based on rational subgroups. This means that samples should be taken in such a way that the variability within them would only reflect the chance variability of the process at the exclusion of the assignable variability.

If the samples taken include both assignable and common causes' observations, the standard deviation used to calculate the control limits would not be the reflection of only the chance variation. The range of the control limits would end up being too wide to detect signal factors.

When items in the same sample are taken within a short time frame, they are more likely to come from the same lot of chance causes. But if the time frame that it takes to complete a subgroup is too wide, the probability for observations reflecting assignable causes of variation being included in the samples would be greater. If the time frame is short, the probability for including special causes of variation within samples will be minimized, while the ability to detect the variations between samples due to those causes will be maximized.

Probability for misinterpreting control charts

The concept of building a control chart is close to performing a hypothesis testing with the null hypothesis suggesting that the process is stable and in control; therefore, all the dots are within the control limits.

$$H_0 : \ Process \ is \ in \ control$$

$$H_a : \ Process \ is \ out \ of \ control$$

The null hypothesis is not rejected as long as all the plotted dots are between the upper and lower control limits. If a dot is outside the control limits, the null hypothesis is rejected and the process is considered out of control. As in the case of any hypothesis testing, opportunities for making errors are also present when interpreting control charts.

Type I error α

One of the purposes of control charts is to reduce the variations to an insignificant level; therefore, the control limits are set to levels where only common causes of variations are tolerated and at the same time the probability for committing a Type I error is minimized.

Based on the rules of hypothesis testing, a Type I error α is founded on rejecting the null hypothesis when in fact it is true, on concluding that the process is out of control when in fact it is in control. The Type I error α is therefore tied to the control limits and its probability can be estimated. The Shewhart control charts are the most commonly used. They are built by using the process mean as the center line and the upper and lower control limits are three standard deviations away from the center line.

If the actual process mean μ and standard deviation σ are known, based on the central limit theorem, the control limits and the center line would be

$$\text{UCL} = \mu + 3 \frac{\sigma}{\sqrt{n}}$$

$$\text{CL} = \mu$$

$$\text{LCL} = \mu - 3 \frac{\sigma}{\sqrt{n}}$$

where n is the number of observations in a subgroup.

If the value of μ is unknown, the mean of the samples' means will be used instead and $\overline{\overline{X}}$ will be used instead of μ.

$$UCL = \overline{\overline{X}} + 3\frac{\sigma}{\sqrt{n}}$$

$$CL = \overline{\overline{X}}$$

$$LCL = \overline{\overline{X}} - 3\frac{\sigma}{\sqrt{n}}$$

where $\overline{\overline{X}} = \frac{\sum \overline{x}}{k}$ for k number of samples.

If the output is normally distributed and a 3-sigma control chart is used, the probability for having a dot within the control limits is 0.9974 and the probability for having a dot outside the control limits is $1 - 0.9974 = 0.0026$ (Fig. 5.3).

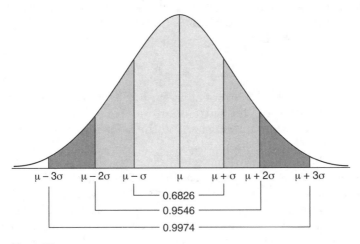

Figure 5.3

The probability of making a Type I error α is therefore 0.0026.

Example 5.2 The weight of the pistons in the previous example is the CTQ characteristic being monitored. The process mean has been determined to be equal to $\mu = 9.25$ and the process standard deviation is $\sigma = 0.05$. A sample of $n = 5$ parts that weigh 10.05, 10, 9.75, 10.05, and 9.89 is taken; based on the central limit theorem, we can derive $\sigma_{\overline{x}}$, the sample's standard deviation.

$$\sigma_{\overline{x}} = \frac{\sigma}{\sqrt{n}} = \frac{0.05}{\sqrt{5}} = \frac{0.05}{2.236} = 0.0224$$

The UCL and the LCL are

$$UCL = 9.25 + 3(0.0224) = 9.3172$$

$$CL = 9.25$$

$$LCL = 9.25 - 3(0.0224) = 9.1828$$

The probability for making a Type I error for this chart is the probability that a dot is plotted outside the control limits. It is

$$1 - P(9.1828 \leq \overline{X} \leq 9.3172)$$

where $\overline{X}$ is the sample mean.

If the scale is 3 sigma and $\overline{X}$ is normally distributed, then $1 - P(9.1828 \leq \overline{X} \leq 9.3172)$ can be estimated as

$$1 - P(-3 \leq Z \leq 3)$$

$$1 - 0.9974 = 0.0026$$

In other words, there is 0.26% chance that a supervisor will be alerted for a process going out of control when in fact the process is performing just fine.

Type II error β

The Type II error (or β error) occurs when we fail to reject the null hypothesis when in actuality it is false. In other words, the control chart indicates that the process is in control when in fact it is not and a supervisor fails to make adjustments on a process when it is warranted.

The β error can be estimated based on the power of a test statistic. The power is interpreted as *correctly* rejecting a false null hypothesis. The power is the probability of indicating that the process is out of control when it really is in control.

$$Power = P(rejecting\ H_0\ when\ H_0\ is\ false)$$

There is a special relationship between the Type II error β and the power.

$$Power = 1 - \beta$$

Example 5.3 The diameter of an O-ring is critical to quality; the production process for manufacturing the O-rings has yielded a mean of 5 cm with a standard deviation of 0.05 cm. The process mean of 5 cm meets engineered standards. After oil was changed on the machine that produces the O-rings, the new process mean has been 5.05 cm with the standard deviation unchanged. A sample of five O-rings is taken.

1. What is the probability that the supervisor should be rightfully alerted to the process going out of control?

2. What is the probability that the supervisor would not be alerted that the process is going out of control when in fact it is?

3. How long should it take the supervisor to realize that the process is out of control?

Solution:

1. This is a case for a power probability, the case where the null hypothesis is correctly rejected when it is false (Fig. 5.4).

$$n = 5 \quad \text{and} \quad \sigma = 0.05$$

$$\sigma_{\bar{x}} = \frac{\sigma}{\sqrt{n}} = \frac{0.05}{\sqrt{5}} = \frac{0.05}{2.236} = 0.0224$$

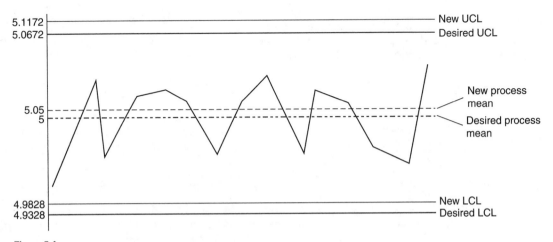

Figure 5.4

The new control limits would be

$$\text{UCL} = 5.05 + 3(0.0224) = 5.1172$$

$$\text{CL} = 5.05$$

$$\text{LCL} = 5.05 - 3(0.0224) = 4.9828$$

The desired control limits will be

$$\text{UCL} = 5 + 3(0.0224) = 5.0672$$

$$\text{CL} = 5$$

$$\text{LCL} = 5 - 3(0.0224) = 4.9328$$

What we are looking for is the probability for a point to fall between the desired UCL and LCL. That probability can be written as

$$p(4.9328 \leq \overline{X} \leq 5.0672)$$

which translates into

$$p\left(\frac{Desired\ LCL - new\ \mu}{\sigma_{\bar{x}}} \le Z \le \frac{Desired\ UCL - new\ \mu}{\sigma_{\bar{x}}}\right)$$

$$p\left(\frac{4.9328 - 5.05}{0.0224} \le Z \le \frac{5.0672 - 5.05}{0.0224}\right)$$

Based on the normal distribution theory, this is equivalent to

$$p(-5.232 \le Z \le 0.768) = \Phi(0.767) - \Phi(-5.323) = 0.777 - 0 = 0.777$$

$\Phi(0.767) = 0.777$ and $\Phi(-5.323) = 0$ are found on the Z table.

The probability that a point would be plotted between the desired control limits is 0.777.

The probability that a point would be plotted outside the desired control limits is $1 - 0.777 = 0.223$. So the power probability, the probability that the supervisor would be rightfully alerted that the process is going out of control, the probability that the null hypothesis is correctly rejected when it is actually true, is 0.223.

2. This is an example of a Type II error β. A false null hypothesis fails to be rejected, the process is believed to be in control when in fact it is not.

Based on the power probability $1 - \beta = 1 - 0.777 = 0.223$, it follows that the probability that the null hypothesis fails to be rejected when it is actually false, in other words, the probability that the supervisor would not be alerted that the process is going out of control when in fact it is $\beta = 0.777$.

3. The time that it takes a control chart to run before an out of control process is detected can also be estimated. The average run length (ARL) measures the average number of samples taken to monitor a process until the process signals that it is operating at a different level from the one it started with. The ARL is based on the probability that a point would be plotted outside the control limits. The probability that the null hypothesis would be correctly rejected when it is actually false is the power probability and it is equal to $1 - \beta$. ARL is the inverse of the power probability and it is, consequently,

$$ARL = \frac{1}{1-\beta} = \frac{1}{0.223} = 4.484$$

It would take 4.484 time measurement units to realize that the process is out of control.

How to determine if the process is out of control—WECO rules

The assumptions thus far have been based on the fact that a process is in control if all the dots are within the control limits and they are randomly spread. Under some circumstances, all the dots can be within the control limits and not be randomly spread. In that case, a close observation of the control chart can reveal that the process is following a pattern that shows that it is about to go out of control.

Figure 5.5 shows a chart with all the dots being within the control limits. However, the quality controller should be alarmed by the trend. If the chart

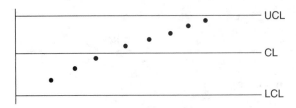

Figure 5.5

is used to monitor the CTQ of a product generated by a machine, this trend could very well be the result of a gradually deteriorating adjustment on the machine.

The patterns of the plots on a chart can be very revealing of a process going out of control, which is why the Western Electric Company (WECO) set out some stringent rules to determine when adjustments would need to be considered based on how the dots are plotted on the control charts. Those rules go beyond just considering whether the dots are within or outside the control limits (Fig. 5.6).

1. Every time a dot is plotted outside the control limits

2. When 2 out of 3 dots are plotted between the control limits and are 2 sigma away from the process mean

3. When 4 out of 5 dots are plotted between 1 and 2 sigma away from the process mean

Any dot above 3 sigma away from the center line

$$UCL = \mu + 3\frac{\sigma}{\sqrt{n}}$$

2 out of the last 3 dots between 2 and 3 sigma away from the center line

$$UCL = \mu + 2\frac{\sigma}{\sqrt{n}}$$

4 out of the last 5 dots between 1 and 2 sigma away from the center line

$$UCL = \mu + \frac{\sigma}{\sqrt{n}}$$

8 consecutive dots on the side of the center line

8 consecutive dots on the side of the center line

$$CL = \mu$$

$$LCL = \mu - \frac{\sigma}{\sqrt{n}}$$

4 out of the last 5 dots between 1 and 2 sigma away from the center line

$$LCL = \mu - 2\frac{\sigma}{\sqrt{n}}$$

2 out of the last 3 dots between 2 and 3 sigma away from the center line

$$LCL = \mu - 3\frac{\sigma}{\sqrt{n}}$$

Any dot below 3 sigma away from the center line

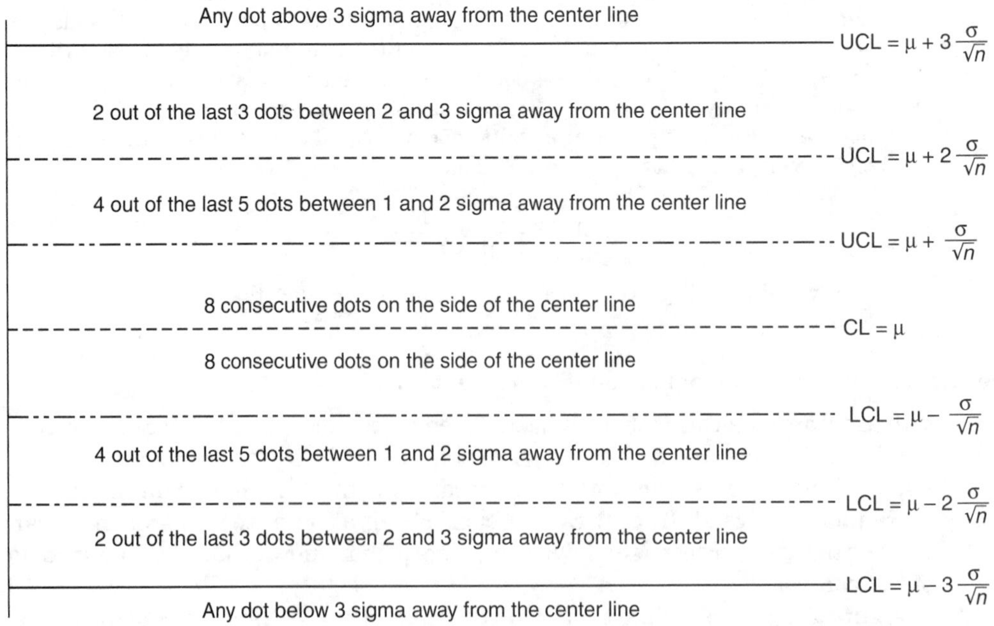

Figure 5.6

4. When 8 consecutive dots are plotted between the center line and 1 sigma away from the center line

5. When 6 dots in a row trend go upward or downward

6. When 14 dots in a row alternating up and down

The WECO rules make the control charts more sensitive to shifts in the process mean and the probability for a Type II error is highly increased. If a process is considered to be out of control only when a dot is plotted outside the control limits, the probability for the process to be out of control would be 0.0026; therefore, the average time before the process signals that it is out of control would be

$$ARL = \frac{1}{0.0026} = 384.6$$

If a sample were taken every hour, a supervisor would be falsely alerted for the process going out of control every 384.6 h.

If the WECO rules were followed, that frequency would be down to 91.75 h. Every 91.75 h, a supervisor would be falsely alerted to the process going out of control.

Categories of Control Charts

The types of control charts used will determine how μ and σ are calculated. There are two main categories of control charts: the univariate control chart, which monitors one CTQ characteristic, and the multivariate control chart, which monitors several CTQ characteristics. Univariate charts are, in general, categorized into attribute and variable control charts.

Variable control charts

The CTQs of most products are measurable. When the quality characteristic being monitored is expressed in continuous measurements, variable control charts are used. Examples of situations where variable control charts are used would be the time that it takes an engine to start running, the weight of an object, the that time it takes an employee to complete a task, or the dimension of a part. Variable control charts are usually paired—a control chart monitoring a process location is associated with one monitoring process variation.

The goal of pairing these charts is to have a simultaneous view of both the changes in the mean (location) and in the standard deviation (the spread) of the process. When a sample is taken from a processing line, the mean of the sample alone is not sufficient to determine the piece-to-piece variations. The spread, the range (the difference between the highest and the lowest values), and the standard deviation within the sample are better estimates of variations within a subgroup.

The most commonly associated control charts are the $\overline{X}$ and R charts and the $\overline{X}$ and s charts.

The mean and range charts— $\overline{X}$ and R charts

The $\overline{X}$ and R charts are paired charts created from CTQ measurements of a process yield to monitor the process central tendency and the process variability. To create $\overline{X}$ and R charts, small samples of the same sizes (usually between 2 and 5) of successive parts are taken at preset time intervals.

The samples are taken, and their mean and range are calculated and plotted on their respective charts. The key factors in creating $\overline{X}$ and R charts are the determination of the sample sizes, the time to order, and the number of samples.

The sample sizes are determined based on rational subgrouping. The samples are chosen in such a way that the probability for having special causes of variation within samples is minimized. In so doing, the range within samples is smaller and only reflects common causes of variation. If only background noises are present within subgroups, then any signal factor that arises between samples would rightfully alert to changes in the process, and prompt for a production stoppage and an investigation for suitable actions. To reduce the opportunities of having mixed common and special causes of variations within the same sample, the pieces of the sample need to be taken at the same time or within a very short time frame. If, for instance, some pieces of a sample were taken at the end of a shift and some pieces at the beginning of the next shift or some before lunch and some after, opportunity for including special causes of variations would more likely be greater. A number of at least 25 samples should be taken initially to provide opportunities for main sources of variations to appear.

The time to order (the frequency of the sampling) should be determined in such a way that it effectively detects signal factors at a reasonable cost and soon enough so that appropriate actions can be taken to bring the process back in control. In the initial phases of the building of the control charts, samples should be taken at shorter time intervals to see how the process behaves. If the process is deemed stable and in control, the time to order (time between two samples) can be gradually increased until it becomes cost effective. Depending on the processes, the time to order can be anywhere between twice per shift to every 30 min.

Calculating the sample statistics to be plotted

The sample statistics plotted on the two charts are $\overline{X}$ and R, with $\overline{X}$ measuring the samples' means (measure of location) and R measuring the samples' ranges (measure of dispersion).

If $X_1, X_2, ..., X_n$ are the pieces contained in a sample being considered, then

$$\overline{X} = \frac{X_1 + X_2 + \cdots + X_n}{n}$$

and

$$R = X_{highest} - X_{lowest}$$

where $X_{highest}$ and X_{lowest} are the highest and the lowest observations, respectively, in the sample.

Example 5.4 A sample of five pieces that weigh 12, 12.03, 12.07, 11.09, and 12.01 are taken from a production line. What are $\overline{X}$ and R?

$$\overline{X} = \frac{X_1 + X_2 + \cdots + X_n}{n} = \frac{12 + 12.03 + 12.07 + 11.09 + 12.01}{5} = \frac{59.2}{5} = 11.84$$

$$R = X_{highest} - X_{lowest} = 12.07 - 11.09 = 0.98$$

Calculating the center line and control limits

To calculate the control limit and the center line, it is customary to start with the range chart first and then the process average chart. If the R chart displays a process out of control due to high variability, there is no need to proceed with the $\overline{X}$ until corrective actions are taken.

Center lines. The center line for the R chart is the mean range and it is obtained by adding the values of the ranges of the samples and dividing them by the number of their scores. For k number of samples,

$$\overline{R} = \frac{R_1 + R_2 + \cdots + R_k}{k}$$

The center line for $\overline{X}$ chart is the mean of the sample means and it is obtained by adding the sample means and then dividing the obtained value by the number of scores.

$$\overline{\overline{X}} = \frac{\overline{X}_1 + \overline{X}_2 + \cdots + \overline{X}_k}{k}$$

Example 5.5 Based on the data in Table 5.1, find the center lines for the $\overline{X}$ and the R charts.

TABLE 5.1

$\overline{X}$	4.99	5.1	5	5	5.1	4.97	5.02	5.01	5.01	4.99	5.01	5.02	5.02	5.1	5.01
R	0.3	0.3	0.3	0.3	0.3	0.29	0.3	0.3	0.3	0.32	0.3	0.3	0.3	0.3	0.3
$\overline{X}$	4.96	5.02	5.03	5	5.03	4.99	4.96	5	5.02	5.02	5.02	5.01			
R	0.31	0.3	0.3	0.3	0.31	0.31	0.31	0.3	0.29	0.31	0.3	0.31			

Solution: The center line for the $\overline{X}$ chart is

$$\overline{\overline{X}} = \frac{\overline{X}_1 + \overline{X}_2 + \cdots + \overline{X}_{27}}{k} = \frac{135.41}{27} = 5.02$$

The center line for the R chart is

$$\overline{R} = \frac{R_1 + R_2 + \cdots + R_{27}}{k} = \frac{8.15}{27} = 0.3$$

Control limits.   Control limits are calculated in such a way that they will contain the process variations if only common causes are present. The upper and lower control limits are three standard deviations away from the center lines.

Control limits for $\overline{X}$ chart

The UCL and LCL for the $\overline{X}$ chart are

$$\text{UCL} = \overline{\overline{X}} + 3\sigma_{\overline{X}}$$

$$\text{CL} = \overline{\overline{X}}$$

$$\text{LCL} = \overline{\overline{X}} - 3\sigma_{\overline{X}}$$

The biggest unknown here is the $\sigma_{\overline{X}}$, the standard deviation of the samples. Its value can be determined in several ways. One way to do it would be through the use of the standard error estimate $\frac{\sigma}{\sqrt{n}}$ based on the central limit theorem and another one would be the use of the mean range.

There is a constant relationship between the mean range and the standard deviation for normally distributed data.

$$\overline{\sigma} = \frac{\overline{R}}{d_2}$$

where the constant d_2 is a function of n.

Standard-error-based $\overline{X}$ chart

The standard-error-based chart is straightforward. Based on the central limit theorem, the standard deviation used for the control limits is nothing but the standard deviation of the process divided by the square root of the sample size. Therefore, we obtain:

$$UCL = \overline{\overline{X}} + 3\left(\frac{\sigma}{\sqrt{n}}\right)$$

$$CL = \overline{\overline{X}}$$

$$LCL = \overline{\overline{X}} - 3\left(\frac{\sigma}{\sqrt{n}}\right)$$

Example 5.6 The weight of washers is critical to quality. The process mean for the washers has been determined to be 12 mm with a standard deviation of 0.03 mm. A sample of five rings is selected for inspection every hour, and the mean of the sample is plotted on an $\overline{X}$ chart. Find the center line, the UCL, and the LCL of the control chart.

$$UCL = 12 + 3\left(\frac{0.03}{\sqrt{5}}\right) = 12 + 3(0.013) = 12.04$$

$$CL = 12$$

$$LCL = 12 - 3\left(\frac{0.03}{\sqrt{5}}\right) = 12 - 3(0.013) = 11.96$$

Since the process standard deviation is seldom known, in theory, these formulas make sense but in actuality, they are impractical. The alternative to that is the use of the mean range.

Mean-range-based $\overline{X}$ control charts

When the sample sizes are relatively small ($n \leq 10$) as it is often the case in SPC, the variations within samples are likely to be small so the range (the difference between the highest and the lowest observed values) can be used to estimate the standard deviation when constructing a control chart.

$$\vec{\sigma} = \frac{\overline{R}}{d_2} \qquad \text{or} \qquad \overline{R} = d_2\vec{\sigma}$$

where $\overline{R}$ is called the relative range and d_2 is a constant that only depends on n. It is found in the control charts constant in Table 5.2.

The mean range is

$$\overline{R} = \frac{R_1 + R_2 + \cdots + R_k}{k}$$

where R_k is the range of the kth sample.

Therefore, the estimator of $\vec{\sigma}$ is $\vec{\sigma} = \dfrac{\overline{R}}{d_2}$ and the estimator of $\dfrac{\vec{\sigma}}{\sqrt{n}} = \dfrac{\overline{R}}{d_2\sqrt{n}}$.

TABLE 5.2   Control Chart Constant

n	A_2	A_3	d_2	D_3	D_4	B_3	B_4
2	1.880	2.659	1.128	–	3.267	–	3.267
3	1.023	1.954	1.693	–	2.574	–	2.568
4	0.729	1.628	2.059	–	2.282	–	2.266
5	0.577	1.427	2.326	–	2.114	–	2.089
6	0.483	1.287	2.534	–	2.004	0.030	1.970
7	0.419	1.182	2.704	0.076	1.924	0.118	1.882
8	0.373	1.099	2.847	0.136	1.864	0.185	1.815
9	0.337	1.032	2.970	0.184	1.816	0.239	1.761
10	0.308	0.975	3.078	0.223	1.777	0.284	1.716
11	0.285	0.927	3.173	0.256	1.744	0.321	1.679
12	0.266	0.886	3.258	0.283	1.717	0.354	1.646
13	0.249	0.850	3.336	0.307	1.693	0.382	1.618
14	0.235	0.817	3.407	0.328	1.672	0.406	1.594
15	0.223	0.789	3.472	0.347	1.653	0.428	1.572
16	0.212	0.763	3.532	0.363	1.637	0.448	1.552
17	0.203	0.739	3.588	0.378	1.622	0.466	1.534
18	0.194	0.718	3.640	0.391	1.608	0.482	1.518
19	0.187	0.698	3.689	0.403	1.597	0.497	1.503
20	0.180	0.680	3.735	0.415	1.585	0.510	1.490
21	0.173	0.663	3.778	0.425	1.575	0.523	1.477
22	0.167	0.647	3.819	0.434	1.566	0.534	1.466
23	0.162	0.633	3.858	0.443	1.557	0.545	1.455
24	0.157	0.619	3.895	0.451	1.548	0.555	1.445
25	0.153	0.606	3.931	0.459	1.541	0.565	1.435

Therefore, the control limits become

$$UCL = \overline{\overline{X}} + \frac{3\overline{R}}{d_2\sqrt{n}}$$

$$CL = \overline{\overline{X}}$$

$$LCL = \overline{\overline{X}} - \frac{3\overline{R}}{d_2\sqrt{n}}$$

These equations can be simplified. If

$$A_2 = \frac{3}{d_2\sqrt{n}}$$

then the formulae for the control limits become

$$UCL = \overline{\overline{X}} + A_2\overline{R}$$

$$CL = \overline{\overline{X}}$$

$$LCL = \overline{\overline{X}} - A_2\overline{R}$$

A_2 is a constant that is function of n and it is found in Table 5.2.

Example 5.7 A sample of five ball joints is taken from a production line for inspection. The weight of the ball joint is the CTQ characteristic of interest. At the end of a day, the measurements obtained are summarized in Table 5.3. Use the data to build an X control chart. The data is in the files *Balljoints.xls* and *balljoint.mtb*.

TABLE 5.3

| Sample numbers | Sample measurements | | | | | $\overline{X}$ | R |
	M1	M2	M3	M4	M5		
1	9.065	10.13	10.343	9.876	9.196	9.722	1.278
2	9.949	10.208	10.349	9.443	10.489	10.088	1.046
3	9.417	9.623	10.267	9.788	10.78	9.975	1.363
4	9.988	9.906	9.656	8.976	10.167	9.739	1.191
5	10.591	9.987	10.266	9.737	9.172	9.951	1.419
6	9.352	9.85	9.909	9.739	9.887	9.747	0.557
7	9.467	10.162	10.271	11.453	9.763	10.223	1.986
8	10.206	10.322	9.752	9.677	10.887	10.169	1.210
9	10.554	10.403	10.345	9.635	9.52	10.091	1.034
10	9.674	9.709	10.532	10.754	11.456	10.425	1.782
11	10.086	9.906	10.733	10.841	10.26	10.365	0.935
12	9.848	8.985	10.295	10.513	9.579	9.844	1.528
13	9.57	10.372	10.572	10.105	10.257	10.175	1.002
14	10.214	10.292	10.459	9.745	10.927	10.327	1.182
15	9.685	10.056	9.462	10.539	9.369	9.822	1.170
16	10.223	9.73	9.94	10.513	10.286	10.138	0.783
17	10.724	10.477	9.762	9.855	9.919	10.147	0.962
18	10.081	10.844	10.256	9.424	10.208	10.163	1.420
19	9.508	9.823	9.741	9.592	10.216	9.776	0.708
20	9.205	10.181	9.772	10.218	10.072	9.890	1.013
21	9.796	9.721	11.01	10.424	9.403	10.071	1.607
22	10.804	10.876	10.679	11.057	9.217	10.527	1.840
23	9.607	9.472	9.497	9.909	9.735	9.644	0.437
24	10.713	10.107	10.112	10.224	9.942	10.220	0.771
Mean						10.052	1.176

Solution: Based on the data, we can determine that

$$\overline{\overline{X}} = 10.052$$

$$\overline{R} = 1.176$$

Since $n = 5$, $A_2 = 0.577$. Therefore,

$$\text{UCL} = 10.052 + 0.577(1.176) = 10.73$$

$$\text{CL} = 10.052$$

$$\text{LCL} = 10.052 - 0.577(1.176) = 9.37$$

SigmaXL can be used to verify the results.

Open the file *Balljoint.xls* (SigmaXL must be opened first). From the menu bar click on SigmaXL and from the SigmaXL menu, click on **Control Charts**. Then from the drop down list, select **X-Bar & R** (Fig. 5.7).

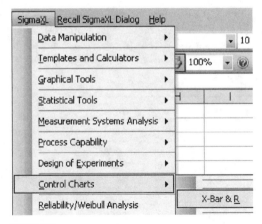

Figure 5.7

The **X-Bar & R** dialog box appears. Select the area that contains the data in the **Please select your data** field including the labels. SigmaXL requires that data labels are used. (Fig. 5.8)

Figure 5.8

Click on the **Next>>** button. The **X-Bar and Range Chart** dialog box appears (Fig. 5.9).

Select the columns one at a time and copy them to the **Numeric Data Variable (Y)>>** field using its button. Then press the **OK>>** button to get the results shown in Fig. 5.10.

The UCL = 10.73, the CL = 10.05, and the LCL = 9.37, which matches what we had found.

Control limits for *R* chart

The R chart is used to monitor changes in the variability of a CTQ of interest. For an R chart, the center line will be $\bar{R}$ and the estimator of sigma is given as $\sigma_R = d_3\sigma$. Since $\sigma = \frac{\bar{R}}{d_2}$, we can replace σ with its value in $\sigma_R = d_3\sigma$ and therefore obtain $\sigma_R = \frac{d_3\bar{R}}{d_2}$.

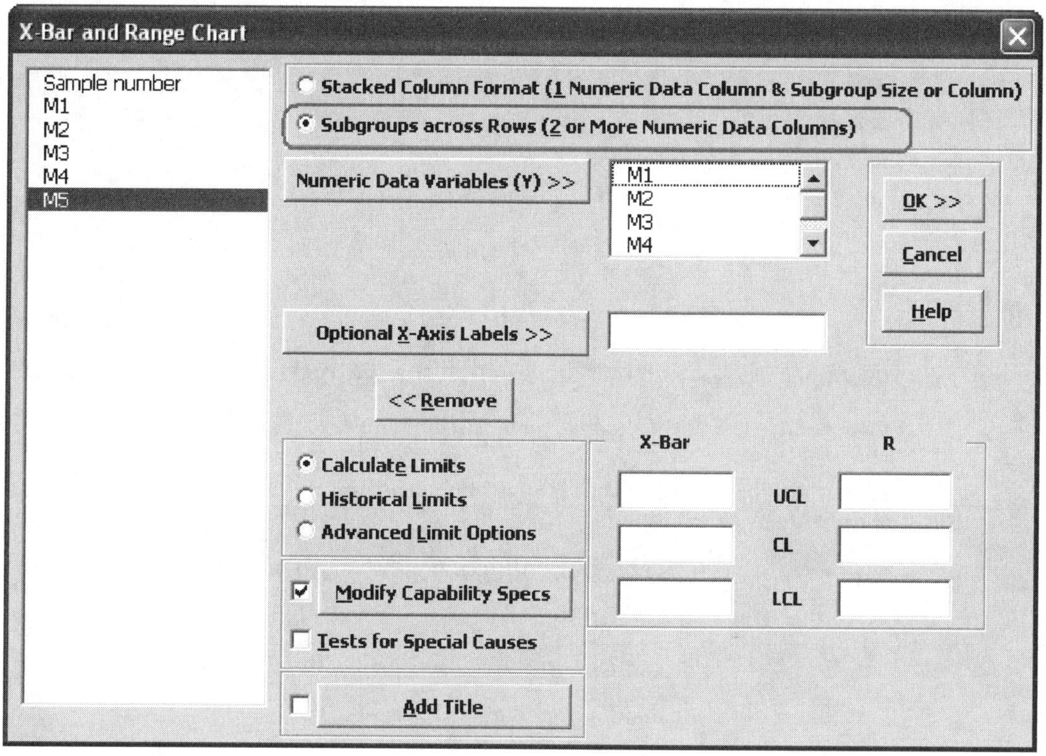

X-Bar and Range Chart

Sample number
M1
M2
M3
M4
M5

○ Stacked Column Format (**1** Numeric Data Column & Subgroup Size or Column)

● Subgroups across Rows (**2** or More Numeric Data Columns)

Numeric Data Variables (Y) >>

M1
M2
M3
M4

OK >>

Cancel

Help

Optional X-Axis Labels >>

<< Remove

● Calculate Limits
○ Historical Limits
○ Advanced Limit Options

☑ **Modify Capability Specs**

☐ **Tests for Special Causes**

☐ **Add Title**

X-Bar   R

UCL

CL

LCL

Figure 5.9

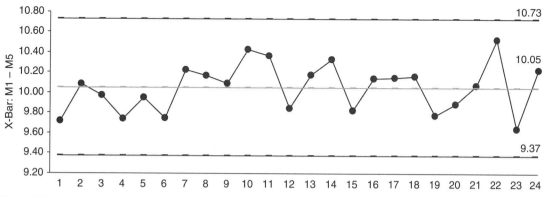

Figure 5.10

Let
$$D_3 = \left(1 - 3\frac{d_3}{d_2}\right) \quad \text{and} \quad D_4 = \left(1 + 3\frac{d_3}{d_2}\right)$$

Therefore, the control limits become

$$\text{UCL} = D_4 \overline{R}$$

$$\text{CL} = \overline{R}$$

$$\text{LCL} = D_3 \overline{R}$$

The values of D_3 and D_4 can be found in the control charts constant table.

Example 5.8   Using the data in Table 5.3, build an R control chart.

Solution: We already know that $\overline{R} = 1.176$ and the sample size is 5; therefore, D_3 does not exist and D_4 is 2.114.

$$\text{UCL} = D_4 \overline{R} = 1.176(2.114) = 2.486$$

$$\text{CL} = 1.176$$

$$\text{LCL} = 0$$

We obtain the same results from SigmaXL with Fig. 5.11.

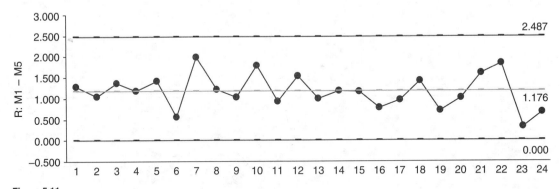

Figure 5.11

To use Minitab, open the file *Balljoint.mtb*. From the menu bar click on **Stat**, then select **Control Charts** and from the submenu, select **Variables Charts for Subgroups** and then **Xbar-R** (Fig. 5.12).

From the Xbar-R chart, select **Observations for a subgroup are in one row of columns,** then select Columns C2 through C6 (Fig. 5.13).

Then click on the **OK** button. Minitab generates the same results with minor differences due to rounding effects (Fig. 5.14).

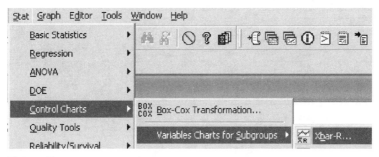

Figure 5.12

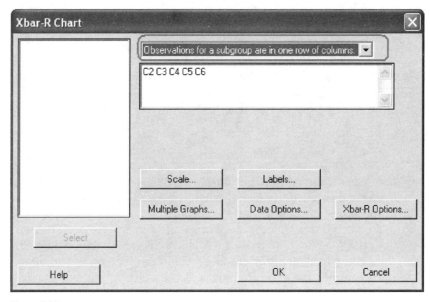

Figure 5.13

We can see that all the dots in both charts are within the control limits and are randomly spread; therefore, we conclude that the process is stable and in control. Note: The computation of R charts can lead to an LCL being negative; in that case, the LCL should be considered equal to zero.

The mean and standard deviation charts $\overline{X}$ and s charts

When the sample sizes are relatively small, the R chart is affective at detecting the presence of special causes of variation. However, one of its inconveniences is that when the samples are big, the R chart still only considers two observations—the highest and the lowest—and it does not account for the ones in between. Consequently, it becomes less affective at assessing variability because the proportion of the observations taken into account by the R chart

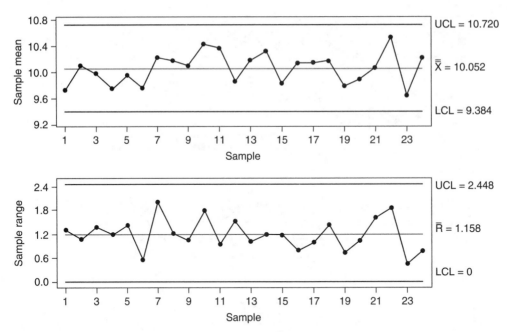

Figure 5.14

becomes smaller. When the sample size is 3, two-thirds of the sample would have been used to calculate R, but when the sample size becomes 20, only one-tenth of the sample would have been considered.

So, when the sample sizes are large, the s charts are usually used in conjunction with the $\overline{X}$ chart instead of the R charts. The R chart measures the range while the s chart measures the standard deviation. The computation of the s charts is more complex than the R chart and the s charts are less sensitive to detecting special causes of variations for small samples. Because the range is the difference between the highest and the lowest observations, any presence of a special cause of variation is easily detected. On the other hand, because the s chart is based on the standard deviation, the presence of a special cause in a sample is less visible in small samples. However, as the sample sizes increase, the s charts become a better estimate than the range. The s charts are usually used to estimate the process variations when the data are recorded using computers because of the complexity involved in the calculations of the standard deviation for each sample.

As in the case of the R chart, when the s chart is used in pair with the $\overline{X}$ chart, the control limits and the center line for $\overline{X}$ are estimated based on the standard deviation s.

Since σ^2, the variance of the process is unknown; it needs to be estimated using the variance of the samples s^2.

$$s^2 = \frac{\sum (x_i - \overline{x})^2}{n-1}$$

Therefore,

$$s = \sqrt{\frac{\sum (x_i - \bar{x})^2}{n-1}}$$

where $\bar{s} = \frac{1}{k} \sum s_i$ is for a number of k samples.

However, using s as an estimator for σ would lead to a biased result. Instead, $c_4 \sigma$ is used, where c_4 is a constant that depends only on the sample size n.

The formula for calculating c_4 is

$$c_4 = \sqrt{\frac{2}{n-1}} \left(\frac{\left(\frac{n-2}{2} \right)!}{\left(\frac{n-3}{2} \right)!} \right)$$

Therefore, c_4 is always a number slightly less than 1.

The expected mean derived from the standard deviation (which is also the center line) will be

$$\bar{s} = E(s) = c_4 \sigma$$

Therefore,

$$\sigma = \frac{\bar{s}}{c_4}$$

and the standard deviation s is

$$\sigma \sqrt{1 - c_4^2}$$

Therefore, the control limits will be

$$\text{UCL} = \bar{S} + 3 \frac{\bar{S}}{c_4} \sqrt{1 - c_4^2}$$

$$\text{CL} = \bar{S}$$

$$\text{LCL} = \bar{S} - 3 \frac{\bar{S}}{c_4} \sqrt{1 - c_4^2}$$

These equations can be simplified by factoring $\bar{s}$. We obtain

$$B_3 = 1 - \frac{3}{c_4}\sqrt{1 - c_4^2}$$

$$B_4 = 1 + \frac{3}{c_4}\sqrt{1 - c_4^2}$$

Therefore,

$$\text{UCL} = B_4 \bar{S}$$

$$\text{CL} = \bar{S}$$

$$\text{LCL} = B_3 \bar{S}$$

The values of B_3 and B_4 are found in the control chart constant table in Table 5.2.

$\bar{S}$ can be used to calculate the control limits for the $\overline{\overline{X}}$ chart paired with the s chart.

The center line will still be $\overline{\overline{X}}$ and the control limits

$$\text{UCL} = \overline{\overline{X}} + 3\frac{\bar{s}}{c_4\sqrt{n}} \qquad \text{and} \qquad \text{LCL} = \overline{\overline{X}} - 3\frac{\bar{s}}{c_4\sqrt{n}}$$

These equalities can also be further simplified. If

$$A_3 = \frac{3}{c_4\sqrt{n}}$$

then

$$\text{UCL} = \overline{\overline{X}} + \bar{S}A_3$$

$$\text{CL} = \overline{\overline{X}}$$

$$\text{LCL} = \overline{\overline{X}} - \bar{S}A_3$$

Example 5.9 The length of a gasket is critical to quality. Every hour, 10 gaskets inspected. The data in Table 5.4 summarize the observations obtained at the end of the first shift; build an s chart and an $\overline{X}$ chart.

TABLE 5.4

Sample ID	M1	M2	M3	M4	M5	M6	M7	M8	M9	M10	s bar	X bar
1	4.732	4.537	5.098	5.041	4.728	4.631	5.042	4.785	5.100	5.289	0.246	4.898
2	4.945	5.262	4.833	5.324	4.905	5.054	5.176	4.853	4.820	5.166	0.188	5.034
3	5.326	5.251	5.554	5.127	4.887	5.133	5.235	5.223	4.777	4.742	0.255	5.125
4	5.019	5.255	4.887	4.950	4.866	5.468	5.214	5.236	5.045	5.116	0.189	5.105
5	5.301	4.526	5.070	4.981	4.629	4.942	5.166	5.100	5.148	5.003	0.240	4.987
6	4.654	4.957	4.775	4.908	5.099	4.703	4.961	4.677	4.856	5.084	0.162	4.867
7	4.886	4.930	4.765	4.913	5.020	5.003	4.957	4.969	5.020	5.300	0.137	4.976
8	5.067	5.259	4.892	5.061	5.161	4.866	5.128	5.058	4.929	4.864	0.136	5.029
Mean											**0.194**	**5.003**

Control limits for the *s* chart   Since $n = 10$,

$$B_3 = 0.284$$

$$B_4 = 1.716$$

$$\bar{s} = 0.194$$

Consequently, the control limits for the *s* chart are

$$\text{UCL} = B_4 \bar{s} = 1.716(0.194) = 0.333$$

$$\text{CL} = 0.194$$

$$\text{LCL} = B_3 \bar{s} = 0.284(0.194) = 0.055$$

Control limits for the $\overline{X}$ chart   The control limits for the $\overline{X}$ chart are

$$A_3 = 0.975$$

$$\overline{\overline{X}} = 5.003$$

$$\text{UCL} = \overline{\overline{X}} + \bar{s}A_3 = 5.003 + 0.194(0.975) = 5.192$$

$$\text{CL} = \overline{\overline{X}} = 5.003$$

$$\text{LCL} = \overline{\overline{X}} - \bar{s}A_3 = 5.003 - 0.194(0.975) = 4.813$$

We can verify the results using SigmaXL. We follow the same process as the one for finding the *R* chart.

The SigmaXL output shows that all the dots in both charts are well within the control limits and the process is in control (Fig. 5.15 and Fig. 5.16).

Individual Values Control Charts

Under some circumstances, subgroups with several observations for control charts are not possible. If we want to track the daily productivity of a shift supervisor on a control chart, it would not be practical to take samples made

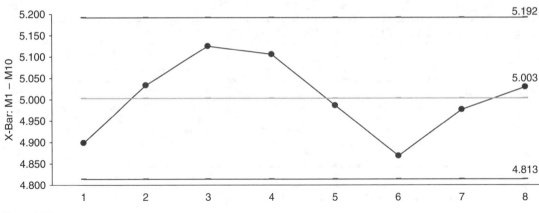

Figure 5.15

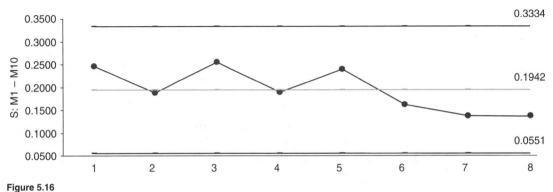

Figure 5.16

up of days of work. Instead, the daily productivities should be tracked on an individual basis. If a quality controller wants to test the chemical reaction to fire for an expensive product, taking more than one piece at a time for testing might end up being exorbitant; in that case, it may be more practical to use samples made up of individual units in order to reduce cost.

Since the samples are made up of one observation, finding the range or the standard deviation to assess the variability within subgroups is excluded. In this case, the variability can be assessed based on the variations between every two individual samples or it can be assessed base on the standard deviation of a set of samples.

Individual Moving Range Charts

Individual moving range (IMR) monitors the variability between two consecutive observations. The variability between two readings is the absolute value of their difference.

$$MR = |X_{i+1} - X_i|$$

where X_i is the initial reading and X_{i+1} is the subsequent reading.

The center line will be the mean of the obtained MR values. For k number of observations,

$$CL = \overline{MR} = \frac{\sum MR}{k-1}$$

The upper and lower control limits are also three standard deviations away from the center line. Since the range is always a difference between two observations, the control limits are estimated using the constants D_3 and D_4.

$$UCL = D_4 \overline{MR}$$

$$CL = \overline{MR}$$

$$LCL = D_3 \overline{MR}$$

Since D_3 and D_4 that depend on $n = 2$ are known constants, we can replace them by their values.

$$D_3 = 0$$

$$UCL = 3.267 \overline{MR}$$

$$CL = \overline{MR}$$

$$LCL = 0$$

Individual Value Chart

For individual value charts, the center line is represented by $\overline{X}$ (not $\overline{\overline{X}}$), the process average.

$$\overline{X} = \frac{\sum X_i}{k}$$

The control limits are still 3 sigma away from the center line. Here again, the $3s$ are estimated from $\overline{MR}/d_2$.

$$UCL = \overline{X} + 3\frac{\overline{MR}}{d_2}$$

$$CL = \overline{X}$$

$$LCL = \overline{X} - 3\frac{\overline{MR}}{d_2}$$

Since for $n = 2$, d_2 is a known constant equal to 1.128, $\frac{3}{d_2}$ is a constant, represented by E_2 on the constant table, equal to 2.66. Thus E_2 can be used to estimate the control limits.

$$UCL = \overline{X} + 2.66\overline{MR}$$

$$CL = \overline{X}$$

$$LCL = \overline{X} - 2.66\overline{MR}$$

Example 5.10   The resistance of a flexible alloy is critical to quality. The production of the alloy is very expensive. Therefore, when monitoring the production process, only samples of one unit are taken every hour for a destructive testing. The data in Table 5.5 summarize the observations obtained in one day of process monitoring. Find the UCL, the CL, and the LCL based on the data. The tables are contained in the files *Strength.xls* and *Strength.mtb*.

TABLE 5.5

Sample ID	Strength in PSI	Ranges
1	149.83	
2	152.002	2.172
3	152.183	0.181
4	147.914	4.269
5	153.958	6.044
6	148.682	5.276
7	151.101	2.419
8	149.361	1.74
9	150.899	1.538
10	148.862	2.037
11	149.623	0.761
12	151.892	2.269
13	151.444	0.448
14	151.481	0.037
15	153.278	1.797
16	152.396	0.882
17	149.542	2.854
18	147.514	2.028
19	151.254	3.74
20	154.396	3.142
21	148.058	6.338
22	149.598	1.54
23	151.435	1.837
24	150.248	1.187
Means	**150.7062917**	**2.37113**

Solution: For the IMR

$$UCL = 3.267\overline{MR} = 3.26(2.37113) = 7.73$$

$$CL = \overline{MR} = 2.37113$$

$$LCL = 0$$

For the individual value chart

$$\mathrm{UCL} = \overline{X} + 2.66\overline{\mathrm{MR}} = 150.7063 + 2.66(2.37113) = 157.014$$

$$\mathrm{CL} = \overline{X} = 150.7063$$

$$\mathrm{LCL} = \overline{X} - 2.66\overline{\mathrm{MR}} = 150.7063 - 2.66(2.37113) = 144.4$$

SigmaXL output shows that both charts (Fig. 5.17*a* and *b*) display a stable and in control process.

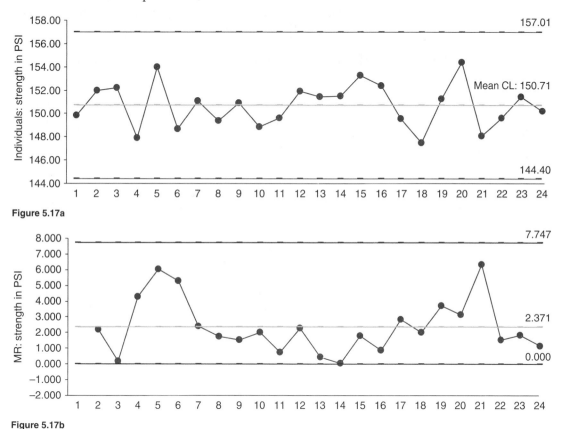

Figure 5.17a

Figure 5.17b

Monitoring Shifts in the Process Mean

Thus far, unless the WECO rules are applied, the decision to determine if a process is in control or not has been solely based on whether the last dot plotted on the control chart is within the control limits. The last dot plotted on the chart is independent from the previous ones and is consequently not affected by previous observations.

Small incremental changes in the distribution of the plotted CTQ that eventually lead to a drift in the process mean (of less than 1.5σ) may go unnoticed when the Shewhart charts are used. In order to add sensitivity to small shifts in the process mean to the control charts, two types of charts are used: the cumulative sum (CUSUM) and the exponentially weighted moving average (EWMA). These two types of charts take into account the recent past data.

CUSUM

The CUSUM charts consider the samples mean deviations from the process target. It plots the cumulative sum of the deviations. Two approaches to CUSUM charts are generally considered: the graphical procedure (V-mask) and the computational procedure. The two are actually equivalent with the graphical version being a visual representation of the computational version.

The V-mask approach. Let us suppose that μ_0 is the process target and $\overline{X}_i$ is the ith sample mean.

$$s_1 = \overline{X}_1 - \mu_0$$

$$s_2 = \sum (\overline{X}_i - \mu_0) = (\overline{X}_1 - \mu_0) + (\overline{X}_2 - \mu_0)$$

$$\vdots$$

$$s_j = \sum (\overline{X}_i - \mu_0) = (\overline{X}_1 - \mu_0) + (\overline{X}_2 - \mu_0) + \cdots + (\overline{X}_j - \mu_0)$$

s_j is the cumulative sum up to the jth sample and that is what is plotted on the CUSUM chart at time t. The cumulative sum s_j includes not only the last sample's deviation from the target but also the cumulative deviations of all the previous ones. For that reason, it is a better tool for estimating the overall deviation from the process target. If the process is in control, if it <u>is</u> not deviating from its target, the cumulative sum should be close to zero ($\overline{X}_i \approx \mu_0$). The process is considered as going out of control if a drift in the cumulative sums drives to an upward or downward process shift.

   Thus far, we have not mentioned any control limits; it is because the CUSUM chart does not have control limits. One way to determine if the process is in control or not is through the use of the V-mask. A V-mask is a transparent plastic overlay with a V shape drawn on it (Fig. 5.18).

   Every time a new dot is plotted on the chart, the mask is laid on it with the point on the V-mask applied on top of the last plotted s_j dot on the chart. The

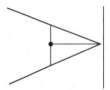

Figure 5.18

line on the mask that goes from the point to the tip of the mask should be parallel to the horizontal line of the chart. The process is deemed to be in control if all the previous dots on the chart are in-between the two arms of the V (Fig. 5.19). If a dot appears to be outside the two arms, the process is considered out of control (Figs. 5.20 and 5.21).

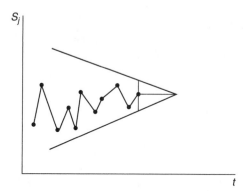

Figure 5.19 Process in control.

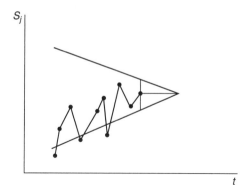

Figure 5.20 Upward drift.

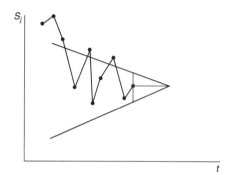

Figure 5.21 Downward drift.

Example 5.11   The amount of sugar contained in a bottle of Welayarde, a fruit juice, is critical to quality. The target has been set at 11.75 mg. The data contained in Table 5.6 represent samples that have been taken over two 10-h shifts. The quality controller wants to determine if the process is in control and at the same time ensure that it has not deviated from the preset target. She chooses to use CUSUM charts to monitor the process mean.

Solution: The first step in creating a CUSUM chart is to determine the mean for each sample, then determine each sample's deviation from the target of 11.75. Then the accumulation of the deviations is determined in the last column.

TABLE 5.6

Sample ID	M1	M2	M3	M4	M5	$\bar{X}$	$\bar{X} - 11.75$	$\sum(\bar{X} - 11.75)$
1	11.532	11.535	11.979	12.641	13.134	12.164	0.414	0.414
2	10.784	10.23	12.39	11.078	12.546	11.406	−0.344	0.07
3	12.449	11.032	10.823	11.999	11.241	11.509	−0.241	−0.171
4	12.005	10.339	11.232	12.794	10.353	11.344	−0.406	−0.577
5	11.743	10.209	12.655	11.739	11.133	11.496	−0.254	−0.831
6	10.398	10.649	11.334	10.569	12.821	11.154	−0.596	−1.427
7	11.901	11.432	11.148	12.231	11.342	11.611	−0.139	−1.566
8	11.454	11.552	10.568	10.756	11.363	11.139	−0.611	−2.177
9	12.985	10.341	10.434	13.154	11.543	11.691	−0.059	−2.236
10	12.836	12.112	12.772	11.827	10.792	12.068	0.318	−1.918
11	12.64	11.596	11.067	12.786	10.399	11.697	−0.053	−1.971
12	11.864	11.511	11.551	10.81	11.529	11.453	−0.297	−2.268
13	9.702	12.588	12.062	10.605	11.42	11.275	−0.475	−2.742
14	13.885	14.241	13.274	12.152	13.319	13.374	1.624	−1.118
15	10.727	13.623	13.852	12.198	10.686	12.217	0.467	−0.651
16	11.403	10.593	11.348	11.648	13.118	11.622	−0.128	−0.779
17	10.221	11.577	12.83	13.243	11.35	11.844	0.094	−0.685
18	12.296	11.358	12.078	12.324	10.618	11.735	−0.015	−0.7
19	12.61	12.064	13.524	12.599	11.557	12.471	0.721	0.021
20	10.937	13.499	11.273	12.37	13.259	12.268	0.518	0.538

The CUSUM chart in Fig. 5.22 is obtained from Microsoft Excel chart wizard by simply charting the last column in Table 5.6. We applied the V-mask to it.

Using Minitab   Open the file *Welayarde.mtb*, then from the menu bar, click on **Stat**, then choose **Control Charts** from the menu, and select **Time-Weighted Charts** from the submenu, and then **CUSUM** from the drop down list (Fig. 5.23).
   The **CUSUM Chart** dialog box appears. Make sure that the **Observations for a subgroup are in row of columns** is selected. Type 11.75 in the **Target** field before pressing on the **CUSUM Options** button (Fig. 5.24).
   The **CUSUM Chart–Options** dialog box appears. Select the **Plan/Type** tab and select the **Two-sided (V-mask)** option (Fig. 5.25).
   Press **OK** and then **OK** again to generate Fig. 5.26.
   The graph matches the one we created with Excel and it shows that the process is barely out of control with one dot slightly on the bottom arm of the V.

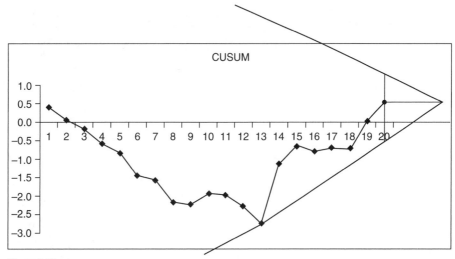

Figure 5.22

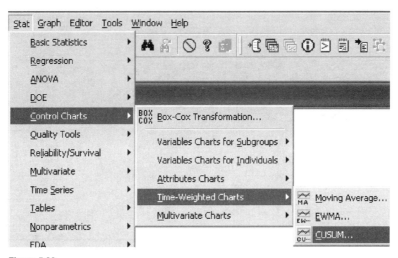

Figure 5.23

Computational Approach

The advantage of using the V-mask is that it allows a visual representation of the trends just as the Shewhart control charts. With the use of some software, another way of assessing CUSUM charts has been made a lot easier. The use of the computational method is better done using a computer because of the complexity involved in manipulating the formula. Three output indices are

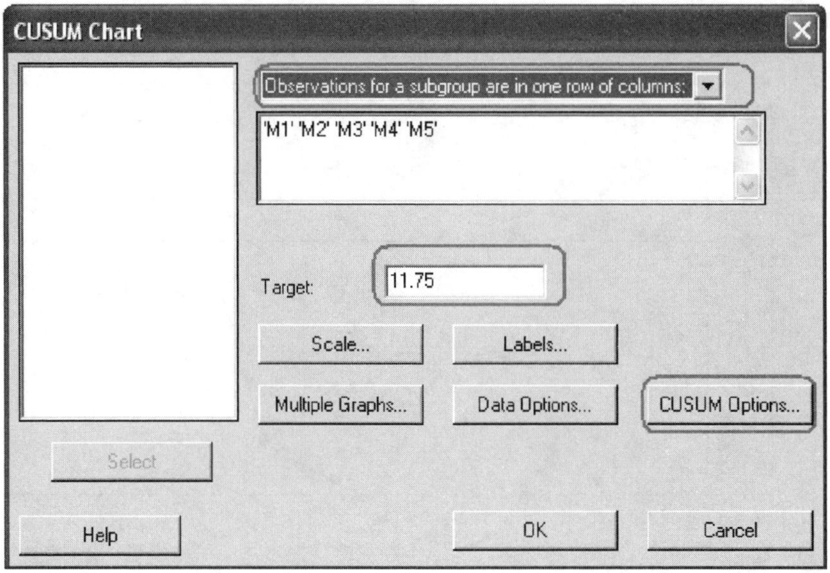

Figure 5.24

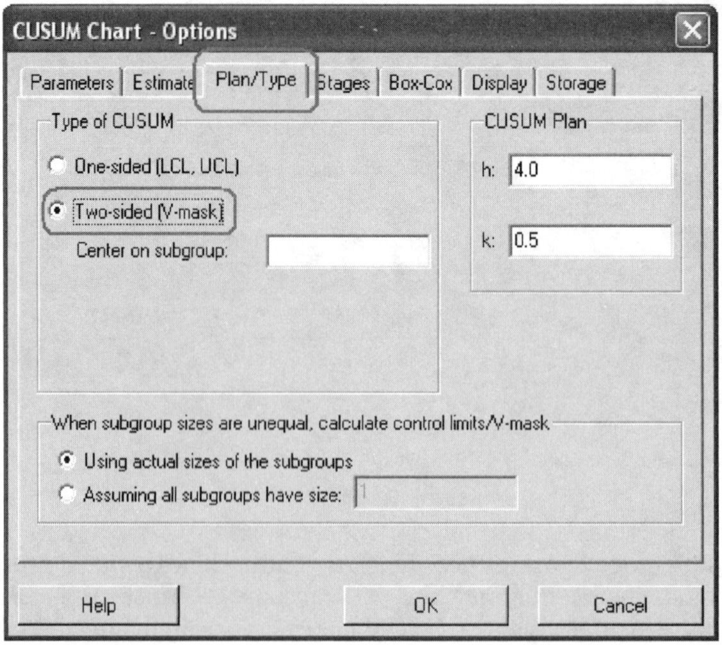

Figure 5.25

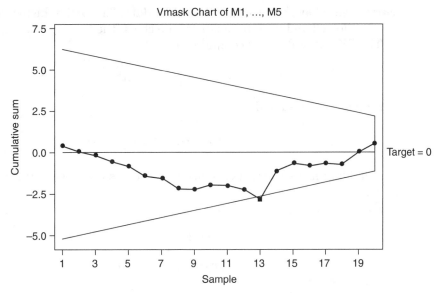

Figure 5.26

of interest when using the computational (or tabular) method for making a CUSUM charts,

$$V_u, V_l, \text{ and } H$$

where V_u is the upper one-sided CUSUM, V_l is the lower one-sided CUSUM, and H is the decision interval.

The process is deemed out of control when V_u or V_l is greater than H.

$$V_u(i) = \max(0, V_{u(i-1)} + (\bar{x}_i - \mu_0 - k))$$

$$V_l(i) = \max(0, V_{l(i-1)} - (\bar{x}_i - \mu_0 + k))$$

k is called the reference value. It is the slope of the lower arm of the V-mask.

$$k = \frac{\Delta}{2} = \frac{\mu_1 - \mu_0}{2}$$

μ_0 is the target and μ_1 indicates a state of out-of-control. The value of K therefore represents the midrange of an out-of-control shift.

Exponentially Weighted Moving Average

The exponentially weighted moving average (EWMA) is another way of monitoring small shifts in the process mean away from its target. EWMA statistics monitor a process mean by averaging the data in such a way that the most current data have more weight than older data. The further historically the data were collected and used to construct the chart, the less significant they are.

The EWMA statistic is computed as

$$w_i = \lambda \overline{X}_i + (1-\lambda)w_{i-1} \qquad \text{with} \qquad 0 > \alpha > 1$$

if samples are used and

$$w_i = \lambda x_i + (1-\lambda)w_{i-1}$$

if individual measurements are used, where λ is the weight (it is generally set at 0.2).

For

$$i = 1$$

$$w_1 = \lambda \overline{X}_1 + (1-\lambda)w_0$$

For

$$i = 2$$

$$w_2 = \lambda \overline{X}_2 + (1-\lambda)w_1 = \lambda \overline{X}_2 + (1-\lambda)[\lambda \overline{X}_1 + (1-\lambda)w_0]$$

Therefore

$$w_2 = \lambda \overline{X}_2 + \lambda \overline{X}_1(1-\lambda) + (1-\lambda)^2 w_0$$

Because $0 < \lambda < 1$, the further the data, the smaller n the exponent of $(1-\lambda)^n$ and, consequently, the smaller their coefficient.

If individual measurements are used and the process target is unknown, then $\overline{X}$ the mean of the initial data can be used as the first value of w_i and when samples are used, $\overline{\overline{X}}$ can be considered.

The standard deviation is

$$\sigma = \sqrt{\left(\frac{\lambda}{2-\lambda}\right)[1-(1-\lambda)^{2i}]}$$

As i increases, $(1-\lambda)^{2i}$ becomes smaller, when i reaches 5, $(1-\lambda)^{2i}$ becomes insignificant, and

$$\sigma = \sqrt{\left(\frac{\lambda}{2-\lambda}\right)}$$

When $i < 5$, the upper and lower control limits are

$$\text{UCL} = \overline{\overline{X}} + 3\frac{\overline{s}}{c_4\sqrt{n}}\sqrt{\left(\frac{\lambda}{2-\lambda}\right)[1-(1-\lambda)^{2i}]}$$

$$\text{LCL} = \overline{\overline{X}} - 3\frac{\overline{s}}{c_4\sqrt{n}}\sqrt{\left(\frac{\lambda}{2-\lambda}\right)[1-(1-\lambda)^{2i}]}$$

When $i > 5$

$$\text{UCL} = \overline{\overline{X}} + 3\frac{\overline{s}}{c_4\sqrt{n}}\sqrt{\left(\frac{\lambda}{2-\lambda}\right)}$$

$$\text{LCL} = \overline{\overline{X}} - 3\frac{\overline{s}}{c_4\sqrt{n}}\sqrt{\left(\frac{\lambda}{2-\lambda}\right)}$$

Example 5.12   The amount of fat is critical to quality for each bar of soap produced at Sogui-Sabunde. The data contained in Table 5.7 were collected over two shifts of production. The data in the table can be found in the file *Sogui sabunde.mtb*.

1. Using EWMA, determine if the process is in control using the samples' historic mean as a process target.

2. What would we say about the process if the process target were 50.5 g of fat per bar?

TABLE 5.7

Sample ID	M1	M2	M3
1	49.08	51.4749	49.4695
2	50.871	50.5511	50.524
3	49.368	49.5386	50.694
4	48.882	50.5051	50.9733
5	50.112	50.0821	50.7661
6	48.921	49.7529	50.8386
7	50.862	48.8734	49.6406
8	50.324	49.5912	51.0464
9	48.566	49.6555	49.3142
10	49.505	50.5136	47.6921
11	49.363	51.0993	51.9579
12	49.447	47.848	48.9557
13	49.156	50.271	49.3338
14	50.062	50.0566	49.7153
15	51.044	51.5204	51.4028
16	50.5	48.2541	49.7105

Using Minitab

1. Open the file *Sogui Sabunde.MTB*, then from the menu bar, click on **Stat**, then choose **Control Charts** from the menu, select **Time-Weighted Charts** from the submenu, and then **EWMA** from the drop down list.

   The **EWMA Chart** dialog box appears. Make sure that the **Observations for a subgroup are in row of columns** is selected. Select M1, M2, and M3 for the Observations' field. The **Weight of EWMA** should already be defaulted to "0.2." Press the **OK** button to get Fig. 5.27.

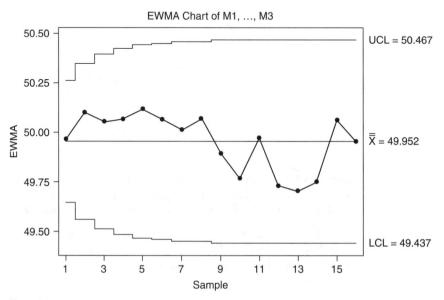

Figure 5.27

The graph shows that the process mean is $\overline{\overline{X}} = 49.952$ if the samples' averages are used to estimate it. The UCL and the LCL are 50.467 and 49.437, respectively. The process is in control and stable.

2. The same procedure is used with Minitab except that in the **EWMA Chart** dialog box, the EWMA Options button is pressed (Fig. 5.28) and "50.5" is typed in the **Mean** field (Fig. 5.29).

   Figure 5.30 shows that the process is out of control if the mean is set at 50.5 with the process mean shifting downward.

Attribute Control Charts

The variable control charts are used to monitor continuous measurements but all process yields are not always expressed in the form of continuous data. Some data used to monitor production processes take the form of attribute data.

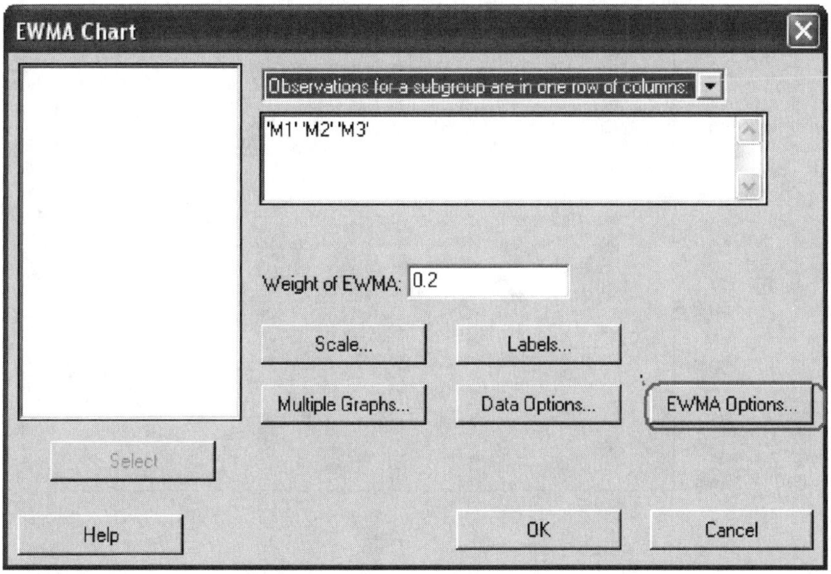

Figure 5.28

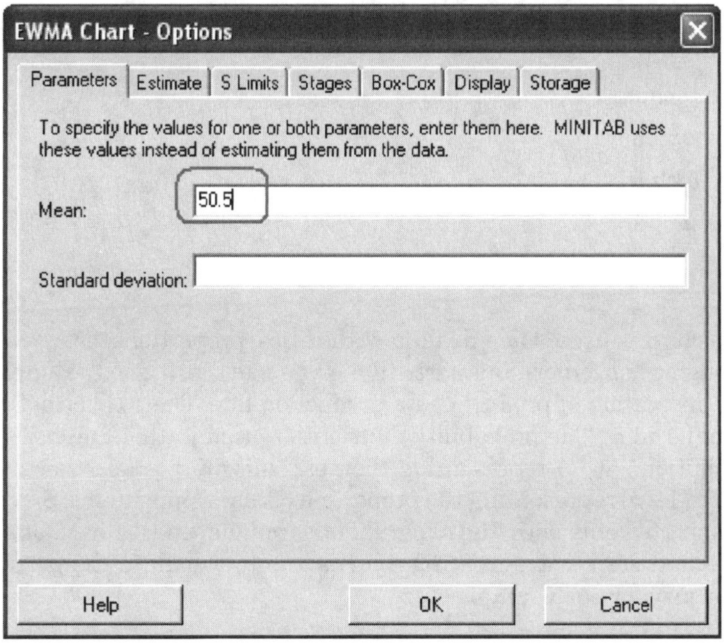

Figure 5.29

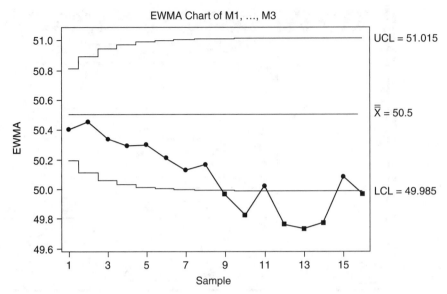

Figure 5.30

Attribute characteristics resemble binary data—they can only take one of two given forms. In quality control, the most common attribute characteristics used are "conforming" or "not conforming," "good" or "bad." Attribute data need to be transformed into discrete data to be meaningful.

The types of charts used for attribute data are

- The **p** chart
- The **np** chart
- The **c** chart
- The **u** chart

The *p* chart

The *p* chart is used when dealing with ratios, proportions, or percentages of conforming or nonconforming parts in a given sample. A good example of a *p* chart is the inspection of products on a production line. They are either conforming or nonconforming. The probability distribution used in this context is the binomial distribution with *p* representing the nonconforming proportion and *q* (which is equal to $1 - p$) representing the proportion of conforming items. Since the products are inspected only once, the experiments are independent from one another.

The first step when creating a *p* chart is to calculate the proportion of nonconformity for each sample.

$$p = \frac{m}{b}$$

where m represents the number of nonconforming items, b is the number of items in the sample, and p is the proportion of nonconformity.

$$\overline{p} = \frac{p_1 + p_2 + \cdots + p_k}{k}$$

where $\overline{p}$ is the mean proportion, k is the number of samples audited, and p is the kth proportion obtained.

The standard deviation of the distribution is

$$\sqrt{\frac{\overline{p}(1 - \overline{p})}{n}}$$

The control limits of a p chart are

$$\text{UCL} = \overline{p} + 3\sqrt{\frac{\overline{p}(1 - \overline{p})}{n}}$$

$$\text{CL} = \overline{p}$$

$$\text{LCL} = \overline{p} - 3\sqrt{\frac{\overline{p}(1 - \overline{p})}{n}}$$

and $\overline{p}$ represents the center line.

Example 5.13   Table 5.8 contains the number of defects found on 27 lots taken from a production line over a period at Podor Tires. We want to build a control chart that monitors the proportions of defects found in each sample taken. The data is contained in the files *Podor.xls* and *Podor.mtb*.

TABLE 5.8

Defects found	Lots inspected	Defects found	Lots inspected	Defects found	Lots inspected
1	25	1	28	0	24
2	21	2	24	2	29
2	19	1	29	2	20
2	25	2	23	2	17
2	24	3	23	2	20
3	26	3	23		
3	19	1	32		
3	24	2	19		
1	21	3	20		
1	27	3	20		
2	26	2	20		

Using SigmaXL, open the file *Podor.xls* and select all the fields containing the data. Then from the menu bar, click on *SigmaXL*, then select *Control Charts*, and then *P*. The **p Chart** dialog box should appear with the selected area in the **Please select your data field**. Press the **Next** button. Fill out the next dialog box as indicated in Fig. 5.31 and then press **OK**.

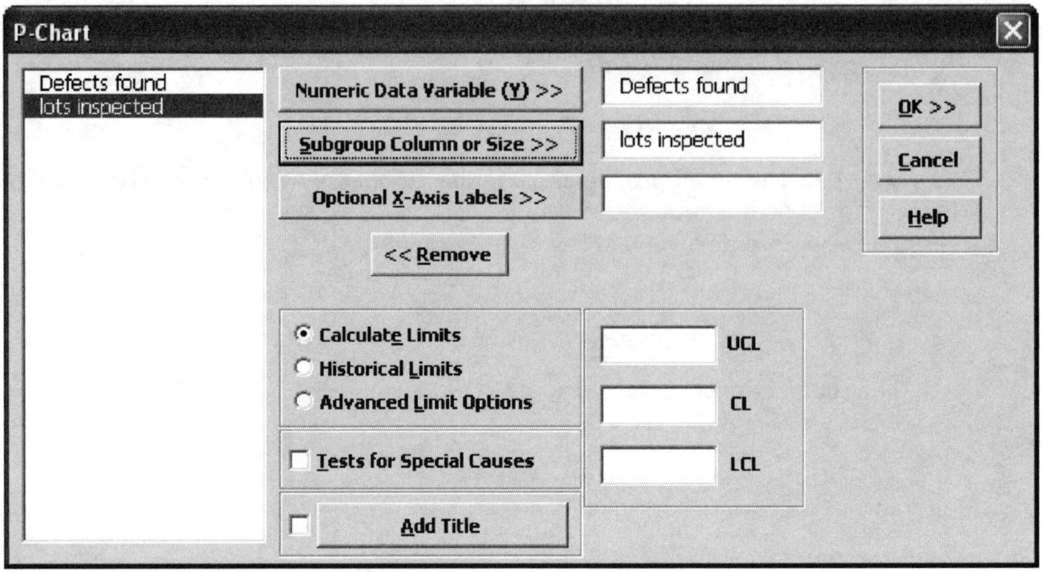

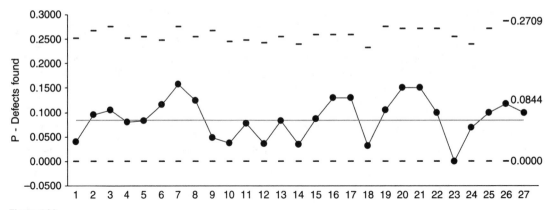

Figure 5.31

Figure 5.32

The control chart in Fig. 5.32 should show.

Using Minitab   Open the file *Podor tire.MPJ*. From the menu bar, click on **Stat,** then select **Control Charts,** then select **Attributes charts,** and click on **P.** Select Defects for the **Variables** and Lots inspected for the **Subgroup Sizes** and then click on the **OK** button to obtain Fig. 5.33.

What are plotted on the chart are not the defects or the sample sizes but rather the proportions of defects found on the samples taken. In this case, we can say that the process is stable and under control since all the plots are

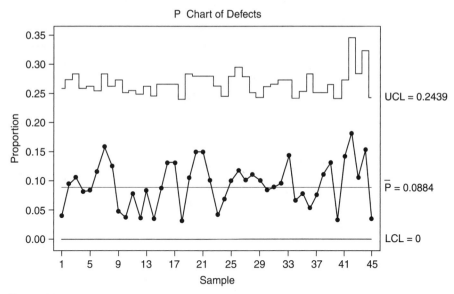

Figure 5.33

within the control limits and the variation exhibits a random pattern around the mean.

One of the advantages of using the p chart is that the variations of the process change with the sizes of the samples and the defects found on each sample.

The *np* chart

The np chart is one of the easiest to build. While the p chart tracks the proportion of nonconformities per sample, the np chart plots the number of nonconforming items per sample. The expected outcome is "good" or "bad," and therefore the mean number of success is np. The control limits for an np chart are as follows:

$$UCL = n\overline{p} + 3\sqrt{n\overline{p}(1 - \overline{p})}$$

$$CL = n\overline{p}$$

$$LCL = n\overline{p} - 3\sqrt{n\overline{p}(1 - \overline{p})}$$

Using the same data on the file *Podor Tires.MPJ* and the same process that was used to build the p chart above, we can construct the np control chart in Fig. 5.34.

Note that the pattern of the chart does not take into account the sample sizes; it just shows how many defects there are on a sample. Sample 2 was of size 21 and had 2 defects and sample 34 was of size 31 and had 2 defects and they are both plotted at the same level on the chart. The chart does not plot the defects

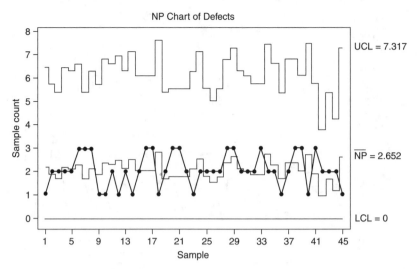

Figure 5.34

relative to the sizes of the samples from which they are taken. For that reason, the *p* chart has superiority over the *np* chart.

Let us consider the same data used to build the chart in Fig. 5.34 with all the samples being equal to 5. We obtain the chart in Fig. 5.35.

These two charts are patterned the same way with two minor differences being the UCL and the center line. If the sample size for the *p* chart is a constant,

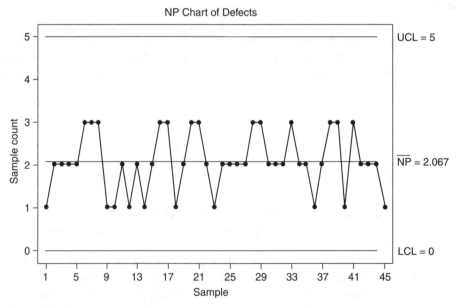

Figure 5.35

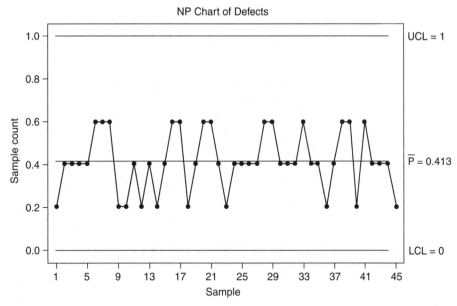

Figure 5.36

the trends for the p chart and the np chart would be identical but the control limits would be different. The p chart in Fig. 5.36 depicts the same data used previously with all the sample sizes being equal to 5.

The c chart

The c chart monitors the process variations due to the fluctuations of defects per item or group of items. The c chart is useful for the process engineer to know not just how many items are not conforming but how many defects there are per item. Knowing how many defects there are on a given part produced on a line might, in some cases, be as important as knowing how many parts are defective. Here, nonconformance must be distinguished from defective items because there can be several nonconformances on a single defective item.

The probability for a nonconformance to be found on an item in this case follows a Poisson distribution. If the sample size does not change and the defects on the items are easy to count, the c chart becomes an effective tool to monitor the quality of the production process. If c is the average nonconformity on a sample, the UCL and the LCL limits will be given as follows for a k-sigma control chart:

$$\text{UCL} = \bar{c} + 3\sqrt{\bar{c}}$$

$$\text{CL} = \bar{c}$$

$$\text{LCL} = \bar{c} - 3\sqrt{\bar{c}}$$

where

$$\bar{c} = \frac{c_1 + c_2 + \cdots + c_k}{k}$$

Example 5.14 Saloum Electrical makes circuit boards for TVs. Each board has 3542 parts. The engineered specification is to have no more than five cosmetic defects per board. The table in the worksheet in the file *Saloum Electrical.xls* contains samples of boards taken for inspection and the number of defects found on them. We want to build a control chart to monitor the production process and determine if it is stable and under control.

Solution: The SigmaXL graph in Fig. 5.37 shows a stable and in control process up to Sample 65. Sample 65 is beyond 3 standard deviations from the mean. Something special must have happened that caused it to be so far out of the control limits. The process needs to be investigated to determine the causes of that deviation and corrective actions taken to bring the process back under control.

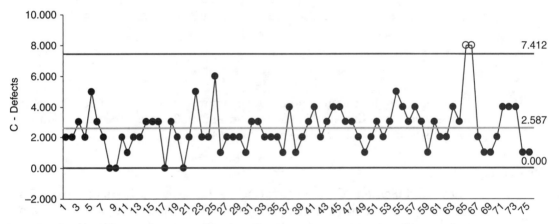

Figure 5.37

The *u* chart

One of the premises for a c chart was that the sample sizes had to be the same. The sample sizes can vary when the u chart is being used to monitor the quality of the production process and the u chart does not require any limit to the number of potential defects. Furthermore, for a p chart or an np chart the number of nonconformances cannot exceed the number of items on a sample. However, for a u chart, it is conceivable because what is being addressed is not the number of defective items but the number of defects on the sample.

The first step in creating a u chart is to calculate the number of defects per unit for each sample.

$$u = \frac{c}{n}$$

where u represents the average defect per sample, c is the total number of defects, and n is the sample size.

Once all the averages are determined, a distribution of the means is created and the next step will be to find the mean of the distribution, in other words, the grand mean.

$$\bar{u} = \frac{u_1 + u_2 + \cdots + u_k}{k}$$

where k is the number of samples.

The control limits are determined based on $\bar{u}$ and the mean of the samples n.

$$\text{UCL} = \bar{u} + 3\sqrt{\frac{\bar{u}}{n}}$$

$$\text{CL} = \bar{u}$$

$$\text{LCL} = \bar{u} - 3\sqrt{\frac{\bar{u}}{n}}$$

Example 5.15   Medina P&L manufactures pistons and liners for diesel engines. The products are assembled in kits of 70 per unit before they are sent to the customers. The quality manager wants to create a control chart to monitor the quality level of the products. He audits 28 units and summarizes the results in the files *medina.xls* and *Medina.MPJ.*

Solution: Notice from the SigmaXL output in Fig. 5.38 that the UCL is not a straight line. This is because the sample sizes are not equal and every time a sample statistic is plotted, adjustments are made to the control limits. The process has shown stability until Sample 27 is plotted. That sample is out of control.

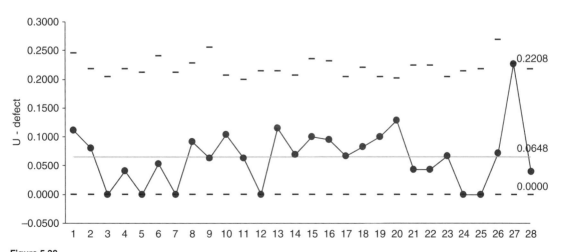

Figure 5.38

Tables

TABLE A.1 Binomial

| | | p | | | | | | | | | | | | | | | | | | | |
n	s	0.01	0.02	0.04	0.05	0.06	0.08	0.1	0.12	0.14	0.15	0.16	0.18	0.2	0.22	0.24	0.25	0.3	0.35	0.4	0.45	0.5
2	0	0.98	0.96	0.922	0.903	0.884	0.846	0.81	0.774	0.74	0.723	0.706	0.672	0.64	0.608	0.578	0.563	0.49	0.423	0.36	0.303	0.25
2	1	0.02	0.039	0.077	0.095	0.113	0.147	0.18	0.211	0.241	0.255	0.269	0.295	0.32	0.343	0.365	0.375	0.42	0.455	0.48	0.495	0.5
2	2			0.002	0.003	0.004	0.006	0.01	0.014	0.02	0.023	0.026	0.032	0.04	0.048	0.058	0.063	0.09	0.123	0.16	0.203	0.25
3	0	0.97	0.941	0.885	0.857	0.831	0.779	0.729	0.681	0.636	0.614	0.593	0.551	0.512	0.475	0.439	0.422	0.343	0.275	0.216	0.166	0.125
3	1	0.029	0.058	0.111	0.135	0.159	0.203	0.243	0.279	0.311	0.325	0.339	0.363	0.384	0.402	0.416	0.422	0.441	0.444	0.432	0.408	0.375
3	2		0.001	0.005	0.007	0.01	0.018	0.027	0.038	0.051	0.057	0.065	0.08	0.096	0.113	0.131	0.141	0.189	0.239	0.288	0.334	0.375
3	3						0.001	0.001	0.002	0.003	0.003	0.004	0.006	0.008	0.011	0.014	0.016	0.027	0.043	0.064	0.091	0.125
4	0	0.961	0.922	0.849	0.815	0.781	0.716	0.656	0.6	0.547	0.522	0.498	0.452	0.41	0.37	0.334	0.316	0.24	0.179	0.13	0.092	0.063
4	1	0.039	0.075	0.142	0.171	0.199	0.249	0.292	0.327	0.356	0.368	0.379	0.397	0.41	0.418	0.421	0.422	0.412	0.384	0.346	0.299	0.25
4	2	0.001	0.002	0.009	0.014	0.019	0.033	0.049	0.067	0.087	0.098	0.108	0.131	0.154	0.177	0.2	0.211	0.265	0.311	0.346	0.368	0.375
4	3					0.001	0.002	0.004	0.006	0.009	0.011	0.014	0.019	0.026	0.033	0.042	0.047	0.076	0.111	0.154	0.2	0.25
4	4										0.001	0.001	0.001	0.002	0.002	0.003	0.004	0.008	0.015	0.026	0.041	0.063
5	0	0.951	0.904	0.815	0.774	0.734	0.659	0.59	0.528	0.47	0.444	0.418	0.371	0.328	0.289	0.254	0.237	0.168	0.116	0.078	0.05	0.031
5	1	0.048	0.092	0.17	0.204	0.234	0.287	0.328	0.36	0.383	0.392	0.398	0.407	0.41	0.407	0.4	0.396	0.36	0.312	0.259	0.206	0.156
5	2	0.001	0.004	0.014	0.021	0.03	0.05	0.073	0.098	0.125	0.138	0.152	0.179	0.205	0.23	0.253	0.264	0.309	0.336	0.346	0.337	0.313
5	3			0.001	0.001	0.002	0.004	0.008	0.013	0.02	0.024	0.029	0.039	0.051	0.065	0.08	0.088	0.132	0.181	0.23	0.276	0.313
5	4							0.001	0.002	0.002	0.003	0.004	0.006	0.009	0.013	0.015	0.028	0.049	0.077	0.113	0.156	
5	5													0.001	0.001	0.001	0.002	0.005	0.01	0.018	0.031	
6	0	0.941	0.886	0.783	0.735	0.69	0.606	0.531	0.464	0.405	0.377	0.351	0.304	0.262	0.225	0.193	0.178	0.118	0.075	0.047	0.028	0.016
6	1	0.057	0.108	0.196	0.232	0.264	0.316	0.354	0.38	0.395	0.399	0.401	0.4	0.393	0.381	0.365	0.356	0.303	0.244	0.187	0.136	0.094
6	2	0.001	0.006	0.02	0.031	0.042	0.069	0.098	0.13	0.161	0.176	0.191	0.22	0.246	0.269	0.288	0.297	0.324	0.328	0.311	0.278	0.234
6	3			0.001	0.002	0.004	0.008	0.015	0.024	0.035	0.041	0.049	0.064	0.082	0.101	0.121	0.132	0.185	0.235	0.276	0.303	0.313
6	4						0.001	0.001	0.002	0.004	0.005	0.007	0.011	0.015	0.021	0.029	0.033	0.06	0.095	0.138	0.186	0.234
6	5											0.001	0.001	0.002	0.002	0.004	0.004	0.01	0.02	0.037	0.061	0.094
6	6																	0.001	0.002	0.004	0.008	0.016
7	0	0.932	0.868	0.751	0.698	0.648	0.558	0.478	0.409	0.348	0.321	0.295	0.249	0.21	0.176	0.146	0.133	0.082	0.049	0.028	0.015	0.008
7	1	0.066	0.124	0.219	0.257	0.29	0.34	0.372	0.39	0.396	0.396	0.393	0.383	0.367	0.347	0.324	0.311	0.247	0.185	0.131	0.087	0.055
7	2	0.002	0.008	0.027	0.041	0.055	0.089	0.124	0.16	0.194	0.21	0.225	0.252	0.275	0.293	0.307	0.311	0.318	0.298	0.261	0.214	0.164
7	3			0.002	0.004	0.006	0.013	0.023	0.036	0.053	0.062	0.071	0.092	0.115	0.138	0.161	0.173	0.227	0.268	0.29	0.292	0.273
7	4						0.001	0.003	0.005	0.009	0.011	0.014	0.02	0.029	0.039	0.051	0.058	0.097	0.144	0.194	0.239	0.273
7	5									0.001	0.001	0.002	0.003	0.004	0.007	0.01	0.012	0.025	0.047	0.077	0.117	0.164
7	6															0.001	0.001	0.004	0.008	0.017	0.032	0.055

7	7																		0.001	0.002	0.004	0.008
8	0	0.923	0.851	0.721	0.663	0.61	0.513	0.43	0.36	0.299	0.272	0.248	0.204	0.168	0.137	0.111	0.1	0.058	0.032	0.017	0.008	0.004
8	1	0.075	0.139	0.24	0.279	0.311	0.357	0.383	0.392	0.39	0.385	0.378	0.359	0.336	0.309	0.281	0.267	0.198	0.137	0.09	0.055	0.031
8	2	0.003	0.01	0.035	0.051	0.07	0.109	0.149	0.187	0.222	0.238	0.252	0.276	0.294	0.305	0.311	0.311	0.296	0.259	0.209	0.157	0.109
8	3			0.003	0.005	0.009	0.019	0.033	0.051	0.072	0.084	0.096	0.121	0.147	0.172	0.196	0.208	0.254	0.279	0.279	0.257	0.219
8	4					0.001	0.002	0.005	0.009	0.015	0.018	0.023	0.033	0.046	0.061	0.077	0.087	0.136	0.188	0.232	0.263	0.273
8	5								0.001	0.002	0.003	0.003	0.006	0.009	0.014	0.02	0.023	0.047	0.081	0.124	0.172	0.219
8	6												0.001	0.001	0.002	0.003	0.004	0.01	0.022	0.041	0.07	0.109
8	7																	0.001	0.003	0.008	0.016	0.031
8	8																			0.001	0.002	0.004
9	0	0.914	0.834	0.693	0.63	0.573	0.472	0.387	0.316	0.257	0.232	0.208	0.168	0.134	0.107	0.085	0.075	0.04	0.021	0.01	0.005	0.002
9	1	0.083	0.153	0.26	0.299	0.329	0.37	0.387	0.388	0.377	0.368	0.357	0.331	0.302	0.271	0.24	0.225	0.156	0.1	0.06	0.034	0.018
9	2	0.003	0.013	0.043	0.063	0.084	0.129	0.172	0.212	0.245	0.26	0.272	0.291	0.302	0.306	0.304	0.3	0.267	0.216	0.161	0.111	0.07
9	3		0.001	0.004	0.008	0.013	0.026	0.045	0.067	0.093	0.107	0.121	0.149	0.176	0.201	0.224	0.234	0.267	0.272	0.251	0.212	0.164
9	4				0.001	0.001	0.003	0.007	0.014	0.023	0.028	0.035	0.049	0.066	0.085	0.106	0.117	0.172	0.219	0.251	0.26	0.246
9	5							0.001	0.002	0.004	0.005	0.007	0.011	0.017	0.024	0.033	0.039	0.074	0.118	0.167	0.213	0.246
9	6										0.001	0.001	0.002	0.003	0.005	0.007	0.009	0.021	0.042	0.074	0.116	0.164
9	7														0.001	0.001	0.001	0.004	0.01	0.021	0.041	0.07
9	8																		0.001	0.004	0.008	0.018
9	9																				0.001	0.002
10	0	0.904	0.817	0.665	0.599	0.539	0.434	0.349	0.279	0.221	0.197	0.175	0.137	0.107	0.083	0.064	0.056	0.028	0.013	0.006	0.003	0.001
10	1	0.091	0.167	0.277	0.315	0.344	0.378	0.387	0.38	0.36	0.347	0.333	0.302	0.268	0.235	0.203	0.188	0.121	0.072	0.04	0.021	0.01
10	2	0.004	0.015	0.052	0.075	0.099	0.148	0.194	0.233	0.264	0.276	0.286	0.298	0.302	0.298	0.288	0.282	0.233	0.176	0.121	0.076	0.044
10	3		0.001	0.006	0.01	0.017	0.034	0.057	0.085	0.115	0.13	0.145	0.174	0.201	0.224	0.243	0.25	0.267	0.252	0.215	0.166	0.117
10	4				0.001	0.002	0.005	0.011	0.02	0.033	0.04	0.048	0.067	0.088	0.111	0.134	0.146	0.2	0.238	0.251	0.238	0.205
10	5						0.001	0.001	0.003	0.006	0.008	0.011	0.018	0.026	0.037	0.051	0.058	0.103	0.154	0.201	0.234	0.246
10	6									0.001	0.001	0.002	0.003	0.006	0.009	0.013	0.016	0.037	0.069	0.111	0.16	0.205
10	7													0.001	0.001	0.002	0.003	0.009	0.021	0.042	0.075	0.117
10	8																	0.001	0.004	0.011	0.023	0.044
10	9																		0.001	0.002	0.004	0.01
10	10																				0.001	0.001

(Continued)

TABLE A.1 Binomial (*Continued*)

n	s	0.01	0.02	0.04	0.05	0.06	0.08	0.1	0.12	0.14	0.15	0.16	0.18	0.2	0.22	0.24	0.25	0.3	0.35	0.4	0.45	0.5
11	0	0.895	0.801	0.638	0.569	0.506	0.4	0.314	0.245	0.19	0.167	0.147	0.113	0.086	0.065	0.049	0.042	0.02	0.009	0.004	0.001	
11	1	0.099	0.18	0.293	0.329	0.355	0.382	0.384	0.368	0.341	0.325	0.308	0.272	0.236	0.202	0.17	0.155	0.093	0.052	0.027	0.013	0.005
11	2	0.005	0.018	0.061	0.087	0.113	0.166	0.213	0.251	0.277	0.287	0.293	0.299	0.295	0.284	0.268	0.258	0.2	0.14	0.089	0.051	0.027
11	3		0.001	0.008	0.014	0.022	0.043	0.071	0.103	0.135	0.152	0.168	0.197	0.221	0.241	0.254	0.258	0.257	0.225	0.177	0.126	0.081
11	4			0.001	0.001	0.003	0.008	0.016	0.028	0.044	0.054	0.064	0.086	0.111	0.136	0.16	0.172	0.22	0.243	0.236	0.206	0.161
11	5						0.001	0.002	0.005	0.01	0.013	0.017	0.027	0.039	0.054	0.071	0.08	0.132	0.183	0.221	0.236	0.226
11	6								0.001	0.002	0.002	0.003	0.006	0.01	0.015	0.022	0.027	0.057	0.099	0.147	0.193	0.226
11	7												0.001	0.002	0.003	0.005	0.006	0.017	0.038	0.07	0.113	0.161
11	8															0.001	0.001	0.004	0.01	0.023	0.046	0.081
11	9																	0.001	0.002	0.005	0.013	0.027
11	10																			0.001	0.002	0.005
11	11																					
12	0	0.886	0.785	0.613	0.54	0.476	0.368	0.282	0.216	0.164	0.142	0.123	0.092	0.069	0.051	0.037	0.032	0.014	0.006	0.002	0.001	
12	1	0.107	0.192	0.306	0.341	0.365	0.384	0.377	0.353	0.32	0.301	0.282	0.243	0.206	0.172	0.141	0.127	0.071	0.037	0.017	0.008	0.003
12	2	0.006	0.022	0.07	0.099	0.128	0.183	0.23	0.265	0.286	0.292	0.296	0.294	0.283	0.266	0.244	0.232	0.168	0.109	0.064	0.034	0.016
12	3		0.001	0.01	0.017	0.027	0.053	0.085	0.12	0.155	0.172	0.188	0.215	0.236	0.25	0.257	0.258	0.24	0.195	0.142	0.092	0.054
12	4			0.001	0.002	0.004	0.01	0.021	0.037	0.057	0.068	0.08	0.106	0.133	0.159	0.183	0.194	0.231	0.237	0.213	0.17	0.121
12	5						0.001	0.004	0.008	0.015	0.019	0.025	0.037	0.053	0.072	0.092	0.103	0.158	0.204	0.227	0.222	0.193
12	6								0.001	0.003	0.004	0.005	0.01	0.016	0.024	0.034	0.04	0.079	0.128	0.177	0.212	0.226
12	7										0.001	0.001	0.002	0.003	0.006	0.009	0.011	0.029	0.059	0.101	0.149	0.193
12	8													0.001	0.001	0.002	0.002	0.008	0.02	0.042	0.076	0.121
12	9																	0.001	0.005	0.012	0.028	0.054
12	10																		0.001	0.002	0.007	0.016
12	11																				0.001	0.003
12	12																					

n	x																					
13	0	0.878	0.769	0.588	0.513	0.447	0.338	0.254	0.19	0.141	0.121	0.104	0.076	0.055	0.04	0.028	0.024	0.01	0.004	0.001		
13	1	0.115	0.204	0.319	0.351	0.371	0.382	0.367	0.336	0.298	0.277	0.257	0.216	0.179	0.145	0.116	0.103	0.054	0.026	0.011	0.004	0.002
13	2	0.007	0.025	0.08	0.111	0.142	0.199	0.245	0.275	0.291	0.294	0.293	0.285	0.268	0.245	0.22	0.206	0.139	0.084	0.045	0.022	0.01
13	3		0.002	0.012	0.021	0.033	0.064	0.1	0.138	0.174	0.19	0.205	0.229	0.246	0.254	0.254	0.252	0.218	0.165	0.111	0.066	0.035
13	4			0.001	0.003	0.005	0.014	0.028	0.047	0.071	0.084	0.098	0.126	0.154	0.179	0.201	0.21	0.234	0.222	0.184	0.135	0.087
13	5					0.001	0.002	0.006	0.012	0.021	0.027	0.033	0.05	0.069	0.091	0.114	0.126	0.18	0.215	0.221	0.199	0.157
13	6							0.001	0.002	0.004	0.006	0.008	0.015	0.023	0.034	0.048	0.056	0.103	0.155	0.197	0.217	0.209
13	7									0.001	0.001	0.002	0.003	0.006	0.01	0.015	0.019	0.044	0.083	0.131	0.177	0.209
13	8													0.001	0.002	0.004	0.005	0.014	0.034	0.066	0.109	0.157
13	9															0.001	0.001	0.003	0.01	0.024	0.05	0.087
13	10																	0.001	0.002	0.006	0.016	0.035
13	11																		0.001	0.001	0.004	0.01
13	12																					0.002
13	13																					
14	0	0.869	0.754	0.565	0.488	0.421	0.311	0.229	0.167	0.121	0.103	0.087	0.062	0.044	0.031	0.021	0.018	0.007	0.002	0.001		
14	1	0.123	0.215	0.329	0.359	0.376	0.379	0.356	0.319	0.276	0.254	0.232	0.191	0.154	0.122	0.095	0.083	0.041	0.018	0.007	0.003	0.001
14	2	0.008	0.029	0.089	0.123	0.156	0.214	0.257	0.283	0.292	0.291	0.287	0.272	0.25	0.223	0.195	0.18	0.113	0.063	0.032	0.014	0.006
14	3		0.002	0.015	0.026	0.04	0.074	0.114	0.154	0.19	0.206	0.219	0.239	0.25	0.25	0.246	0.24	0.194	0.137	0.085	0.046	0.022
14	4			0.002	0.004	0.007	0.018	0.035	0.058	0.085	0.1	0.115	0.144	0.172	0.195	0.214	0.22	0.229	0.202	0.155	0.104	0.061
14	5					0.001	0.003	0.008	0.016	0.028	0.035	0.044	0.063	0.086	0.11	0.135	0.147	0.196	0.218	0.207	0.17	0.122
14	6							0.001	0.003	0.007	0.009	0.012	0.021	0.032	0.047	0.064	0.073	0.126	0.176	0.207	0.209	0.183
14	7								0.001	0.001	0.002	0.003	0.005	0.009	0.015	0.023	0.028	0.062	0.108	0.157	0.195	0.209
14	8												0.001	0.002	0.004	0.006	0.008	0.023	0.051	0.092	0.14	0.183
14	9														0.001	0.001	0.002	0.007	0.018	0.041	0.076	0.122
14	10																	0.001	0.005	0.014	0.031	0.061
14	11																		0.001	0.003	0.009	0.022
14	12																			0.001	0.002	0.006
14	13																					0.001
14	14																					

(Continued)

TABLE A.1 Binomial (Continued)

n	s												p										
		0.01	0.02	0.04	0.05	0.06	0.08	0.1	0.12	0.14	0.15	0.16	0.18	0.2	0.22	0.24	0.25	0.3	0.35	0.4	0.45	0.5
15	0	0.86	0.739	0.542	0.463	0.395	0.286	0.206	0.147	0.104	0.087	0.073	0.051	0.035	0.024	0.016	0.013	0.005	0.002			
15	1	0.13	0.226	0.339	0.366	0.378	0.373	0.343	0.301	0.254	0.231	0.209	0.168	0.132	0.102	0.077	0.067	0.031	0.013	0.005	0.002	
15	2	0.009	0.032	0.099	0.135	0.169	0.227	0.267	0.287	0.29	0.286	0.279	0.258	0.231	0.201	0.171	0.156	0.092	0.048	0.022	0.009	0.003
15	3		0.003	0.018	0.031	0.047	0.086	0.129	0.17	0.204	0.218	0.23	0.245	0.25	0.246	0.234	0.225	0.17	0.111	0.063	0.032	0.014
15	4			0.002	0.005	0.009	0.022	0.043	0.069	0.1	0.116	0.131	0.162	0.188	0.208	0.221	0.225	0.219	0.179	0.127	0.078	0.042
15	5				0.001	0.001	0.004	0.01	0.021	0.036	0.045	0.055	0.078	0.103	0.129	0.154	0.165	0.206	0.212	0.186	0.14	0.092
15	6						0.001	0.002	0.005	0.01	0.013	0.017	0.029	0.043	0.061	0.081	0.092	0.147	0.191	0.207	0.191	0.153
15	7								0.001	0.002	0.003	0.004	0.008	0.014	0.022	0.033	0.039	0.081	0.132	0.177	0.201	0.196
15	8										0.001	0.001	0.002	0.003	0.006	0.01	0.013	0.035	0.071	0.118	0.165	0.196
15	9													0.001	0.001	0.003	0.003	0.012	0.03	0.061	0.105	0.153
15	10																0.001	0.003	0.01	0.024	0.051	0.092
15	11																	0.001	0.002	0.007	0.019	0.042
15	12																			0.002	0.005	0.014
15	13																				0.001	0.003
15	14																					
15	15																					

TABLE A.2   Poisson

Events	Mean 0.1	0.2	0.3	0.4	0.5	0.6	0.7	0.8	0.9	1
0	0.90484	0.81873	0.74082	0.67032	0.60653	0.54881	0.49659	0.44933	0.40657	0.36788
1	0.09048	0.16375	0.22225	0.26813	0.30327	0.32929	0.34761	0.35946	0.36591	0.36788
2	0.00452	0.01637	0.03334	0.05363	0.07582	0.09879	0.12166	0.14379	0.16466	0.18394
3	0.00015	0.00109	0.00333	0.00715	0.01264	0.01976	0.02839	0.03834	0.0494	0.06131
4	0	0.00005	0.00025	0.00072	0.00158	0.00296	0.00497	0.00767	0.01111	0.01533
5	0	0	0.00002	0.00006	0.00016	0.00036	0.0007	0.00123	0.002	0.00307

Events	1.1	1.2	1.3	1.4	1.5	1.6	1.7	1.8	1.9	2
0	0.33287	0.30119	0.27253	0.2466	0.22313	0.2019	0.18268	0.1653	0.14957	0.13534
1	0.36616	0.36143	0.35429	0.34524	0.3347	0.32303	0.31056	0.29754	0.28418	0.27067
2	0.20139	0.21686	0.23029	0.24167	0.25102	0.25843	0.26398	0.26778	0.26997	0.27067
3	0.07384	0.08674	0.09979	0.11278	0.12551	0.13783	0.14959	0.16067	0.17098	0.18045
4	0.02031	0.02602	0.03243	0.03947	0.04707	0.05513	0.06357	0.0723	0.08122	0.09022
5	0.00447	0.00625	0.00843	0.01105	0.01412	0.01764	0.02162	0.02603	0.03086	0.03609
6	0.00082	0.00125	0.00183	0.00258	0.00353	0.0047	0.00612	0.00781	0.00977	0.01203
7	0.00013	0.00021	0.00034	0.00052	0.00076	0.00108	0.00149	0.00201	0.00265	0.00344
8	0.00002	0.00003	0.00006	0.00009	0.00014	0.00022	0.00032	0.00045	0.00063	0.00086
9	0	0	0.00001	0.00001	0.00002	0.00004	0.00006	0.00009	0.00013	0.00019
10	0	0	0	0	0	0.00001	0.00001	0.00002	0.00003	0.00004

Events	2.1	2.2	2.3	2.4	2.5	2.6	2.7	2.8	2.9	3
0	0.12246	0.1108	0.10026	0.09072	0.08208	0.07427	0.06721	0.06081	0.05502	0.04979
1	0.25716	0.24377	0.2306	0.21772	0.20521	0.19311	0.18145	0.17027	0.15957	0.14936
2	0.27002	0.26814	0.26518	0.26127	0.25652	0.25104	0.24496	0.23838	0.23137	0.22404
3	0.18901	0.19664	0.20331	0.20901	0.21376	0.21757	0.22047	0.22248	0.22366	0.22404
4	0.09923	0.10815	0.1169	0.12541	0.1336	0.14142	0.14882	0.15574	0.16215	0.16803
5	0.04168	0.04759	0.05378	0.0602	0.0668	0.07354	0.08036	0.08721	0.09405	0.10082
6	0.01459	0.01745	0.02061	0.02408	0.02783	0.03187	0.03616	0.0407	0.04546	0.05041
7	0.00438	0.00548	0.00677	0.00826	0.00994	0.01184	0.01395	0.01628	0.01883	0.0216
8	0.00115	0.00151	0.00195	0.00248	0.00311	0.00385	0.00471	0.0057	0.00683	0.0081
9	0.00027	0.00037	0.0005	0.00066	0.00086	0.00111	0.00141	0.00177	0.0022	0.0027
10	0.00006	0.00008	0.00011	0.00016	0.00022	0.00029	0.00038	0.0005	0.00064	0.00081
11	0.00001	0.00002	0.00002	0.00003	0.00005	0.00007	0.00009	0.00013	0.00017	0.00022
12	0	0	0	0.00001	0.00001	0.00001	0.00002	0.00003	0.00004	0.00006
13	0	0	0	0	0	0	0	0.00001	0.00001	0.00001
14	0	0	0	0	0	0	0	0	0	0
15	0	0	0	0	0	0	0	0	0	0

Events	3.1	3.2	3.3	3.4	3.5	3.6	3.7	3.8	3.9	4
0	0.04505	0.04076	0.03688	0.03337	0.0302	0.02732	0.02472	0.02237	0.02024	0.01832
1	0.13965	0.13044	0.12171	0.11347	0.10569	0.09837	0.09148	0.08501	0.07894	0.07326
2	0.21646	0.2087	0.20083	0.1929	0.18496	0.17706	0.16923	0.16152	0.15394	0.14653
3	0.22368	0.22262	0.22091	0.21862	0.21579	0.21247	0.20872	0.20459	0.20012	0.19537
4	0.17335	0.17809	0.18225	0.18582	0.18881	0.19122	0.19307	0.19436	0.19512	0.19537
5	0.10748	0.11398	0.12029	0.12636	0.13217	0.13768	0.14287	0.14771	0.15219	0.15629
6	0.05553	0.06079	0.06616	0.0716	0.0771	0.08261	0.0881	0.09355	0.09893	0.1042
7	0.02459	0.02779	0.03119	0.03478	0.03855	0.04248	0.04657	0.05079	0.05512	0.05954
8	0.00953	0.01112	0.01287	0.01478	0.01687	0.01912	0.02154	0.02412	0.02687	0.02977
9	0.00328	0.00395	0.00472	0.00558	0.00656	0.00765	0.00885	0.01019	0.01164	0.01323
10	0.00102	0.00126	0.00156	0.0019	0.0023	0.00275	0.00328	0.00387	0.00454	0.00529
11	0.00029	0.00037	0.00047	0.00059	0.00073	0.0009	0.0011	0.00134	0.00161	0.00192
12	0.00007	0.0001	0.00013	0.00017	0.00021	0.00027	0.00034	0.00042	0.00052	0.00064
13	0.00002	0.00002	0.00003	0.00004	0.00006	0.00007	0.0001	0.00012	0.00016	0.0002
14	0	0.00001	0.00001	0.00001	0.00001	0.00002	0.00003	0.00003	0.00004	0.00006
15	0	0	0	0	0	0	0.00001	0.00001	0.00001	0.00002

(Continued)

TABLE A.2   Poisson (*Continued*)

	Mean									
Events	4.1	4.2	4.3	4.4	4.5	4.6	4.7	4.8	4.9	5
0	0.01657	0.015	0.01357	0.01228	0.01111	0.01005	0.0091	0.00823	0.00745	0.00674
1	0.06795	0.06298	0.05834	0.05402	0.04999	0.04624	0.04275	0.0395	0.03649	0.03369
2	0.13929	0.13226	0.12544	0.11884	0.11248	0.10635	0.10046	0.09481	0.0894	0.08422
3	0.19037	0.18517	0.1798	0.17431	0.16872	0.16307	0.15738	0.15169	0.14601	0.14037
4	0.19513	0.19442	0.19328	0.19174	0.18981	0.18753	0.18493	0.18203	0.17887	0.17547
5	0.16	0.16332	0.16622	0.16873	0.17083	0.17253	0.17383	0.17475	0.17529	0.17547
6	0.10934	0.11432	0.11913	0.12373	0.12812	0.13227	0.13617	0.1398	0.14315	0.14622
7	0.06404	0.06859	0.07318	0.07778	0.08236	0.08692	0.09143	0.09586	0.10021	0.10444
8	0.03282	0.03601	0.03933	0.04278	0.04633	0.04998	0.05371	0.05752	0.06138	0.06528
9	0.01495	0.01681	0.01879	0.02091	0.02316	0.02554	0.02805	0.03068	0.03342	0.03627
10	0.00613	0.00706	0.00808	0.0092	0.01042	0.01175	0.01318	0.01472	0.01637	0.01813
11	0.00228	0.00269	0.00316	0.00368	0.00426	0.00491	0.00563	0.00643	0.00729	0.00824
12	0.00078	0.00094	0.00113	0.00135	0.0016	0.00188	0.00221	0.00257	0.00298	0.00343
13	0.00025	0.0003	0.00037	0.00046	0.00055	0.00067	0.0008	0.00095	0.00112	0.00132
14	0.00007	0.00009	0.00011	0.00014	0.00018	0.00022	0.00027	0.00033	0.00039	0.00047
15	0.00002	0.00003	0.00003	0.00004	0.00005	0.00007	0.00008	0.0001	0.00013	0.00016

	5.1	5.2	5.3	5.4	5.5	5.6	5.7	5.8	5.9	6
0	0.0061	0.00552	0.00499	0.00452	0.00409	0.0037	0.00335	0.00303	0.00274	0.00248
1	0.03109	0.02869	0.02646	0.02439	0.02248	0.02071	0.01907	0.01756	0.01616	0.01487
2	0.07929	0.07458	0.07011	0.06585	0.06181	0.05798	0.05436	0.05092	0.04768	0.04462
3	0.13479	0.12928	0.12386	0.11853	0.11332	0.10823	0.10327	0.09845	0.09377	0.08924
4	0.17186	0.16806	0.16411	0.16002	0.15582	0.15153	0.14717	0.14276	0.13831	0.13385
5	0.17529	0.17479	0.17396	0.17282	0.1714	0.16971	0.16777	0.1656	0.16321	0.16062
6	0.149	0.15148	0.15366	0.15554	0.15712	0.1584	0.15938	0.16008	0.16049	0.16062
7	0.10856	0.11253	0.11634	0.11999	0.12345	0.12672	0.12978	0.13263	0.13527	0.13768
8	0.06921	0.07314	0.07708	0.08099	0.08487	0.0887	0.09247	0.09616	0.09976	0.10326
9	0.03922	0.04226	0.04539	0.04859	0.05187	0.05519	0.05856	0.06197	0.0654	0.06884
10	0.02	0.02198	0.02406	0.02624	0.02853	0.03091	0.03338	0.03594	0.03859	0.0413
11	0.00927	0.01039	0.01159	0.01288	0.01426	0.01573	0.0173	0.01895	0.0207	0.02253
12	0.00394	0.0045	0.00512	0.0058	0.00654	0.00734	0.00822	0.00916	0.01018	0.01126
13	0.00155	0.0018	0.0018	0.00209	0.00241	0.00277	0.00316	0.0036	0.00409	0.00462	0.0052
14	0.00056	0.00067	0.00079	0.00093	0.00109	0.00127	0.00147	0.00169	0.00195	0.00223
15	0.00019	0.00023	0.00028	0.00033	0.0004	0.00047	0.00056	0.00065	0.00077	0.00089

	6.1	6.2	6.3	6.4	6.5	6.6	6.7	6.8	6.9	7
0	0.00224	0.00203	0.00184	0.00166	0.0015	0.00136	0.00123	0.00111	0.00101	0.00091
1	0.01368	0.01258	0.01157	0.01063	0.00977	0.00898	0.00825	0.00757	0.00695	0.00638
2	0.04173	0.03901	0.03644	0.03403	0.03176	0.02963	0.02763	0.02575	0.02399	0.02234
3	0.08485	0.08061	0.07653	0.07259	0.06881	0.06518	0.0617	0.05837	0.05518	0.05213
4	0.12939	0.12495	0.12053	0.11615	0.11182	0.10755	0.10335	0.09923	0.09518	0.09123
5	0.15786	0.15494	0.15187	0.14867	0.14537	0.14197	0.13849	0.13495	0.13135	0.12772
6	0.16049	0.1601	0.15946	0.15859	0.15748	0.15617	0.15465	0.15294	0.15105	0.149
7	0.13986	0.1418	0.14352	0.14499	0.14623	0.14724	0.14802	0.14857	0.1489	0.149
8	0.10664	0.1099	0.11302	0.11599	0.11882	0.12148	0.12397	0.12628	0.12842	0.13038
9	0.07228	0.07571	0.07911	0.08248	0.08581	0.08908	0.09229	0.09541	0.09846	0.1014
10	0.04409	0.04694	0.04984	0.05279	0.05578	0.05879	0.06183	0.06488	0.06794	0.07098
11	0.02445	0.02646	0.02855	0.03071	0.03296	0.03528	0.03766	0.04011	0.04261	0.04517
12	0.01243	0.01367	0.01499	0.01638	0.01785	0.0194	0.02103	0.02273	0.0245	0.02635
13	0.00583	0.00652	0.00726	0.00806	0.00893	0.00985	0.01084	0.01189	0.01301	0.01419
14	0.00254	0.00289	0.00327	0.00369	0.00414	0.00464	0.00519	0.00577	0.00641	0.00709
15	0.00103	0.00119	0.00137	0.00157	0.0018	0.00204	0.00232	0.00262	0.00295	0.00331
16	0.00039	0.00046	0.00054	0.00063	0.00073	0.00084	0.00097	0.00111	0.00127	0.00145
17	0.00014	0.00017	0.0002	0.00024	0.00028	0.00033	0.00038	0.00045	0.00052	0.0006
18	0.00005	0.00006	0.00007	0.00008	0.0001	0.00012	0.00014	0.00017	0.0002	0.00023
19	0.00002	0.00002	0.00002	0.00003	0.00003	0.00004	0.00005	0.00006	0.00007	0.00009
20	0	0.00001	0.00001	0.00001	0.00001	0.00001	0.00002	0.00002	0.00002	0.00003

(Continued)

TABLE A.2 Poisson (*Continued*)

Events	Mean 7.1	7.2	7.3	7.4	7.5	7.6	7.7	7.8	7.9	8
0	0.00083	0.00075	0.00068	0.00061	0.00055	0.0005	0.00045	0.00041	0.00037	0.00034
1	0.00586	0.00538	0.00493	0.00452	0.00415	0.0038	0.00349	0.0032	0.00293	0.00268
2	0.0208	0.01935	0.018	0.01674	0.01556	0.01445	0.01342	0.01246	0.01157	0.01073
3	0.04922	0.04644	0.0438	0.04128	0.03889	0.03661	0.03446	0.03241	0.03047	0.02863
4	0.08736	0.0836	0.07993	0.07637	0.07292	0.06957	0.06633	0.06319	0.06017	0.05725
5	0.12406	0.12038	0.1167	0.11303	0.10937	0.10574	0.10214	0.09858	0.09507	0.0916
6	0.1468	0.14446	0.14199	0.13941	0.13672	0.13394	0.13108	0.12816	0.12517	0.12214
7	0.1489	0.14859	0.14807	0.14737	0.14648	0.14542	0.14419	0.1428	0.14126	0.13959
8	0.13215	0.13373	0.13512	0.13632	0.13733	0.13815	0.13878	0.13923	0.1395	0.13959
9	0.10425	0.10698	0.1096	0.11208	0.11444	0.11666	0.11874	0.12067	0.12245	0.12408
10	0.07402	0.07703	0.08	0.08294	0.08583	0.08866	0.09143	0.09412	0.09673	0.09926
11	0.04777	0.05042	0.05309	0.0558	0.05852	0.06126	0.064	0.06674	0.06947	0.07219
12	0.02827	0.03025	0.0323	0.03441	0.03658	0.0388	0.04107	0.04338	0.04574	0.04813
13	0.01544	0.01675	0.01814	0.01959	0.0211	0.02268	0.02432	0.02603	0.02779	0.02962
14	0.00783	0.00862	0.00946	0.01035	0.0113	0.01231	0.01338	0.0145	0.01568	0.01692
15	0.00371	0.00414	0.0046	0.00511	0.00565	0.00624	0.00687	0.00754	0.00826	0.00903
16	0.00164	0.00186	0.0021	0.00236	0.00265	0.00296	0.0033	0.00368	0.00408	0.00451
17	0.00069	0.00079	0.0009	0.00103	0.00117	0.00132	0.0015	0.00169	0.0019	0.00212
18	0.00027	0.00032	0.00037	0.00042	0.00049	0.00056	0.00064	0.00073	0.00083	0.00094
19	0.0001	0.00012	0.00014	0.00016	0.00019	0.00022	0.00026	0.0003	0.00035	0.0004
20	0.00004	0.00004	0.00005	0.00006	0.00007	0.00009	0.0001	0.00012	0.00014	0.00016

Events	8.1	8.2	8.3	8.4	8.5	8.6	8.7	8.8	8.9	9
0	0.0003	0.00027	0.00025	0.00022	0.0002	0.00018	0.00017	0.00015	0.00014	0.00012
1	0.00246	0.00225	0.00206	0.00189	0.00173	0.00158	0.00145	0.00133	0.00121	0.00111
2	0.00996	0.00923	0.00856	0.00793	0.00735	0.00681	0.0063	0.00584	0.0054	0.005
3	0.02689	0.02524	0.02368	0.02221	0.02083	0.01952	0.01828	0.01712	0.01602	0.01499
4	0.05444	0.05174	0.04914	0.04665	0.04425	0.04196	0.03977	0.03766	0.03566	0.03374
5	0.0882	0.08485	0.08158	0.07837	0.07523	0.07217	0.06919	0.06629	0.06347	0.06073
6	0.11907	0.11597	0.11285	0.10972	0.10658	0.10345	0.10033	0.09722	0.09414	0.09109
7	0.13778	0.13585	0.1338	0.13166	0.12942	0.12709	0.12469	0.12222	0.1197	0.11712
8	0.1395	0.13924	0.13882	0.13824	0.13751	0.13663	0.1356	0.13445	0.13316	0.13176
9	0.12555	0.12687	0.12803	0.12903	0.12987	0.13055	0.13108	0.13146	0.13168	0.13176
10	0.1017	0.10403	0.10626	0.10838	0.11039	0.11228	0.11404	0.11568	0.1172	0.11858
11	0.07488	0.07755	0.08018	0.08276	0.0853	0.08778	0.0902	0.09255	0.09482	0.09702
12	0.05055	0.05299	0.05546	0.05793	0.06042	0.06291	0.06539	0.06787	0.07033	0.07277
13	0.03149	0.03343	0.03541	0.03743	0.03951	0.04162	0.04376	0.04594	0.04815	0.05038
14	0.01822	0.01958	0.02099	0.02246	0.02399	0.02556	0.0272	0.02888	0.03061	0.03238
15	0.00984	0.0107	0.01162	0.01258	0.01359	0.01466	0.01577	0.01694	0.01816	0.01943
16	0.00498	0.00549	0.00603	0.0066	0.00722	0.00788	0.00858	0.00932	0.0101	0.01093
17	0.00237	0.00265	0.00294	0.00326	0.00361	0.00399	0.00439	0.00482	0.00529	0.00579
18	0.00107	0.00121	0.00136	0.00152	0.0017	0.0019	0.00212	0.00236	0.00261	0.00289
19	0.00046	0.00052	0.00059	0.00067	0.00076	0.00086	0.00097	0.00109	0.00122	0.00137
20	0.00018	0.00021	0.00025	0.00028	0.00032	0.00037	0.00042	0.00048	0.00055	0.00062

Events	9.1	9.2	9.3	9.4	9.5	9.6	9.7	9.8	9.9	10
0	0.00011	0.0001	0.00009	0.00008	0.00007	0.00007	0.00006	0.00006	0.00005	0.00005
1	0.00102	0.00093	0.00085	0.00078	0.00071	0.00065	0.00059	0.00054	0.0005	0.00045
2	0.00462	0.00428	0.00395	0.00365	0.00338	0.00312	0.00288	0.00266	0.00246	0.00227
3	0.01402	0.01311	0.01226	0.01145	0.0107	0.00999	0.00932	0.0087	0.00811	0.00757
4	0.03191	0.03016	0.0285	0.02691	0.0254	0.02397	0.02261	0.02131	0.02008	0.01892
5	0.05807	0.05549	0.053	0.05059	0.04827	0.04602	0.04386	0.04177	0.03976	0.03783
6	0.08807	0.08509	0.08215	0.07926	0.07642	0.07363	0.0709	0.06822	0.06561	0.06306
7	0.11449	0.11183	0.10915	0.10644	0.10371	0.10098	0.09825	0.09551	0.09279	0.09008
8	0.13024	0.12861	0.12688	0.12506	0.12316	0.12118	0.11912	0.117	0.11483	0.1126
9	0.13168	0.13147	0.13111	0.13062	0.13	0.12926	0.12839	0.1274	0.12631	0.12511
10	0.11983	0.12095	0.12193	0.12279	0.1235	0.12409	0.12454	0.12486	0.12505	0.12511

(Continued)

TABLE A.2 Poisson (*Continued*)

					Mean					
Events	9.1	9.2	9.3	9.4	9.5	9.6	9.7	9.8	9.9	10
11	0.09913	0.10116	0.10309	0.10493	0.10666	0.10829	0.10982	0.11124	0.11254	0.11374
12	0.07518	0.07755	0.0799	0.08219	0.08444	0.08663	0.08877	0.09084	0.09285	0.09478
13	0.05262	0.05488	0.05716	0.05943	0.06171	0.06398	0.06624	0.06848	0.07071	0.07291
14	0.03421	0.03607	0.03797	0.0399	0.04187	0.04387	0.04589	0.04794	0.05	0.05208
15	0.02075	0.02212	0.02354	0.02501	0.02652	0.02808	0.02968	0.03132	0.033	0.03472
16	0.0118	0.01272	0.01368	0.01469	0.01575	0.01685	0.01799	0.01918	0.02042	0.0217
17	0.00632	0.00688	0.00749	0.00812	0.0088	0.00951	0.01027	0.01106	0.01189	0.01276
18	0.00319	0.00352	0.00387	0.00424	0.00464	0.00507	0.00553	0.00602	0.00654	0.00709
19	0.00153	0.0017	0.00189	0.0021	0.00232	0.00256	0.00282	0.00311	0.00341	0.00373
20	0.0007	0.00078	0.00088	0.00099	0.0011	0.00123	0.00137	0.00152	0.00169	0.00187
	10.1	10.2	10.3	10.4	10.5	10.6	10.7	10.8	10.9	11
0	0.00004	0.00004	0.00003	0.00003	0.00003	0.00002	0.00002	0.00002	0.00002	0.00002
1	0.00041	0.00038	0.00035	0.00032	0.00029	0.00026	0.00024	0.00022	0.0002	0.00018
2	0.0021	0.00193	0.00178	0.00165	0.00152	0.0014	0.00129	0.00119	0.0011	0.00101
3	0.00705	0.00657	0.00613	0.00571	0.00531	0.00495	0.0046	0.00428	0.00398	0.0037
4	0.01781	0.01676	0.01577	0.01483	0.01395	0.01311	0.01231	0.01156	0.01086	0.01019
5	0.03598	0.0342	0.03249	0.03085	0.02929	0.02779	0.02635	0.02498	0.02367	0.02242
6	0.06056	0.05814	0.05578	0.05348	0.05125	0.04909	0.04699	0.04496	0.04299	0.04109
7	0.08739	0.08472	0.08207	0.07946	0.07688	0.07433	0.07183	0.06937	0.06695	0.06458
8	0.11033	0.10801	0.10567	0.1033	0.1009	0.09849	0.09607	0.09365	0.09122	0.08879
9	0.12381	0.12242	0.12093	0.11936	0.11772	0.116	0.11422	0.11238	0.11048	0.10853
10	0.12505	0.12486	0.12456	0.12414	0.12361	0.12296	0.12221	0.12137	0.12042	0.11938
11	0.11482	0.11578	0.11663	0.11737	0.11799	0.11849	0.11888	0.11916	0.11932	0.11938
12	0.09664	0.09842	0.10011	0.10172	0.10324	0.10467	0.106	0.10724	0.10839	0.10943
13	0.07508	0.07722	0.07932	0.08137	0.08339	0.08534	0.08725	0.08909	0.09088	0.09259
14	0.05416	0.05626	0.05836	0.06045	0.06254	0.06462	0.06668	0.06873	0.07075	0.07275
15	0.03647	0.03826	0.04007	0.04191	0.04378	0.04566	0.04757	0.04949	0.05141	0.05335
16	0.02302	0.02439	0.0258	0.02724	0.02873	0.03025	0.03181	0.0334	0.03503	0.03668
17	0.01368	0.01463	0.01563	0.01667	0.01774	0.01886	0.02002	0.02122	0.02246	0.02373
18	0.00767	0.00829	0.00894	0.00963	0.01035	0.01111	0.0119	0.01273	0.0136	0.0145
19	0.00408	0.00445	0.00485	0.00527	0.00572	0.0062	0.0067	0.00724	0.0078	0.0084
20	0.00206	0.00227	0.0025	0.00274	0.003	0.00328	0.00359	0.00391	0.00425	0.00462
21	0.00099	0.0011	0.00122	0.00136	0.0015	0.00166	0.00183	0.00201	0.00221	0.00242
22	0.00045	0.00051	0.00057	0.00064	0.00072	0.0008	0.00089	0.00099	0.00109	0.00121
23	0.0002	0.00023	0.00026	0.00029	0.00033	0.00037	0.00041	0.00046	0.00052	0.00058

TABLE A.3 Chi Square

df\area	0.995	0.99	0.975	0.95	0.9	0.75	0.5	0.25	0.1	0.05	0.025	0.01	0.005
1	0.00004	0.00016	0.00098	0.00393	0.01579	0.10153	0.45494	1.3233	2.70554	3.84146	5.02389	6.6349	7.87944
2	0.01003	0.0201	0.05064	0.10259	0.21072	0.57536	1.38629	2.77259	4.60517	5.99146	7.37776	9.21034	10.59663
3	0.07172	0.11483	0.2158	0.35185	0.58437	1.21253	2.36597	4.10834	6.25139	7.81473	9.3484	11.34487	12.83816
4	0.20699	0.29711	0.48442	0.71072	1.06362	1.92256	3.35669	5.38527	7.77944	9.48773	11.14329	13.2767	14.86026
5	0.41174	0.5543	0.83121	1.14548	1.61031	2.6746	4.35146	6.62568	9.23636	11.0705	12.8325	15.08627	16.7496
6	0.67573	0.87209	1.23734	1.63538	2.20413	3.4546	5.34812	7.8408	10.64464	12.59159	14.44938	16.81189	18.54758
7	0.98926	1.23904	1.68987	2.16735	2.83311	4.25485	6.34581	9.03715	12.01704	14.06714	16.01276	18.47531	20.27774
8	1.34441	1.6465	2.17973	2.73264	3.48954	5.07064	7.34412	10.21885	13.36157	15.50731	17.53455	20.09024	21.95495
9	1.73493	2.0879	2.70039	3.32511	4.16816	5.89883	8.34283	11.38875	14.68366	16.91898	19.02277	21.66599	23.58935
10	2.15586	2.55821	3.24697	3.9403	4.86518	6.7372	9.34182	12.54886	15.98718	18.30704	20.48318	23.20925	25.18818
11	2.60322	3.05348	3.81575	4.57481	5.57778	7.58414	10.341	13.70069	17.27501	19.67514	21.92005	24.72497	26.75685
12	3.07382	3.57057	4.40379	5.22603	6.3038	8.43842	11.34032	14.8454	18.54935	21.02607	23.33666	26.21697	28.29952
13	3.56503	4.10692	5.00875	5.89186	7.0415	9.29907	12.33976	15.98391	19.81193	22.36203	24.7356	27.68825	29.81947
14	4.07467	4.66043	5.62873	6.57063	7.78953	10.16531	13.33927	17.11693	21.06414	23.68479	26.11895	29.14124	31.31935
15	4.60092	5.22935	6.26214	7.26094	8.54676	11.03654	14.33886	18.24509	22.30713	24.99579	27.48839	30.57791	32.80132
16	5.14221	5.81221	6.90766	7.96165	9.31224	11.91222	15.3385	19.36886	23.54183	26.29623	28.84535	31.99993	34.26719
17	5.69722	6.40776	7.56419	8.67176	10.08519	12.79193	16.33818	20.48868	24.76904	27.58711	30.19101	33.40866	35.71847
18	6.2648	7.01491	8.23075	9.39046	10.86494	13.67529	17.3379	21.60489	25.98942	28.8693	31.52638	34.80531	37.15645
19	6.84397	7.63273	8.90652	10.11701	11.65091	14.562	18.33765	22.71781	27.20357	30.14353	32.85233	36.19087	38.58226
20	7.43384	8.2604	9.59078	10.85081	12.44261	15.45177	19.33743	23.82769	28.41198	31.41043	34.16961	37.56623	39.99685
21	8.03365	8.8972	10.2829	11.59131	13.2396	16.34438	20.33723	24.93478	29.61509	32.67057	35.47888	38.93217	41.40106
22	8.64272	9.54249	10.98232	12.33801	14.04149	17.23962	21.33704	26.03927	30.81328	33.92444	36.78071	40.28936	42.79565
23	9.26042	10.19572	11.68855	13.09051	14.84796	18.1373	22.33688	27.14134	32.0069	35.17246	38.07563	41.6384	44.18128
24	9.88623	10.85636	12.40115	13.84843	15.65868	19.03725	23.33673	28.24115	33.19624	36.41503	39.36408	42.97982	45.55851
25	10.51965	11.52398	13.11972	14.61141	16.47341	19.93934	24.33659	29.33885	34.38159	37.65248	40.64647	44.3141	46.92789
26	11.16024	12.19815	13.8439	15.37916	17.29188	20.84343	25.33646	30.43457	35.56317	38.88514	41.92317	45.64168	48.28988
27	11.80759	12.8785	14.57338	16.1514	18.1139	21.7494	26.33634	31.52841	36.74122	40.11327	43.19451	46.96294	49.64492
28	12.46134	13.56471	15.30786	16.92788	18.93924	22.65716	27.33623	32.62049	37.91592	41.33714	44.46079	48.27824	50.99338
29	13.12115	14.25645	16.04707	17.70837	19.76774	23.56659	28.33613	33.71091	39.08747	42.55697	45.72229	49.58788	52.33562
30	13.78672	14.95346	16.79077	18.49266	20.59923	24.47761	29.33603	34.79974	40.25602	43.77297	46.97924	50.89218	53.67196

TABLE A.4   Z Table

	0	0.01	0.02	0.03	0.04	0.05	0.06	0.07	0.08	0.09
0	0	0.004	0.008	0.012	0.016	0.0199	0.0239	0.0279	0.0319	0.0359
0.1	0.0398	0.0438	0.0478	0.0517	0.0557	0.0596	0.0636	0.0675	0.0714	0.0753
0.2	0.0793	0.0832	0.0871	0.091	0.0948	0.0987	0.1026	0.1064	0.1103	0.1141
0.3	0.1179	0.1217	0.1255	0.1293	0.1331	0.1368	0.1406	0.1443	0.148	0.1517
0.4	0.1554	0.1591	0.1628	0.1664	0.17	0.1736	0.1772	0.1808	0.1844	0.1879
0.5	0.1915	0.195	0.1985	0.2019	0.2054	0.2088	0.2123	0.2157	0.219	0.2224
0.6	0.2257	0.2291	0.2324	0.2357	0.2389	0.2422	0.2454	0.2486	0.2517	0.2549
0.7	0.258	0.2611	0.2642	0.2673	0.2704	0.2734	0.2764	0.2794	0.2823	0.2852
0.8	0.2881	0.291	0.2939	0.2967	0.2995	0.3023	0.3051	0.3078	0.3106	0.3133
0.9	0.3159	0.3186	0.3212	0.3238	0.3264	0.3289	0.3315	0.334	0.3365	0.3389
1	0.3413	0.3438	0.3461	0.3485	0.3508	0.3531	0.3554	0.3577	0.3599	0.3621
1.1	0.3643	0.3665	0.3686	0.3708	0.3729	0.3749	0.377	0.379	0.381	0.383
1.2	0.3849	0.3869	0.3888	0.3907	0.3925	0.3944	0.3962	0.398	0.3997	0.4015
1.3	0.4032	0.4049	0.4066	0.4082	0.4099	0.4115	0.4131	0.4147	0.4162	0.4177
1.4	0.4192	0.4207	0.4222	0.4236	0.4251	0.4265	0.4279	0.4292	0.4306	0.4319
1.5	0.4332	0.4345	0.4357	0.437	0.4382	0.4394	0.4406	0.4418	0.4429	0.4441
1.6	0.4452	0.4463	0.4474	0.4484	0.4495	0.4505	0.4515	0.4525	0.4535	0.4545
1.7	0.4554	0.4564	0.4573	0.4582	0.4591	0.4599	0.4608	0.4616	0.4625	0.4633
1.8	0.4641	0.4649	0.4656	0.4664	0.4671	0.4678	0.4686	0.4693	0.4699	0.4706
1.9	0.4713	0.4719	0.4726	0.4732	0.4738	0.4744	0.475	0.4756	0.4761	0.4767
2	0.4772	0.4778	0.4783	0.4788	0.4793	0.4798	0.4803	0.4808	0.4812	0.4817
2.1	0.4821	0.4826	0.483	0.4834	0.4838	0.4842	0.4846	0.485	0.4854	0.4857
2.2	0.4861	0.4864	0.4868	0.4871	0.4875	0.4878	0.4881	0.4884	0.4887	0.489
2.3	0.4893	0.4896	0.4898	0.4901	0.4904	0.4906	0.4909	0.4911	0.4913	0.4916
2.4	0.4918	0.492	0.4922	0.4925	0.4927	0.4929	0.4931	0.4932	0.4934	0.4936
2.5	0.4938	0.494	0.4941	0.4943	0.4945	0.4946	0.4948	0.4949	0.4951	0.4952
2.6	0.4953	0.4955	0.4956	0.4957	0.4959	0.496	0.4961	0.4962	0.4963	0.4964
2.7	0.4965	0.4966	0.4967	0.4968	0.4969	0.497	0.4971	0.4972	0.4973	0.4974
2.8	0.4974	0.4975	0.4976	0.4977	0.4977	0.4978	0.4979	0.4979	0.498	0.4981
2.9	0.4981	0.4982	0.4982	0.4983	0.4984	0.4984	0.4985	0.4985	0.4986	0.4986
3	0.4987	0.4987	0.4987	0.4988	0.4988	0.4989	0.4989	0.4989	0.499	0.499

TABLE A.5   t Table

	0.4	0.25	0.1	0.05	0.025	0.01	0.005	0.0005
1	0.32492	1	3.077684	6.313752	12.7062	31.82052	63.65674	636.6192
2	0.288675	0.816497	1.885618	2.919986	4.30265	6.96456	9.92484	31.5991
3	0.276671	0.764892	1.637744	2.353363	3.18245	4.5407	5.84091	12.924
4	0.270722	0.740697	1.533206	2.131847	2.77645	3.74695	4.60409	8.6103
5	0.267181	0.726687	1.475884	2.015048	2.57058	3.36493	4.03214	6.8688
6	0.264835	0.717558	1.439756	1.94318	2.44691	3.14267	3.70743	5.9588
7	0.263167	0.711142	1.414924	1.894579	2.36462	2.99795	3.49948	5.4079
8	0.261921	0.706387	1.396815	1.859548	2.306	2.89646	3.35539	5.0413
9	0.260955	0.702722	1.383029	1.833113	2.26216	2.82144	3.24984	4.7809
10	0.260185	0.699812	1.372184	1.812461	2.22814	2.76377	3.16927	4.5869
11	0.259556	0.697445	1.36343	1.795885	2.20099	2.71808	3.10581	4.437
12	0.259033	0.695483	1.356217	1.782288	2.17881	2.681	3.05454	4.3178
13	0.258591	0.693829	1.350171	1.770933	2.16037	2.65031	3.01228	4.2208
14	0.258213	0.692417	1.34503	1.76131	2.14479	2.62449	2.97684	4.1405
15	0.257885	0.691197	1.340606	1.75305	2.13145	2.60248	2.94671	4.0728
16	0.257599	0.690132	1.336757	1.745884	2.11991	2.58349	2.92078	4.015
17	0.257347	0.689195	1.333379	1.739607	2.10982	2.56693	2.89823	3.9651
18	0.257123	0.688364	1.330391	1.734064	2.10092	2.55238	2.87844	3.9216
19	0.256923	0.687621	1.327728	1.729133	2.09302	2.53948	2.86093	3.8834
20	0.256743	0.686954	1.325341	1.724718	2.08596	2.52798	2.84534	3.8495
21	0.25658	0.686352	1.323188	1.720743	2.07961	2.51765	2.83136	3.8193
22	0.256432	0.685805	1.321237	1.717144	2.07387	2.50832	2.81876	3.7921
23	0.256297	0.685306	1.31946	1.713872	2.06866	2.49987	2.80734	3.7676
24	0.256173	0.68485	1.317836	1.710882	2.0639	2.49216	2.79694	3.7454
25	0.25606	0.68443	1.316345	1.708141	2.05954	2.48511	2.78744	3.7251
26	0.255955	0.684043	1.314972	1.705618	2.05553	2.47863	2.77871	3.7066
27	0.255858	0.683685	1.313703	1.703288	2.05183	2.47266	2.77068	3.6896
28	0.255768	0.683353	1.312527	1.701131	2.04841	2.46714	2.76326	3.6739
29	0.255684	0.683044	1.311434	1.699127	2.04523	2.46202	2.75639	3.6594
30	0.255605	0.682756	1.310415	1.697261	2.04227	2.45726	2.75	3.646
inf	0.253347	0.67449	1.281552	1.644854	1.95996	2.32635	2.57583	3.2905

TABLE A.6 F Table

df\area	0.995	0.99	0.975	0.95	0.9	0.75	0.5	0.25	0.1	0.05	0.025	0.01	0.005
1	0.00004	0.00016	0.00098	0.00393	0.01579	0.10153	0.45494	1.3233	2.70554	3.84146	5.02389	6.6349	7.87944
2	0.01003	0.0201	0.05064	0.10259	0.21072	0.57536	1.38629	2.77259	4.60517	5.99146	7.37776	9.21034	10.59663
3	0.07172	0.11483	0.2158	0.35185	0.58437	1.21253	2.36597	4.10834	6.25139	7.81473	9.3484	11.34487	12.83816
4	0.20699	0.29711	0.48442	0.71072	1.06362	1.92256	3.35669	5.38527	7.77944	9.48773	11.14329	13.2767	14.86026
5	0.41174	0.5543	0.83121	1.14548	1.61031	2.6746	4.35146	6.62568	9.23636	11.0705	12.8325	15.08627	16.7496
6	0.67573	0.87209	1.23734	1.63538	2.20413	3.4546	5.34812	7.8408	10.64464	12.59159	14.44938	16.81189	18.54758
7	0.98926	1.23904	1.68987	2.16735	2.83311	4.25485	6.34581	9.03715	12.01704	14.06714	16.01276	18.47531	20.27774
8	1.34441	1.6465	2.17973	2.73264	3.48954	5.07064	7.34412	10.21885	13.36157	15.50731	17.53455	20.09024	21.95495
9	1.73493	2.0879	2.70039	3.32511	4.16816	5.89883	8.34283	11.38875	14.68366	16.91898	19.02277	21.66599	23.58935
10	2.15586	2.55821	3.24697	3.9403	4.86518	6.7372	9.34182	12.54886	15.98718	18.30704	20.48318	23.20925	25.18818
11	2.60322	3.05348	3.81575	4.57481	5.57778	7.58414	10.341	13.70069	17.27501	19.67514	21.92005	24.72497	26.75685
12	3.07382	3.57057	4.40379	5.22603	6.3038	8.43842	11.34032	14.8454	18.54935	21.02607	23.33666	26.21697	28.29952
13	3.56503	4.10692	5.00875	5.89186	7.0415	9.29907	12.33976	15.98391	19.81193	22.36203	24.7356	27.68825	29.81947
14	4.07467	4.66043	5.62873	6.57063	7.78953	10.16531	13.33927	17.11693	21.06414	23.68479	26.11895	29.14124	31.31935
15	4.60092	5.22935	6.26214	7.26094	8.54676	11.03654	14.33886	18.24509	22.30713	24.99579	27.48839	30.57791	32.80132
16	5.14221	5.81221	6.90766	7.96165	9.31224	11.91222	15.3385	19.36886	23.54183	26.29623	28.84535	31.99993	34.26719
17	5.69722	6.40776	7.56419	8.67176	10.08519	12.79193	16.33818	20.48868	24.76904	27.58711	30.19101	33.40866	35.71847
18	6.2648	7.01491	8.23075	9.39046	10.86494	13.67529	17.3379	21.60489	25.98942	28.8693	31.52638	34.80531	37.15645
19	6.84397	7.63273	8.90652	10.11701	11.65091	14.562	18.33765	22.71781	27.20357	30.14353	32.85233	36.19087	38.58226
20	7.43384	8.2604	9.59078	10.85081	12.44261	15.45177	19.33743	23.82769	28.41198	31.41043	34.16961	37.56623	39.99685
21	8.03365	8.8972	10.2829	11.59131	13.2396	16.34438	20.33723	24.93478	29.61509	32.67057	35.47888	38.93217	41.40106
22	8.64272	9.54249	10.98232	12.33801	14.04149	17.23962	21.33704	26.03927	30.81328	33.92444	36.78071	40.28936	42.79565
23	9.26042	10.19572	11.68855	13.09051	14.84796	18.1373	22.33688	27.14134	32.0069	35.17246	38.07563	41.6384	44.18128
24	9.88623	10.85636	12.40115	13.84843	15.65868	19.03725	23.33673	28.24115	33.19624	36.41503	39.36408	42.97982	45.55851
25	10.51965	11.52398	13.11972	14.61141	16.47341	19.93934	24.33659	29.33885	34.38159	37.65248	40.64647	44.3141	46.92789
26	11.16024	12.19815	13.8439	15.37916	17.29188	20.84343	25.33646	30.43457	35.56317	38.88514	41.92317	45.64168	48.28988
27	11.80759	12.8785	14.57338	16.1514	18.1139	21.7494	26.33634	31.52841	36.74122	40.11327	43.19451	46.96294	49.64492
28	12.46134	13.56471	15.30786	16.92788	18.93924	22.65716	27.33623	32.62049	37.91592	41.33714	44.46079	48.27824	50.99338
29	13.12115	14.25645	16.04707	17.70837	19.76774	23.56659	28.33613	33.71091	39.08747	42.55697	45.72229	49.58788	52.33562
30	13.78672	14.95346	16.79077	18.49266	20.59923	24.47761	29.33603	34.79974	40.25602	43.77297	46.97924	50.89218	53.67196

Index

Note: Page numbers referencing figures are italicized and followed by an "*f*".